Water, Sanitary and Waste Services for Buildings

Water, Sanitary and Waste Services for Buildings

Fifth Edition

A F E WISE and J A SWAFFIELD

Routledge
Taylor & Francis Group

LONDON AND NEW YORK

First published by The Mitchell Publishing Company Ltd., a subsidiary of
B.T. Batsford 1979
Second edition 1981
Third edition 1986
Fourth edition published by Longman Scientific and Technical 1995
Fifth edition published by Butterworth-Heinemann 2002

This edition published 2011 by Routledge
2 Park Square, Milton Park, Abingdon, Oxon OX14 4RN
52 Vanderbilt Avenue, New York, NY 10017, USA

First issued in paperback 2020

Routledge is an imprint of the Taylor & Francis Group, an informa business

British Library Cataloguing in Publication Data
A catalogue record for this book is available from the British Library

Library of Congress Cataloguing in Publication Data
A catalogue record for this book is available from the Library of Congress

ISBN 13: 978-0-367-57859-6 (pbk)
ISBN 13: 978-0-7506-5255-1 (hbk)

Contents

Preface

The services in buildings represent a substantial proportion of the cost of construction — perhaps 50 per cent in complex buildings such as hospitals, 20–30 per cent in simpler buildings. Water, sanitary and waste services are a significant part of the whole, representing perhaps 10 per cent of the cost of a building. Research on this subject has, nevertheless, been rather neglected and the reports of what has been done are scattered in the literature. In planning this book the aim was to draw together published research, to set out the principles behind design, and to put forward design recommendations within this general framework.

Since the first edition was published in 1979 there have been many developments in codes, standards and regulations, which play such a major part in building services design and use. New Water Regulations to sit alongside Building Regulations have been the most recent change. In England and Wales the Water Supply (Water Fittings) Regulations 1999 were issued by the Secretaries of State — the 'Regulator' — to replace water byelaws. The Water Byelaws 2000 largely mirror the new regulations in Scotland. It is expected that similar regulations will be issued in Northern Ireland in due course. These new requirements affect a range of topics dealt with in this book, including the preservation of water quality in building services, aspects of design and maintenance and water conservation; as a result, new material is incorporated in several chapters.

A further major initiative in this period has been the start of the preparation of harmonized European standards through the European Committee for Standardization (CEN).

This continuing work, which draws on national experience and that of the International Organization for Standardization, is a response to the adoption of various European legislation including the Construction Products and Public Procurement Directives. European standards, with appropriate national provisions and adopted as British Standards, are expected to cover most of the major fields dealt with in this book. They are intended to help ensure compliance with the essential requirements of relevant Directives, and may cover design and installation as well as products. New material relating to these European standards is incorporated in the text. Amendments have also been made in view of new guidance in Approved Document H(2002) related to Building Regulations 2000 for England and Wales.

Whilst the book as a whole has been brought up-to-date as far as possible, chapter 9 looks beyond current design procedures and codification. It outlines advances in computer modelling to describe the unsteady flow conditions which are common in water supply and drainage services, and gives examples of recent applications. These methods have great potential for the future design and codification in building services engineering.

Amidst these changes and developments this fifth edition will, it is hoped, have a part to play by emphasizing the common underlying principles and data relevant to many of the subjects dealt with. Wider interests are covered by reference to research and practice in countries outside Europe.

AFEW and JAS
April 2002

Acknowledgements

The modelling of unsteady flows within building drainage and water supply networks, including rainwater systems, has been supported by a series of research awards from the UK Engineering and Physical Sciences Research Council. This support is gratefully acknowledged. In addition the work of the Drainage Research Group at Heriot-Watt University has been supported by a wide range of UK and international industry and professional institutions, including London Underground Ltd; Hepworth Plumbing Products, UK; Caroma Industries, Australia; Toto, Japan; Studor Ltd, Belgium; the Chartered Institution of Building Services Engineers and the US Association of Plumbing Engineers Research Foundation. In addition to the research funding support mentioned above, JB Engineering, Indiana USA; Nottingham Teaching Hospital; Roodlands Hospital; Heriot-Watt University; Edinburgh City Council and Dundee City Council have either provided data collection facilities to support the research or have contributed data and illustration to this text. All of these contributions are gratefully acknowledged.

The authors are grateful to members of the Drainage Research Group at Heriot-Watt – Dr Arthur, Dr Campbell, Dr Jack, Dr McDougall and Dr Wright for the results of their research, and to Dr Campbell for preparing some of the new illustrations in this edition. Discussions within the Conseil International du Batiment W62 Working Group on Water Supply and Drainage for Buildings have contributed by focusing on some of the research undertaken. The authors are grateful to Tony Gorman of Hepworth Plumbing Products for comments on chapter 14, including the data on PB in table 14.2; and to Glynwed Pipe Systems for the data on PVC-C and to Uponor Ltd. for data on PE-X, both in table 14.2.

Crown copyright illustrations (figures 6.4 and 12.2) have been reproduced from Building Research Establishment publications listed in the references, and appear by permission of the Controller of HMSO.

Figure 13.1 is reproduced from BS 6367 (1983) by permission of BSI. Complete copies of the standard can be obtained from BSI Sales, Linford Wood, Milton Keynes, MK14 6LE, UK.

We are grateful to the following for permission to reproduce copyright material: Elsevier Science Ltd, The Boulevard, Langford Lane, Kidlington, OX5 1GB, UK for figure 1.3; Institution of Water and Environmental Management for table 1.2; Thomas Telford Publications for table 11.3; Wilo Salmson Pumps Ltd for figure 11.3.

Whilst every effort has been made to trace the owners of copyright material, this has not always proved possible, and we take this opportunity to apologize to anyone whose rights we may have unwittingly infringed.

1 Water Use, Load and Storage Estimation

Over half of the potable water supplied by public water undertakings in the UK is used in buildings of various kinds. It is required for what is commonly termed domestic use, including all the various water-using activities in the home and also the supply to sanitary and kitchen accommodation in offices, schools, hospitals and elsewhere. The designer of building services requires a wide range of information on water usage. His or her concern is with the demand in classes of building and in individual buildings, with the requirements of the various water-using activities and with the patterns of use that develop in occupied buildings, all with the objective of estimating the water storage requirements and the flow loads that are likely to arise from the intermittent use of many different sanitary appliances. This requirement is by no means easy to satisfy. It demands both practical information on water usage in occupied buildings related to information about the use of the buildings themselves, and a theoretical framework within which to fit the data so obtained and to provide a basis for a design method. Statistical and probability considerations provide the necessary framework for the theoretical analysis and design procedure, but suitable practical data on water demand and patterns of use are much harder to come by. Experimental work to provide such data must be designed carefully and carried out thoroughly if it is to be at all useful. It may be necessary to monitor such quantities as the total supply to a building and its variation throughout the day, the variable flows in branches within networks of piping, the frequency of use of the

sanitary appliances, and the actual quantities discharged to the drains.

It is not surprising, therefore, that the experimental investigations in this general field have been few and far between. They have been mostly carried out by national institutes concerned with building research and by universities. Designers, therefore, have been obliged to make do with very limited practical information, to use it within as sound a theoretical framework as possible, and to make up for the lack of reliable data by engineering judgement.

Water use

We turn now to some practical data that set the scene for considering load estimation. Tables 1.1 and 1.2 and figure 1.1 (DOE/WO 1992), the latter assuming a total use in England and Wales of 13000 Ml/day, summarize information which is not intended for direct design use but gives useful indications. Thus the consumption of water in WC flushing and for washing and bathing in dwellings suggests that these appliances merit careful attention from a load-producing standpoint. Urinal flushing is seen as a substantial consumer of water in commerce and industry. Changes in the amount of water used, for example a reduction in WC cistern capacity, would affect the overall picture in the course of time — see table 1.2. Water economy is dealt with in chapter 11 which has further information on usage.

From the design standpoint, it is necessary next to consider how the water-using activities and hence the consumption of water are

Table 1.1   Water use as percentage of total use in major sectors (based on DOE/WO 1992)

Activity	Household (55% of total use)	Commerce and service industry (15% of total use)	Industry and agriculture (30% of total use)
WC flushing	31	35	5
Personal washing	26	26	1
Drinking, cooking and food preparation	15	9	13
Washing machines	12	8	—
Washing up	10	2	—
Urinal flushing	—	15	2
External use	5	4	17
Other	1	1	62

Table 1.2   Domestic consumption from studies in south west England (Hall *et al.* 1988)

Activity	Consumption (litres per head per day)	
	1977	1985
WC flushing	32.5	32.6
Personal washing	35.9	41.1
Drinking and cooking	5.0	5.0
Washing and cleaning	12.4	14.0
Laundry	13.4	18.2
Waste disposal units	0.1	0.3
External use	5.1	9.1
Other use	5.6	11.3
Total	110.0	131.6

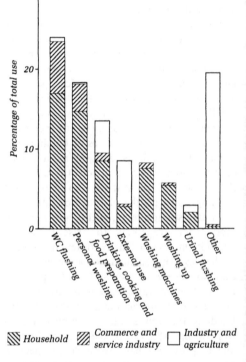

Household ▨   Commerce and service industry ▨   Industry and agriculture ☐

Figure 1.1   Estimated total water use in England and Wales

distributed throughout the day. It is obvious that the influence of the large use of water by WCs is likely to vary depending on whether it is distributed evenly throughout the day or whether it is concentrated in several peak periods. The substantial use of water by urinals occurs at regular intervals throughout the day with automatic flushing. These simple examples serve to illustrate the importance of knowing something of the pattern of water-using activities in different classes of building.

Few data are available and perhaps the most extensive investigation was carried out in a block of flats in London, using instrumentation and also involving the use of questionnaires (Webster 1972). Figure 1.2 illustrates in a simplified form some of the data obtained on the pattern of water-using activity throughout a typical weekday in the larger flats studied. The major activities in this study were washing, using the WC and preparing a drink, and these consistently accounted for about 80 per cent of

the total. Broadly speaking there were three main periods of use of the sanitary appliances, both in the larger flats (two and three bedroom) which were in the main occupied by families with one or two adults working and in some 55 per cent of which there were children, and also in the smaller flats (single bedroom) where many occupants were retired. These periods were from 06.00 to 09.00; from 17.00 to 20.00; and from 21.00 to 24.00. Use of the sink

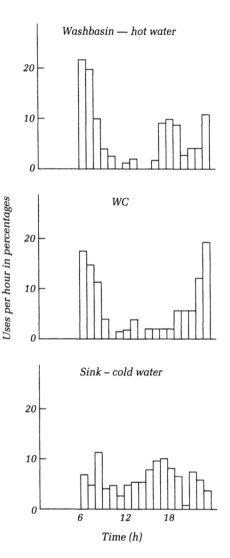

Washbasin — hot water

WC

Sink – cold water

Time (h)

Uses per hour in percentages

Note: Results in terms of average number of uses per hour as percentages of uses for the day

Figure 1.2 Examples of average weekday water use in flats

discussed later. As a detail not recorded in these illustrations, the questionnaire showed that, for this sample, bathing was distributed throughout the week and there was no evidence of a particular period when use of the bath was favoured.

More recently, a small-scale diary survey was carried out by Butler (1991) with 28 households of one to five people living in houses in the south east of England. The sample had a 50 per cent managerial/professional content as against 15 per cent in this group nationally. All had access to a WC, basin, bath and sink, with other appliances available to a less extent. This work, done over seven consecutive days in the middle of December 1987, showed that, as with the study reported above, appliances generally were used most frequently during the weekday morning peak, defined here as 07.30–08.30. The washbasin and the WC in particular were most used in this period, with sink use more uniformly distributed throughout the day. Weekend use generally was more uniformly distributed and less onerous. It was, therefore, recommended that data for a weekday peak should continue to be used for practical design. Table 1.3 summarizes the main numerical values obtained in terms of usage per dwelling (Butler 1993).

The analysis went further and looked at the influence of dwelling occupancy on usage. For the WC, basin, sink, bath and washing machine, frequency of use — uses/dwelling/day — were found to increase directly in proportion to the number of occupants. The data for the study as a whole have been smoothed and plotted on a per capita basis in figure 1.3, which shows both the significance of

was rather more uniformly distributed throughout the day than use of the WC and washbasin. It is to be noted that these records relate to use of the water supply outlets, and that many of the uses would not have involved filling the basin or sink with water — although there is no direct evidence on that aspect. The latter is more significant as far as the estimation of hydraulic loads on the soil and waste pipe system is concerned, a point to be

Table 1.3 Frequencies of appliance weekday usage per dwelling (Butler 1993)

Appliance	Uses per hour per dwelling	
	Average	Peak
WC	0.42	1.20
Basin	0.38	1.53
Bath	0.026	0.14
Shower	0.035	0.32
Sink	0.22	0.76
Washing machine	0.018	0.03

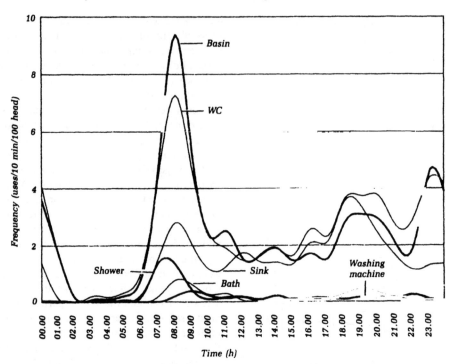

Figure 1.3   Water use from a study of 28 households

the morning peak and provides information on this basis for practical use if desired (multiply the frequency values by 0.06 to obtain frequency/head/hour). For comparison, the author notes that his peak figure for the WC from figure 1.3 amounts to 0.45 uses/head/hour, somewhat less than that deduced from Webster's data, 0.59 uses/head/hour.

The difference may, for instance, be due to different social groups tested and different geographical spread — a range of individual houses as against a block of flats. He also notes an apparent inconsistency in the reported use of hot water in the study by Webster (1972), which suggests that the hot water usage figures may be too high.

Following Butler's methodology, Gatt (1993) has reported a survey of 51 households in Malta which provides extensive data of the kind outlined above. He obtains a figure of 0.39 uses/head/hour as weekday peak for the WC — compare the values in the preceding paragraph.

The 1990s have seen experimental work in this field extended by T.P. Konen and colleagues at the Stevens Institute in New Jersey, S. Murakama and colleagues at the University of Hiroshima, and F. Kiya at the Tokyo Institute of Technology.

Probability of use

Values of the probability of an outlet being in use are required for the analysis to be described later. Such probabilities may be expressed in terms of the average time for which an outlet is in use in relation to the average interval between uses over some period during which there is a series of uses at random. In other words the probability of an outlet being on at any instant is given by the ratio t/T where t is the average time for which an outlet is on and T is the average interval between uses. In the particular study in question probabilities were established by determining the average proportion of each hour for which the individual water outlets were on. Day-long curves obtained in this way take the same general shape as those relating to water-using activities (figure 1.2). For long periods of the day when an appliance is not used the value of the probability that it may be found in use is low; during the main periods of

Table 1.4 Examples of maximum hourly probabilities for water supply points

		Weekday		Saturday		Sunday	
	Water supply point	Maximum probability	Period (hours)	Maximum probability	Period (hours)	Maximum probability	Period (hours)
Flats							
Small, 1.5	WC	0.0155	7–8	0.0133	7–8	0.0116	10–11
occupants on	Washbasin (C)	0.0085	8–9	0.0142	14–15	0.0080	9–10
average	Washbasin (H)	0.0038	8–9	0.0039	9–10	0.0037	8–9
	Sink (C)	0.0034	9–10	0.0055	8–9	0.0152	10–11
	Sink (H)	0.0154	17–18	0.0136	8–9	0.0179	10–11
	Bath (C)	0.0017	7–8	0.0041	14–15	0.0020	13–14
	Bath (H)	0.0059	16–17	0.0427	14–15	0.0042	13–14
Large, 3.2	WC	0.0501	7–8	0.0417	8–9	0.0443	10–11
occupants on	Washbasin (C)	0.0076	7–8	0.0050	9–10	0.0053	9–10
average	Washbasin (H)	0.0108	7–8	0.0080	7–8	0.0085	9–10
	Sink (C)	0.0258	17–18	0.0329	10–11	0.0441	11–12
	Sink (H)	0.0342	18–19	0.0310	9–10	0.0415	10–11
	Bath (C)	0.0030	18–19	0.0038	9–10	0.0035	10–11
	Bath (H)	0.0068	22–23	0.0142	11–12	0.0125	14–15
Hospital ward	Washbasin (C)	0.031					
	Washbasin (H)	0.042					
Office building	WC (Men)	Av. interval between uses: 1200 s. Period: 9–12 h					
		Probability based on 75 s inflow duration: 0.0625					
50 men, four WCs	WC (Women)	Av. interval between uses: 600 s. Period: 11–12 h					
50 women, six WCs		Probability based on 75 s inflow duration: 0.125					

use higher values of probability occur. As an example, the highest average probability reported for the WC for the large flats in this study was about 0.05 for an hour during the morning peak period. This relates to the ballvalve to the flushing cistern and means that the ballvalve was flowing for 0.05 of the hour in question, i.e. for about 3 minutes in the hour.

Table 1.4 gives a summary of the information on probabilities from this study together with the few other data that have been traced in the literature (Crisp and Sobolev 1959, HERU 1961). Figure 1.4 illustrates the distribution of use throughout the day for the study of the office building from which the office data in table 1.4 have been derived.

Whereas the foregoing data relate to the water supply outlets and are applicable to the design of the water supply system, design of the soil and waste drainage system requires values of probability related to the outflows from the sanitary appliances. Early data (table 1.5) were determined in studies in local authority flats and houses (Wise and Croft 1954). This work which has been widely used gave the highest

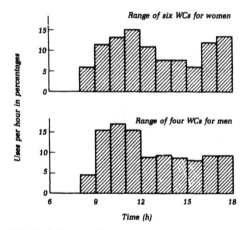

Note: Results in terms of uses per range as percentages of uses for the day

Figure 1.4   Examples of water use in offices

hourly probabilities during the morning peak periods of use set out in table 1.5. The value for the washbasin covers occasions when it was filled and discharged but not occasions when the tap was run without the plug being inserted. The value for the sink included occasions when a bowl or bucket of water was emptied to waste.

Table 1.5   Maximum hourly probabilities for discharge in domestic use

	Duration of discharge, t(s)	Interval between discharges, T(s)	$p = t/T$
WC	5	1140	0.0044
Washbasin	10	1500	0.0067
Sink	25	1500	0.0167

It is worth considering whether the information much more recently reported on water supply outlets can be related in any way to the figures in table 1.4, to support and amplify them or to show changes in patterns of living. From the study of water supply a value of 0.05 might be taken as reasonable (table 1.4) for the probability of use of WCs in a domestic situation during the main period of use. This relates as already noted to the inflowing ballvalve and is, therefore, relevant to the design of the water supply system. For soil and waste pipe systems, however, the relevant quantity is the probability related to the outflow from the pan. This is smaller than that related to the inflowing ballvalve, in proportion to the actual flow time.

In order to convert to an outflow probability it is necessary to know both the inflow time and the outflow time. In the study described it is reported that the inflow time for the WCs studied was mainly between 60 and 120 seconds, and it exceeded 75 seconds for 90 per cent of the time. Taking 75 seconds as a basis of calculation and assuming that the pan outflow duration was 5 seconds, it may be seen

Table 1.6   Conversion of supply probabilities to discharge probabilities

	WC	Washbasin
Supply probability	0.05	0.009
Duration of flow at an individual water supply point (s)	75	12
Duration of discharge from an appliance (s)	5	10
Conversion factor	5/75	10/12
Supply probability multiplied by conversion factor	0.0033	0.0075
Discharge probability (table 1.5)	0.0044	0.0067

that the probability related to the outflow is 5/75 or 1/15 of the probability related to the inflow, i.e. 1/15 of 0.05 or 0.0033. This sequence of calculation is given in table 1.6 which also gives the result of a similar calculation for a washbasin. For the latter, the conversion is more questionable because, for the inflow data, there is no means of knowing whether the supply was used to fill the basin or whether it ran straight through to waste. In the latter case the outflow and inflow probabilities would be more or less the same. Table 1.6 gives tentative assessments of the same order as the probability values previously determined and which are also listed in table 1.6. Observations of discharge parameters from his study of 28 households were reported by Butler (1991).

Data for the office building in table 1.4 can also be interpreted tentatively for discharge. Assuming a discharge time of 5 seconds for the WCs and the given intervals between uses, the outflow probabilities are, for men, 0.0042, and for women, 0.0083. These data relate to populations of 50 per range. Code recommendations may permit higher populations for ranges of this size. For a population of 75, for example, the frequencies and probabilities — for both demand and discharge — are likely to be higher. No data are available for washbasins but it might be assumed that they would be used at the same frequency as the WCs — although people do not always wash after using the WC and when they do it is often with a running tap rather than a filled basin.

For purposes of code writing and standardization it is necessary to reduce the usage data to as simple a form as possible and to give the designer appropriate guidance on their application. Earlier data (table 1.5) have been widely used as a guide to intervals between uses and hence probabilities for domestic use. The interval between uses for the three appliances listed is about 1200 seconds (20 minutes) and the latter figure has often been taken as a value appropriate to domestic installations for both supply and discharge. With offices and factories, however, much heavier use may be expected and an extreme might be conceived where a WC was flushed regularly as soon as its cistern was full — as frequently as once every 120 seconds or so. Data are lacking as already explained but experience suggests that this high frequency is unlikely in practice. Instead a figure of 300 seconds has often been assumed as a time interval representative of 'peak' or 'congested' use, with 600 seconds for 'public' or 'commercial' use, and these values are used later. Designers then use their experience and the sparse, published data in choosing the level of probability likely to be reasonable for particular situations.

Theoretical considerations

If all the appliances available in a building were to be in use at the same time the design of systems to give satisfactory supply and discharge of water would be a matter of fluid mechanics only. Such a condition has already been seen to be far removed from practice; water-using activities are distributed throughout the day and the probability of finding any one appliance in a building in use at any instant is small. Some means must, therefore, be found for determining what combination of appliances may reasonably be assumed to be in use at the same instant as a basis for designing or testing water supply or soil and waste pipe installations. For the former, interest lies in the water supply points at the appliances; for the latter, interest centres on the discharge from the appliances whose probability of occurrence is likely to differ, as already seen, from the supply probability. Once a load has been estimated, the normal procedures of fluid mechanics as outlined in chapter 8 may be used to complete the sizing of the installation.

The basis for the method to be used lies in statistical theory and has been applied to solve problems broadly similar in principle in a range of subjects. Hunter (1940) was one of the first to apply statistical methods to the design of water supply services in buildings. Wise and Croft (1954) introduced a similar approach in research on soil and waste pipe design and the method has since been used widely for these particular systems and further refined and developed.

The idea of probability related to the use of sanitary appliances was introduced in the preceding section. The systems under consideration serve many appliances distributed throughout a building and being used more or less independently of one another. If any one appliance is used on average for a period of t seconds every T seconds, the probability p of finding it in use at any instant is given by $p = t/T$. The appliance is not in use for a time $(T - t)$ every T seconds and the probability of finding it not in use is, therefore, $(T - t)/T$, i.e. $(1 - t/T)$ or $(1 - p)$. As an example, in the preceding section the peak probability of finding a domestic WC discharging was given as 0.0044; the probability of finding it not discharging is much greater, i.e. 0.9956. In other words, each WC was not loading the soil and waste system for 99.56 per cent of the time on average during the peak period of use.

The probability of finding various numbers of appliances of a given type in use simultaneously, or not in use, is found as the product of the individual probabilities. Thus the probability of finding any two in use together is given by p^2, three in use by p^3 and so on. The probability that a given number of appliances and no others will be in use must take account both of the probabilities of some being in use and of others not being in use. Moreover, it is necessary to allow for every combination of appliances possible in estimating a likely load; for example, if there are three appliances A, B and C, A with B, A with C and B with C are all possible combinations of use. These considerations lead to the binomial theorem which expresses the probability P of finding any number r appliances in use out of a total of n of the same sort of appliance:

$$P = c_r^n p^r (1 - p)^{n-r} \qquad (1.1)$$

where c_r^n denotes the number of combinations of n appliances taken r at a time.* This equation reduces to a simpler expression due to Poisson when the probability is small as it is for many of the practical situations under consideration:

$$P = \frac{e^{-\varepsilon}\varepsilon^r}{r!} \qquad (1.2)$$

where e is the base of the natural logarithm and equals 2.7183 and $\varepsilon = np$. Table 1.7 gives some examples of probabilities calculated from the two expressions (Poisson in parentheses). The simple Poisson expression may be useful for preliminary explorations of a problem.

Equation (1.1) provides a basic tool for estimating design and test loadings for water

*$c_r^n = n!/r!\ (n-r)!$ where ! denotes factorial and $n! = n\ (n-1)(n-2)\cdots$

Table 1.7   Examples of calculated probabilities

| | Probability of finding r appliances discharging out of a total of 10 | | |
	$r = 0$	$r = 1$	$r = 2$
WC	0.956	0.0425	0.000841
	(0.956)	(0.0421)	(0.000926)
Washbasin	0.934	0.0630	0.00191
	(0.935)	(0.0628)	(0.00210)
Sink	0.845	0.143	0.0110
	(0.846)	(0.141)	(0.0118)

supply or soil and waste pipe installations. The procedure is to estimate the probability of finding one or more, two or more, three or more appliances in use at the same instant and to compare the result with some agreed criterion or standard or service. By way of illustration consider a group of 10 sinks in a building for which an estimate of a design load is required. The entries in table 1.7 provide the necessary information. The probability of finding one or more sinks in use simultaneously is given by $(0.141 + 0.0118 + \ldots) = 0.1534 \ldots$ and for the example as a whole the probabilities are:

$r = 0 \ldots 10 \; P = 1.00$

$r = 1 \ldots 10 \; P = 0.1534$

$r = 2 \ldots 10 \; P = 0.0124$

$r = 3 \ldots 10 \; P = 0.00068$

$r = 4 \ldots 10 \; P = 0.00002$ etc.

With more than two or three in combination the probabilities are very small. As a practical criterion or standard of service a value of 0.01 has commonly been taken; in other words that actual loadings should exceed the design load for less than 1 per cent of the time. On this basis the design load in the above example would be any two sinks in combination, for the chance of three sinks being in use simultaneously is less than 0.01. In general terms these considerations for the suggested design point are expressed by:

$$\sum_{r=r}^{r=n} c_r^n p^r (1-p)^{n-r} = 0.01$$

where $\sum$ represents the sum of the terms as indicated above. This procedure is the basis of table 4.2 in chapter 4 which can, of course, be extended to larger numbers of appliances if required. Such results may, if required, be expressed as flow rates by multiplying the computed number of appliances by the flow rate appropriate to each appliance, either the supply rate or the discharge rate, depending on the problem under consideration. Figure 1.5 is an example for discharge in domestic use based on the individual flow rates in table 1.8, in which the curves have been produced by rounding and smoothing computed results.

Establishing design procedures

The discussion so far has centred on separate groups of appliances all of the same kind but, in practice, the pipe systems usually serve various mixes of appliances. The method put forward above does not take into account the probability, or rather the improbability, of overlapping between different kinds of appliance in a mixed system and thus tends to overdesign. Hunter (1940) first recognized this deficiency in estimating water supply loads in large buildings and attempted to overcome it by assigning a 'weighting' or 'fixture unit' to each kind of appliance as an expression of its load-producing propensity. His approach has been widely used and has also been applied to soil and waste pipe design. Several such schemes have been proposed over the years, e.g. Griffiths (1962), Burberry and Griffiths (1962). More recently a rigorous application of what is termed the generalized binomial distribution to the design of water supply services for a block of flats has been attempted. This requires a more complex expression than equation (1.1) in which the differing individual probabilities of the various appliances are directly taken into account. Further research on these lines is merited since the study referred to suggested that the method offered scope for economy in design over the more conventional treatment.

Fixture unit methods

Consider first the fixture unit approach which is well established and has been used in various codes. This is based on the fact that a given flow load may be produced by different numbers of appliances depending on their characteristics. Thus figure 1.5 shows that a domestic drainage

Table 1.8   Examples of supply and discharge data

Appliance	Interval between uses (s)	Supply		Discharge		
		Flow rate (l/s)	Demand unit	Discharge rate (l/s)	Discharge unit	
Washdown WC	1200	0.11	0.9	2.3	7(10)
9 litre flush	600	0.11	1.8	2.3	14(20)
12 mm supply (13 litre in parentheses)	300	0.11	3.6	2.3	28(40)
Washbasin	1200	0.15	0.3	0.6	1
12 mm supply, hot or cold	600	0.15	0.8	0.6	3
30 mm trap	300	0.15	1.5	0.6	6
Bath	4500	0.30	1.0	1.1	7
18 mm supply, hot or cold, 40 mm trap	1800	0.30	3.3	1.1	18
Sink	1200	0.30	1.0	0.9	6
18 mm supply, hot or cold	600	0.30	2.5	0.9	14
40 mm trap	300	0.30	5.0	0.9	27
Group — domestic WC, bath and/or shower, basin and sink*	—	—	—	—	14
Urinal (per unit)	1200	0.004	—	0.15	0.3
Shower	—	0.1	—	0.1	—
Spray tap basin	—	0.05	—	0.05	—

* May include washing machine

Notes: For bath and sink, 25 mm supply: multiply demand units in table by 2
For sink, 12 mm supply: multiply demand units in table by 0.7
A volume of 13 litres has been used in Scotland

flow of 10 l/s is likely to be produced by about 160 WCs (13 litre) or 270 sinks; 131 was traditional in Scotland and is taken as an example. Hence the relative loading weights of these appliances are:

$$\frac{1}{160} : \frac{1}{270}$$

Suppose that a 13 litre WC is assigned a numerical value of 10 as an expression of its loading effect. It follows from these figures that the sink has a value of 5.9 relative to that of the WC, i.e. $10 \times 160/270$. We may say that the flow of 10 l/s is the effect of 160 WCs each with a value of 10 or 270 sinks each with a value of 5.9. This idea is the basis of the 'fixture unit' method, the essential parts of which are (i) a carefully chosen set of units to express the load-producing effects of different appliances used at different frequencies and (ii) a relationship between flow rate and the total appliance load expressed in terms of these units. The two components of the method are intimately related.

Table 1.8 sets out supply and discharge unit values for commonly used appliances; also given

are the supply and discharge rate values for the individual appliances which were used in deriving the basic curves such as those in figure 1.5. These

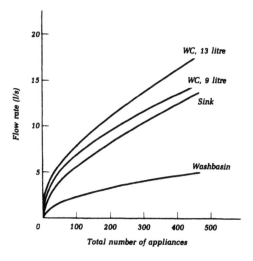

Figure 1.5 Discharge for a given number of appliances with a probability criterion of 0.01. Domestic use

Table 1.9 Examples of discharge unit totals for given flows

	WC 13 litre		Sink	
Flow (l/s)	Number	Discharge units	Number	Discharge units
Domestic use				
5	35	350	70	420
10	160	1600	270	1600
15	330	3300	490	2900
20	530	5300	730	4400
Public/ commercial use				
5	15	300		
10	85	1700	—	—
15	170	3400		
20	260	5200		

units are related as shown to the intervals between use set out in previous paragraphs, with additional figures for baths as a guide. Table 1.9 gives examples of discharge unit totals for given flows. The figures for domestic use are derived from figure 1.5. Thus a flow of 10 l/s is produced as already seen by 160 13 litre WCs or 270 sinks, in each case a total of 1600 discharge units. Similar curves gave the data for public/ commercial use. Plotting the data in table 1.9 gives the curves in figure 1.6 and we have moved from the spread of figure 1.5 to virtually a single curve. Figure 1.7 gives smoothed curves prepared on this basis.

In determining a design flow for a pipe the procedure is to list the sanitary appliances to be connected to it and their unit values, whether demand or discharge, and then to work out the total unit load for all these appliances. Reference to figure 1.7 then permits the design flow to be established. Where it is known that appliances such as showers and spray tap basins may also be in use occasionally with a low rate of flow, the additional flow may be added directly to the rate derived from figure 1.7, or a separate demand assessment may be made. There is, of course, much to be said for exercising judgement about the likelihood of simultaneous use of appliances in carrying out the design work. For example, the information set out in previous paragraphs indicates that the use of baths is not likely to coincide with the morning peak use of WCs, basins and sinks in dwellings. In assessing a load to correspond to a morning peak, therefore, the use of baths may reasonably be neglected. This is the basis for the discharge unit value suggested for a group of appliances — WC, basin, bath and sink — in table 1.8.

Some codified procedures

The British Standard on sanitary pipework, BS 5572 (see Appendix 1), contains the discharge curves given in figure 1.7 together with discharge unit values for individual appliances. In the first edition (1978) the units were those given in table 1.8. The 1994 edition contains the wider range of units illustrated in table 1.10, to reflect the introduction of new appliances and flushing volumes. The curve for mixed appliances is also given in BS 8301.

The original British Code of Practice CP310 on water supply to buildings contained a simple fixture unit method which was carried forward into BS 6700 (1987 and 1997 editions). In addition to design flow rates the Standard now includes minimum flow rates, exemplified in table 1.11, given also in the Water Regulations 1999. It is suggested there that the flow rate from an appliance should not fall below the minimum when simultaneous discharge occurs. The basis is given in table 1.11 and figure 1.8. It does not offer a means to allow for differing intervals between uses, except for the suggestion that the loading unit for a washbasin should be increased from 1.5 to 3 where there is a peak period of use. Large load values are assigned to baths, 10 or 22 depending

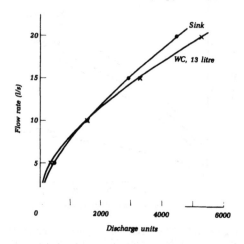

Figure 1.6 Flow rate related to total discharge units

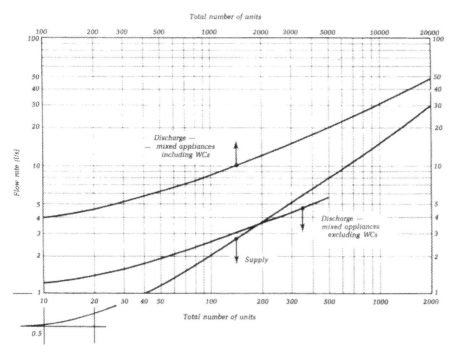

Figure 1.7   Design curves for supply and discharge related to tables 1.8 and 1.10

Table 1.10   Examples of discharge units from BS 5572 (1994)

Appliance	Interval between uses (s)	Discharge unit
Washdown WC	1200	7
9 l	600	14
	300	27
Washdown WC	1200	6
7.5 l	600	12
	300	24
Washdown WC	1200	6
6 l	600	11
	300	23
Washbasin	1200	1
32 mm branch	600	3
	300	5
Sink	1200	7
40 mm branch	600	13
	300	26
Bath	4500	7
40 mm branch	1800	17
Washing machine	240	18
	Domestic	3
Group: WC, bath, sink, one or two basins, washing machine		14

Notes: WCs with high- or low-level cistern
Shower — add 0.07–0.35 l/s
Dishwasher — add 0.25 l/s

on pipe size. Table 1.11 and figure 1.8 also give details of the method recommended by the Chartered Institution of Building Services Engineers (1986). The design flows given by the three methods, i.e. table 1.8 and figure 1.7 as against table 1.11 and figure 1.8, are of much the same order in some situations, diverging more in others, e.g. table 1.12. In small systems, with all such procedures the calculated flow may turn out to be less than the minimum required for one of the appliances which then becomes the determining factor (e.g. see appendix 2).

With the move towards harmonization of standards in Europe, the methods of countries across the Union are of interest. Procedures used in Germany and repeated in several other countries rely as an approximation for both water supply and drainage on simple power law relationships between the design flow rates and total numbers of units. The 'unit' is unique to a sanitary appliance, and allowance for frequency of use and type of occupancy is made through factors related to type of building. In American and UK practice, on the other hand, the 'unit' usually varies for a given appliance with the assumed interval

Table 1.11   Examples of supply units and flow rates related to figure 1.8

Supply fitting at appliance	BS 6700			CIBSE Demand unit		
	Flow rate (l/s)		Loading unit	Private (1200 s)	Public (600 s)	Congested (300 s)
	Design	Minimum				
Basin	0.15*	0.1	1.5 to 3	3	5	10
Bath						
18 mm	0.3*	0.2	10	12	25	47
25 mm	0.6*	0.4	22			
Sink						
15 mm	0.2*	0.1	3	11	22	43
20 mm	0.3*	0.2	5			
WC	0.13*	0.05	2	5	10	22
Bidet	0.2	0.1	1	—	—	—
Washing machine	0.2	0.15	3	—	—	—
Dishwasher	0.15	0.1		—	—	—
Shower head	0.2	0.1	3	0.1 l/s continuous (minimum)	—	—

* CIBSE flow rates as for BS 6700
For washing machines and dishwashers consult manufacturers on flow rates for other than single dwellings
For showers, flow rate depends on type fitted — consult manufacturer

between uses, e.g. table 1.8. The basic form of equation to give the expected peak flow rate Q in the former procedures is:

$$Q = a\left(\sum u\right)^n + b$$

where u is the design flow rate (unit) value for an individual appliance, $\sum$ denotes a summation for all the appliances concerned, and a, b and n are constants that depend on building type and loading. With the German standards, the discharge constants given (DIN 1986) are described as a rough guide; those for water supply (DIN 1988) are more extensive and detailed and said to be based on numerous measurements. The design flows obtained from the UK and German procedures differ somewhat according to circumstances and are reflected in some differences in pipe sizes obtained.

Proposals for a European standard for drainage above ground have, for simplicity, followed the general form of the above relationship (De Cuyper 1993), with ranges of the constants given in an attempt to allow for variations in appliances, usage patterns and building types in different countries of the Union. Designers can then exercise a choice in producing designs for particular local conditions. Proposals for discharge tests of completed systems, on the other hand, have been based on the probability method set out earlier in the present chapter, from which table 4.2 is derived. Now BSEN 12056:2000 with National Annex brings together the relevant information.

The European standard for drains and sewers outside buildings (EN 752) gives the equation for design flow rate above but assigns priority to the use of national guidelines, including BS 8301. The latter, with discharge curves based on BS 5572 (mixed flows in figure 1.7), therefore remains as a recommended procedure.

Once the design flows have been established it is a matter of hydraulic calculations to establish pipe sizes — see chapters 5, 8 and 12.

More advanced methods of load estimation

In an attempt to improve on the empirical fixture unit methods, a few investigators have sought to make a more rigorous theoretical analysis of field data. They have made use of the so-called generalized binomial distribution and the multinomial distribution, and used the computer to assist in the extensive arithmetic required. The aim of such work is a greater

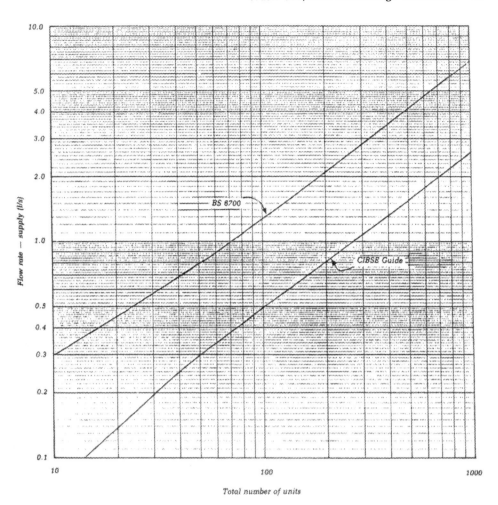

Figure 1.8   Design flow rate as a function of unit totals related to table 1.11

Table 1.12   Comparison of supply rates (l/s) from different procedures

Example	Book (figure 1.7)	BS 6700	CIBSE	
Appendix 2	0.56	0.75	0.59	as in
figure A.2.4	0.48	0.58	0.49	the
	0.41	0.50	0.42	figure
Office building				
600 s interval				
50 WCs and basins	2.6	2.4	2.1	
150 WCs and basins	6.2	5.6	5.1	
Domestic building				
1200/1800 s interval				
20 WCs, basins, baths, sinks	2.4	3.0	1.8	
100 WCs, basins, baths, sinks	8.8	9.8	6.6	

precision in design, but if a more rigorous analysis is to be used to produce standard design information it must be proved over a suitable range of building types and occupancies. If the brief descriptions that follow stimulate more work of this kind they will have served their purpose.

The generalized binomial distribution provides a means of treating directly a system serving a variety of sanitary appliances having different probabilities of use. In the work already outlined (Webster 1972) the peak hours of flow in the systems in the large flats were found from a detailed analysis to be:

Hot water — Monday 18.00–19.00
Cold water — Sunday 10.00–11.00
Drinking water — Sunday 11.00–12.00

The generalized binomial distribution was used to determine design flow rates for the three individual systems based on the observed probabilities for these hours. The basic principles of the binomial distribution have been set out in previous paragraphs. The method may be generalized as follows: if a supply system serves three independent outlets, e.g. one each of types A, B and C, the distribution of flow rates in the common supply pipe is given by the expression:

$$(p_A + q_A)(p_B + q_B)(p_C + q_C)$$

where p and q are the probabilities that an outlet is 'on' or 'off' respectively. Expansion of this expression gives various terms, each of which corresponds to a particular combination of outlets being 'on', e.g. $p_B p_C q_A$ is the probability that outlets B and C are on simultaneously, when the flow rate is the sum of that due to B plus that due to C. When a pipe is serving n independent groups of outlets the general expression for the distribution of flow rates is:

$$[(p_A + q_A)(p_B + q_B)(p_C + q_C)]^n$$

which, when expanded, gives the necessary information for the assessment of a design load against some chosen criterion of service. In the study described the author used for each type of outlet the actual mean flow rates observed, which are summarized in table 1.13. The observed mean flow rates are mostly less than the data (table 1.8) given as design values, suggesting that the latter are on the safe side. This method gave results which suggested considerably lower design flow rates than would be obtained by fixture unit methods of the type already described.

Both the fixture unit method and that described in the previous paragraph have made use of probabilities based on hourly intervals, but it may be envisaged that shorter-term random peaks will occur within any hour. Moreover, certain uses of appliances may not be independent as has been assumed; for example, use of a washbasin may follow use of a WC, and the ending of popular television programmes may give rise to correlations between water uses in different flats as occupants take the opportunity to prepare a drink or visit the WC. This thinking led to a

further computer analysis of the data from the block of flats already described in which the combinations of outlets simultaneously in use in each of the 30 flats studied were established for each 300 second period of the week. Table 1.14 has been reproduced from the report of this work (Courtney 1976) to illustrate the kind of detail examined for just one 300 second period. The multinominal distribution was then used with the computer to combine flows from different flats as a basis for determining design flows against some chosen criterion of service, selected as the flow rate exceeded for 1 per cent of the peak demand interval. Again the design recommendations were substantially less than those obtained by a fixture unit method although, as a result of the examination of random peaks in shorter time intervals, not generally as low as those

Table 1.13 Mean flow rates used

Outlet	Mean flow rate (l/s)
Washbasin	
Hot	0.087
Cold	0.074
Bath	
Hot	0.15
Cold	0.13
WC	0.080
Sink	
Hot	0.098
Cold	0.098

Table 1.14 Combinations of outlets in use Monday, 18.25–18.30

Outlets in use	Mean flow rate (l/s)	% of time
None	0	89.09
WBC (washbasin cold)	0.074	0.14
WC	0.080	3.92
WBH (washbasin hot)	0.087	0.18
SH (sink hot)	0.098	2.31
SC (sink cold)	0.098	1.92
BC (bath cold)	0.130	0.56
BH (bath hot)	0.150	0.11
WC, WBC	0.154	0.03
WBC, WBH	0.161	0.03
WC, WBH	0.167	0.18
WC, SH	0.178	0.49
SC, SH	0.196	0.49
WC, WBC, WBH	0.241	0.03
BC, BH	0.280	0.31
BC, BH, SH	0.378	0.21

Note: Mean flow during the period 0.0124 l/s

obtained by using hourly probabilities with the generalized binomial distribution.

Sanitary accommodation in large buildings

Operational research of the type outlined in the foregoing paragraphs is equally applicable to the study of sanitary accommodation needs in large buildings. Accommodation requirements in different classes of buildings have been established over many years from experience and were not originally the subject of research. It is, therefore, of interest to consider a more rigorous approach to this matter as has been done in a few studies on offices, a shopping centre and schools.

Sanitary provision in office buildings is controlled by the Workplace Regulations published by the Health and Safety Executive (1992) and is also dealt with in BS 6465 (1995) which contains minimum scales of provision based on those of each type of appliance in the Regulations. The number is linearly related to the population served, with differences in WC provision for men depending on whether or not urinals are provided. To establish whether this provision is realistic or represents an over- or underprovision requires (i) information on the usage of sanitary accommodation in occupied buildings and (ii) a computer simulation based on the field data obtained. The purpose of the simulation is to investigate the probability of queuing and the waiting times for various practical situations, and hence to establish realistic scales of provision. The latter may be regarded as widely applicable if a sufficient range of situations has been studied.

Data were obtained on the occupation times for sanitary fitments from studies in UK office buildings (Davidson and Courtney 1976), in an American office building (Konen 1989) and in a Canadian shopping centre (Flenning 1977). The mean times of occupation are summarized in table 1.15, based on observations ranging from 200 to 600 in number. The times, in fact, varied widely, ranging from some 20 to 250 or more seconds, for example, with women's WCs. There is a reasonable measure of agreement in spite of the differences in type of occupancy, whilst the difference for the men's WCs may perhaps be partly explained in terms of the relative numbers of WCs and urinals, full information not being available.

The demand on sanitary accommodation was studied in seven UK office buildings and it was found that the peak demand generally occurred near the lunch period, but with men's WCs being most in demand near the start of the working day (cf. figure 1.4). The peak demands were analysed to give the number of arrivals at the accommodation in 300 second intervals. The probability that r users arrive in any such interval was found to follow the Poisson law (equation 1.2) where ϵ is here the mean number of arrivals in any 300 second period. The mean arrival rates found in the field investigations are given in table 1.16. The averages for male and female are 3 per cent and 3.5 per cent of the population, e.g. those proportions of the population may be assumed to arrive to use the sanitary accommodation each 300 seconds during the peak period. A figure of 1 per cent was found to apply to the male population

Table 1.15   Mean occupation times of sanitary accommodation

	Mean occupation time (s)	
Situation	Male	Female
Office building, UK		
WC	267	80
Urinal	39	—
Washbasin	18	19
Office building, USA		
WC	260	88
Urinal	36	—
Washbasin	15.5	23
Shopping centre, Canada		
WC	184 (54*)	92
Urinal	35	—
Washbasin	14	20

* Often used as a urinal

Table 1.16   Mean arrivals in 300 second periods at sanitary accommodation in seven offices during peak use

Percentage of male population	Percentage of female population
1.8	3.6
2.5	2.9
4.0	4.4
3.3	4.6
3.1	2.9
4.0	2.7
2.9	2.6

Table 1.17   Relative usage in percentage terms of sanitary appliances in an office during peak periods

	Male	Female
Use only the washbasin	10	13
Use the WC first	11	87
Use the urinal first	79	—

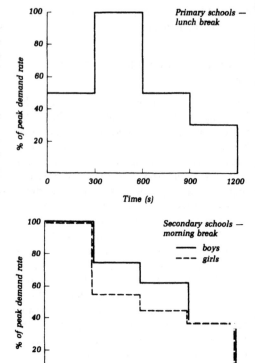

using the WCs earlier in the day. The populations ranged from less than 10 to over 250. The Canadian study also gave such data relating to the large shopping centre for which the overall population was determined by survey, amounting to some 2000 over the peak shopping times. Typically some 8 per cent of the population was found to use the sanitary accommodation in a peak hour, either on a Friday evening or on a Saturday afternoon. In corresponding quarter-hour periods some 3 per cent (female) and 2 per cent (male) used the accommodation. The study on offices, furthermore, gave information about the relative usage of appliances during peak periods, e.g. those wishing to use the washbasin only, those using the WC first, and so on. In summary the results were as in table 1.17.

Similar studies (Davidson and Courtney 1980) have also been carried out in seven primary and six secondary schools in the UK with the results in tables 1.18 to 1.20 and figures 1.9 and 1.10. The primary schools varied as regards the provision of sanitary facilities, most having sets of WC cubicles and washbasins in or adjacent to classrooms and separate additional cloakrooms. One school was 'open plan' and

Figure 1.10   Peak demand patterns based on observations in schools

the two girls' and two boys' cloakrooms each served about 90 pupils. As a result the accommodation was classified in the study as in table 1.18, with the infants and junior mixed sets assumed to serve both sexes as occurred with the younger children in these examples.

In the primary schools the distribution of occupation times as, for example, in figure 1.9 did not differ significantly from one school to another. Times varied up to 170 seconds or more with the mean values given in table 1.18. The demand pattern, on the other hand, varied considerably depending on the school layout and the teaching methods and policies adopted by the staff, and this is indicated by the summary of arrival rates in table 1.18. It was, moreover, not always clear whether users of the washbasins had previously used the WCs. For the purposes of simulation it was decided, to be on the safe side, to assume a value of 20 per cent as the proportion of the population

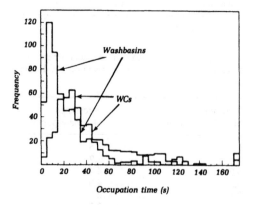

Figure 1.9   Examples of the distribution of occupation times for mixed juniors in four primary schools

Table 1.18 Mean occupation times and arrivals at sanitary accommodation in seven primary schools

	Infants		Junior mixed		Junior girls		Junior boys		
	WC	Basin	WC	Basin	WC	Basin	WC	Urinal	Basin
Mean occupation time (s)	50	28	43	25	42	23	71	22	20
Arrivals as percentage of population served in 300 second period of peak use	3	6	2	2	—	—	—	—	—
	(12)	(10)	(10)	(10)					
	9	5	—	—	—	—	—	—	—
	(22)	(53)		(10)		(12)	(10)		
	—	—	—	—	2	14	13	—	18
					(8)	(23)	(22)		(27)
	16	16	5	4	—	—	—	—	—
	(24)	(26)	(7)	(11)					
	—	—	—	—	—	3	—	—	—
						(8)			
	—	—	—	—	8	12	8	—	8
					(10)	(15)	(11)		(16)
	—	—	—	—	8	10	14	—	8

Note: Percentage of population served is given as mean and maximum recorded (in parentheses)

served by the cloakroom during the peak 300 second period of the school day, occurring at the beginning of lunch. With accommodation covered by the first three columns of table 1.18, half the users were assumed to require the washbasins alone and half the WCs and then the washbasins; for the boys alone, the ratio of requirements for WCs, urinals and basins was taken to be 2:25:15.

With the secondary schools studied the peak rate of demand occurred during the morning break rather than at lunch time and varied much less than with primary schools, presumably because the children were discouraged from visiting the cloakrooms during lessons. Unlike primary schools, washbasins were hardly used at all by the boys and there was a much reduced use by the girls. Few uses of WCs by boys were recorded. Table 1.19 summarizes mean occupation times for the sanitary fitments and table 1.20 the demand rates observed. For simulation 12 per cent and 9 per cent were assumed for girls and boys, respectively, as demand rates during a 300 second peak. First requirement was assumed to be on the basis of 5:1 (WC:basin) for girls and 1:16:1 (WC:urinal:basin) for boys. On the basis of the observations the demand patterns shown in figure 1.10 were assumed in the simulation.

Both the UK and the Canadian studies incorporated such data into models which were run on a computer to examine various practical situations. A major output of such computer

runs is the proportion of time for which all the sanitary appliances provided are occupied, hence giving the probability that a potential user has to wait before gaining access. As a

Table 1.19 Mean occupation times, in seconds, for sanitary accommodation in six secondary schools

Boys		Girls	
WC	Urinal	WC	Basin
190	22	71	20

Table 1.20 Percentage of population served in each 300 second period of morning break in six secondary schools

Boys' urinals				Girls' WCs			
300 second periods				300 second periods			
1	2	3	4	1	2	3	4
8	7	6	2	—	—	—	—
5	4	2	—	11	5	5	—
—	—	—	—	6	5	2	1
—	—	—	—	Mean = 6			
7	3	0	1	—	—	—	—
8	4	2	3	11	6	4	5
3	6	8	5 ←	Lunch break			
8	6	5	3	11	6	5	4
							Assumed

Note: Lunch break in one school is included. 'Assumed' refers to the distribution adopted for simulation

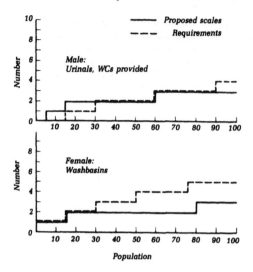

Figure 1.11 Comparison of recommendations with requirements for offices

general criterion for establishing a reasonable level of provision, a probability of 0.01 was used (cf. the criterion for fixture units) in the offices study. In other words, a potential user will find, for example, the urinals occupied on 1 per cent of the occasions that he or she wishes to use them during a peak period. On this basis scales of provision were recommended of which figure 1.11 illustrates examples in comparison with BS recommendations. It is interesting to note that, although the results generally suggest economies could be made compared with present requirements, in certain instances there is underprovision on the criteria used in this study. The authors note that their model did not take into account the compensating behaviour of users when facilities are under peak demand. Economies in provision were also suggested as a result of the study in schools, for which a 5 per cent criterion of service was used. It is to be hoped that more studies of this nature will be undertaken in order to confirm the general applicability of such information from research and to broaden the range of reliable data available. In the 1990s, O.M. Goncalves at the University of Sao Paolo, Brazil, has applied these methods to facility provision and to the estimation of flow rates.

An updated version of BS 6465 (1995) repeated in Part 1 many of the long-standing requirements on scales of provision but draws on experience and research to recommend some new data for several types of accommodation.

Storage requirements

Several authors have sought to establish a rational basis for the quantity of cold and hot water to be stored in buildings. They have relied on a 'design day' approach in which information obtained from a series of demand metering studies in buildings was reduced to data for a single day assumed to be representative of the particular building type chosen. The design curve so obtained was selected on a trial and error basis so as to cover expected worst cases. Typically the investigations (e.g. Spielvogel 1969) determined in this way which of the various parameters of site size was most closely related to hot water consumption for each size of building that was metered. The results were then expressed in terms of this parameter to give per unit data, e.g. quantity per meal served for a restaurant. These data were scanned to find the maximum 1, 2, 3, etc. hours consumption, and the results combined to give a 'design day' for the site. It is assumed that the relationship between water consumption and the chosen parameter of site size remains constant for all buildings of the type considered.

Whilst this kind of approach was an advance on traditional methods, it lacked a sound statistical basis. In a series of studies on hospital wards carried out by the Building Services Research Unit at Glasgow University (HERU 1961, Maver 1964) statistical methods and computer simulation were used, and for the first time quantified the relation between storage volume and rates of heating for hot water and between storage volume and rates of supply for cold water. Account was taken of the magnitude and time of occurrence of demands during each day studied by using the metered data as input to a digital computer simulation of the supply/storage/demand system. By running the demand data through the model using different supply rates it was possible to determine, for each day studied, the maximum storage requirements for that day. Hence a storage/supply design curve was obtained.

More recent work (Gibson 1978) has built on these studies by improving on some of the assumptions made, e.g. by incorporating in the simulation an improved ballvalve operating characteristic. The work has also distinguished between the requirements for 'reserve' storage (to provide a reserve of water in the event of failure of the mains supply) and 'buffer' storage (to limit the maximum rate of draw-off from the mains). The former is intended to take account of the variability between days, whilst the latter is a function of the method of supply and the variability of demand within days and may be termed the 'within day' requirement. The buffer storage is the additional storage required in order that the supply rate can be limited to some value less than the maximum demand in the building whilst maintaining the reserve storage at its required level under all conditions. The work has not provided much data for practical use but provides a theoretical framework as a starting point for the further practical investigations needed on various building types. Appendix 2 gives an example of storage calculations.

2 Requirements and Regulations for Water

'Bad water' had been suspected for centuries as a cause of illness, and in 1845 cholera was, for the first time, convincingly linked with the drinking of contaminated water. John Snow brought the epidemic to an end by removing the handle of the Broad Street pump in a London neighbourhood where over 500 people had died of the disease; the well supplying this pump was close to a sewer. Investigations of the spread of cholera, typhoid fever and dysentery led to the characterization of the organisms causing such diseases and showed their presence in contaminated water. By the end of the nineteenth century the requirements for providing microbiologically safe drinking water were clear: a matter of excluding excremental contamination from the supply and preventing contamination of the treated water. In more recent years chemical contamination has been the subject of extensive investigation, with a view to establishing proper control measures.

Water companies in the UK have applied recommendations of the World Health Organization (WHO) (1984) and DHSS and DOE (1983) to control the microbiological and chemical quality of water they supply. This ensures that the supply to buildings is normally fit for human consumption, i.e. potable, and also palatable, although the quality may sometimes deteriorate between the point of treatment and the point of use. An EC (1980) Directive on water quality superseded these recommendations in 1985, the Directive being intended to apply at a point where the water is made available to the consumer. Regulations have been introduced in the UK (DOE 1989) as a basis for the application of the Directive and require the water companies to take samples at the consumers' water outlets as a check that standards are being met. The Drinking Water Inspectorate, reporting for the first time in 1990 (Department of the Environment/Welsh Office 1990), showed a high degree of compliance with the requirements, the results gradually improving with standards being met in 99.82 per cent of some 3 million tests carried out by water companies in 1999. A desire for continuing improvement in the quality of the drinking water has seen some updating of the EC Directive in 1998, and a revision of UK regulations, the latest to be introduced from 2004.

In recent decades more attention has been focused on the extent to which potable and palatable water deteriorates in pipework and systems within buildings. In the past, Byelaws made under the Water Acts, 1945 to 1989 and most recently 1991, have helped to prevent such deterioration, having the aim also of preventing misuse, waste and undue consumption and the contamination of the supply mains themselves. The Byelaws were replaced in 1999 for England and Wales by the Water Supply (Water Fittings) Regulations, having broadly the same purposes as the Byelaws, being concerned to control supplies in domestic and commercial plumbing installations. As with the Byelaws, the Water Undertakers have the duty to enforce the Regulations. In Scotland the Water Byelaws 2000 in large part mirror the 1999 Regulations for England and Wales.

The present chapter deals with some of the main considerations as regards water quality and the formulation of regulations, leading up to the 1999 Regulations and Byelaws 2000 for control within buildings. The requirements of the user of such installations are generally wider than those factors covered by regulation, and are summarized at the beginning of chapter 3. That chapter also deals with the practical aspects of preserving water quality through design and maintenance of installations in buildings.

Aesthetic qualities

The EC (1980) Directive listed and attempted a quantification of 'organoleptic parameters' including colour, turbidity, odour and taste, recognizing their importance to the consumer. Most complaints by consumers of contamination have indeed related to aesthetic qualities — whether the water is palatable or not. A study in London (Colbourne 1981) showed that growth in the plumbing installation of the micro-organisms known as fungi was the most common cause of complaints about taste or odour. The origin is thought to be metabolic products released into the water by fungi multiplying on materials used in the installation. A combination of warming of the water and nutrients from materials, especially those based on natural ingredients, encourages fungal growth. Other complaints included 'metallic taste' apparently derived from the leaching of copper from new pipework (which diminishes as a film of carbonate builds up on the pipe surface), and 'plastic taste' associated with some polythene pipes.

An extreme case of copper contamination, found in a newly constructed hospital building, was reported by Felstead (1974). The staff complained of the taste and colour of the drinking water and sampling revealed copper contamination at about 300 mg/l, compared with 3 mg/l maximum after 16 hours contact recommended by WHO (the EC 1980 Directive specifies 3 mg/l after 12 hours where the water is made available to the consumer). This installation contained long lateral mains to supply drinking water points, and the inner surfaces of these pipes were found on investigation to be corroded and covered in part by a black film of oxide. The explanation lay in the fact that the pipework had been incompletely drained after pressure testing. Some of the laterals had been left partly filled for some time at temperatures of between 25 and 30 °C and it seemed that no protective carbonate film had

developed under these conditions. Replacement of the pipework was mooted but the cost and the disruption of the hospital that would result was unacceptable. Instead the pipework was cleaned chemically at about one-tenth of the replacement cost.

Drinks delivered by vending machines have sometimes been the subject of complaint caused by contamination from particulate matter or slimes (Colbourne 1981). In most of these the contaminant has been microbial in origin. If the internal components of machines are not readily cleaned, the powdered ingredients may remain lodged in crevices for some time and become a centre for microbial action. The resulting odour may taint the drinks. Materials used in vending machines are, therefore, being evaluated against approved specifications. A machine may also become contaminated if unsuitable tubing is used to connect it to the water supply. An example quoted was the growth of yeasts in a short length of clear, plasticized PVC tubing not supplied with the machine but used by the company fitting out the restaurant where the machine was located. Bacterial colonization of dental equipment has sometimes followed a similar pattern, whilst the growth of bacteria and fungi has also been observed in the pipework and water softeners attached to renal dialysis equipment in the home.

Discoloration accounts for a substantial proportion of the complaints about quality received by a water company. Iron or manganese is usually the cause, arising from inadequate removal in treatment works or, if it is iron, due to corrosion of the mains as internal coatings deteriorate. Plumbing systems are not usually the prime cause, therefore, of any discoloration.

Health aspects

Specific and precise compared with earlier regulations, the EC (1980) Directive listed some 60 parameters for which limits were given. These included values for various toxic metals including lead, cadmium and arsenic, phenolic and other compounds, nitrates and so on. The publication prompted surveys of drinking water quality in many EC member states, as a result of which various actions were required to ensure compliance with EC law. In the UK attention was particularly focused on lead and nitrates. It is the past use of lead pipework to supply buildings that causes most concern, because lead may have an effect on the mental development of children and be a factor in behavioural problems. Lead concentrations in excess of levels recommended by WHO had been found in some areas of the UK in a 1975 survey; the variation depends on the extent to which the water is plumbosolvent, with soft water being of most concern. In a recent survey (DOE/WO 1990), of 200 000 samples tested, around 1 per cent exceeded the 50 micrograms per litre limit specified in the EC (1980) Directive, with rather more in 1991. Lead can also occur in a particulate (insoluble) form of varying size, and this may depend on water quality and on disturbances due to hydraulic conditions such as waterhammer or high flow velocity (Hulsmann 1990). In view of health concerns, especially regarding children, WHO proposed a reduction of the limit from 50 to 10 micrograms per litre. The EC 1998 Directive, therefore, tightens the standard for lead from the present 50 to 25 micrograms per litre by 2004 and to 10 by 2014, the values to be applied 'as a weekly average value ingested by consumers'. Such matters as the sampling required and the practicality and cost of sampling and remedial measures, are under consideration. The steps needed eventually to achieve compliance will rest with the water undertakings. An evaluation of the problems involved in meeting the higher standard has been made by Hayes *et al.* (1997).

The microbiological aspects of water quality and health have been reviewed (Dadswell 1990), drawing on an extensive examination of waterborne diseases in the UK between 1937 and 1986 (Galbraith *et al.* 1987). The year 1937 was of particular significance, involving a notable typhoid outbreak in Croydon. The reviews record 34 outbreaks of several different kinds in that period, of which 21 were due to public water supplies; 11 of the latter were due to contamination at source, 10 to contamination within the water distribution system. Dadswell (1990) has noted that with ageing water mains and sewers and the inevitably extensive repair work, there is a greater risk of the ingress of waterborne pathogens into treated water. Maintaining the integrity of the distribution system is, therefore, a vital factor

in the supply of wholesome water. Proper design, operation and maintenance of plumbing systems in buildings is one important aspect of the overall requirements for such supply.

The micro-organisms found in water include bacteria, viruses and intestinal parasites. With plumbing systems bacteria are of most interest. As with fungi, bacteria depend for their growth on such factors as temperature, time, the presence of nutrients, and pH (intensity of acidity or alkalinity). Temperature is a critical factor — below 10 °C many bacteria cease to function or multiply; at 50 °C, or above, bacteria begin to die owing to thermal damage, taking perhaps an hour at 50 °C and a minute or so at 70 °C. The optimum temperature for multiplication usually lies between 25 and 40 °C, with the more pathogenic species at the upper end of the range. The effect is further illustrated in table 2.1.

Bacteria found in treated water in the UK are not, in general, a significant hazard to public health, but some may present a hazard to the debilitated patient in hospitals. *Pseudonymous aeruginosa* is one of those principally responsible for cross-infection (Colbourne 1981). The organism has been found in tap water in association with some construction materials, and laboratory tests suggest that some washers, gaskets, paints and lubricants are capable of supporting growth at favourable temperatures.

Legionella pneumophila causes a type of pneumonia known as legionnaires' disease and may also lead to a less serious flu-like illness known as Pontiac fever. A major outbreak

Table 2.1 Effect of temperature and heating system on bacterial growth, based on Meers (1981)

Hospital	Location	Bacteria/ml
A	Mains water	25
	After softener	487
	In cold storage tank	1384
	At 60 °C from taps	13
B	In cold storage tank	190
	From calorifiers storing water at 43 °C nominal	11 840
C	In cold storage tank	130
	At 41 °C from taps — provided by water from calorifiers at 65 °C mixed with water at 15 °C from storage tank	46

Table 2.2   Categories of risk, based on the 1999 Water Regulations

Fluid category	Description	Examples
1	Wholesome water	Water direct from a water undertaker's main
2	Wholesome water become less palatable owing to: [a] a change in temperature, or [b] a change in taste, colour or appearance i.e. aesthetic impairment	Hot water mixed with cold water Domestic water softeners (common salt regeneration) Ice making machines Drink vending machines with no ingredients or carbon dioxide added
3	Water with a slight health hazard due to the presence of substances of low toxicity including: [a] ethylene glycol, copper sulphate solution of similar chemical additives, or [b] sodium hypochlorite (chloros and common disinfectants)	Water in heating systems in a house Washbasins, baths and showers Domestic clothes and dishwasher machines Drink vending machines with added ingredients or carbon dioxide Home dialysing machines Domestic hand-held hoses
4	Water with a significant health hazard due to the presence of toxic substances including: [a] chemical, carcinogenic substances or pesticides [b] environmental substances of potential health significance	Water in heating systems in other than a house Fire sprinkler systems using anti-freeze Commercial dishwashers; bottle washers Mini-irrigation systems for gardens, with pop-up sprinkers or permeable hoses Some food processing equipment Various industrial and commercial equipment
5	Water with a serious health hazard due to the presence of pathogenic organisms, radioactive or very toxic substances including: [a] faecal material or other human waste [b] butchery or other animal waste, or [c] pathogens from any other source	Sinks, urinals, WC pans and bidets Industrial cisterns, various plant and equipment Grey water recycling systems A variety of medical equipment; food processing and catering equipment; various agricultural equipment including commercial irrigation outlets at or below ground level and/or permeable pipes

of this pneumonia — the first recorded — occurred in 1976 amongst members of the American Legion attending their annual convention at a hotel in Philadelphia. The organism responsible for the outbreak was identified in the following year and named *Legionella pneumophila*. Since then many more outbreaks have been reported from different parts of the world, in which the risk of infection was greatest in buildings and locations used by the general public or workers for significant periods of time, including hotels, hospitals, business premises and residential establishments. The UK Public Health Laboratory service has provided extensive data, including some reported by the Institute of Plumbing (1990).

Infection is due to inhaling fine droplets of water (about 5 micrometres in size) containing legionella. Certain groups of people are more likely to be affected than others — men three

times more likely than women, those over the age of 40, heavy smokers or drinkers, people whose immune system has been reduced by the medication they are taking or chronic disease. In some outbreaks about 10 per cent of people who caught pneumonia died. An example of a serious outbreak was at Stafford in 1985 when more than 20 people were known to have died.

For an outbreak to occur it is necessary for there to be a reserve of water, containing legionella, with conditions right for growth to high concentrations. There must be a means for creating a spray or aerosol effect, and atmospheric conditions such as to permit the organism to remain active. A person susceptible to the infection must inhale legionella. Modern plumbing and air-conditioning installations often make good breeding grounds for the organism, which probably gains access to such systems in several ways. It may be found in low numbers in mains water supplies to buildings, i.e. treated water; legionella may also gain access to buildings via open access points such as vents and tanks or during construction or repair of pipes, the organism being present widely in the environment (Colbourne and Dennis 1989).

Once present in building service systems, growth may occur. Legionella grows well at water temperatures between 20 and 45 °C, with an optimum for growth of about 36 °C. It may lie dormant in water below 20 °C. Whilst time is important for growth, water may, of course, lie stagnant for considerable periods in pipe and tank installations. Furthermore, sediment, algae and slime present may provide nutrients for the growth, facilitated by the presence of some kinds of constructional materials (Bartlett 1981). In building services, too, the means of forming sprays or aerosols are often present with, for example, shower heads, spas, cooling towers and evaporative condensers; the impact of water on a hard surface or turning on a tap can give rise to water droplets. Ellis (1993) has provided a comprehensive review of the subject with many references to field studies and research.

The bacteria will grow in systems supplied with mains water chlorinated to 0.5 ppm but are susceptible to higher concentrations of free residual chlorine. Treatment to eliminate the possible recurrence of legionnaires' disease in a building thus includes the regular cleaning of systems and chlorination of water, other than drinking water, to ensure the delivery to cold water outlets of water containing 1–2 ppm of free residual chlorine. Disinfection methods using chlorine are described in BS 6700 and in HS (G)70 issued by the Health and Safety Executive in the UK. The latter, in its 2001 version, requires risk assessments and a management plan for all non-domestic water systems, and places great emphasis on thorough building maintenance.

Classifying risk

Over many years, how best to preserve quality once water enters a building, and thus to provide a sound basis for the design and installation of pipework and fittings, has been a major consideration for those drafting regulations and standards. The matter has been approached by examining and classifying the risks involved.

To this end, early studies in the UK and continental Europe formed the basis for tripartite discussions between the UK, France and Germany and then in the standards organization CEN. A suggestion by the tripartite working group was for a two-level classification:

(i) permanent or occasional possibility of contamination which may make the water harmful to health; and

(ii) possibility of contamination which is not harmful to health but which may give rise to complaints by consumers.

The UK report of 1974 (DOE 1974) went further in recognizing two classes within (i), and this triple classification formed the basis for the development of the subject in the UK model bylaws and the code on water services in buildings, BS 6700 (1987).

The UK triple classification (termed method 1 here), in effect, recognized four qualities of water: potable water for human consumption and water contaminated at three levels of risk. The latter, in descending order of importance, are:

1. Risk of serious contamination which is likely to be harmful to health (continuing or frequent).

    This might arise, for example, from WC pans and urinals; various medical equipment; various industrial processes, and research; automatic car washing; launderettes; some agricultural processes; drinking troughs; tankers; fire fighting supplies; ship–shore connections; gully emptiers; hose union taps.

2. Risk of contamination by a substance not continually or frequently present which may be harmful to health.

    This might arise, for example, in the home from taps at sinks, baths and basins, washing machines, flushing cisterns, flexible shower fittings with bath or basin, hose union taps, heating systems (primary circuits and indirect central heating systems including feed cisterns); also from some drink vending machines; water softeners (non-domestic); heating circuits in laundries.

3. Risk of contamination by a substance which is not harmful to health but which could give cause for complaint by consumers.

    This might arise, for example, in the home from flexible shower fittings in a cabinet, mixing and combination taps in certain situations, domestic water softeners, water storage cisterns for gravity discharge.

Whilst this classification was taken up in the UK, there were continuing attempts, notably in France, to make the approach less subjective, for instance, by introducing the idea of testing for contamination against a specified standard.

An alternative approach (method 2) might thus recognize five qualities of water on the following lines:

(i)   potable water for human consumption;
(ii) water fit for human consumption, including that from a potable distribution system, which has deteriorated in taste, odour or colour or become warmer or cooler (compares with 3 above);
(iii) water with a human health hazard due to the presence of one or more toxic substances; a lethal dose might be in the range of, say, LD50 more than 200 mg per kg of bodyweight;
(iv) as (iii) but with LD50 $\leq$ 200 mg per kg or with one or more radioactive, carcinogenic or mutagenic substances;
(v)   water with a human health hazard due to the presence of harmful bacteria or viruses.

The term LD50 is widely recognized and means the oral dose required to kill 50 per cent of a target population, here of rats.

Subjective elements remain in this second system of classification. The question of frequency and duration of contamination may present difficulties in practice. Uncertainty about some aspects — e.g. an incident causing brief contamination which then clears — is inevitable. It remains to be seen, therefore, whether such a system offers worthwhile practical advantages in the long term.

Fluid categories in the 1999 Regulations

Risk in the 1999 Regulations is dealt with by recognizing five qualities of water as in the example (method 2) described above. The approach does not go as far, however, as defining a contaminated fluid by means of an objective test. The subjective element is retained, and this appears simpler and more practical. The five categories are described in table 2.2, based on the Regulations. The prime requirement is for wholesome water — potable and palatable — supplied by a water undertaker and complying with the regulations made under section 67 of the Water Act, 1991. Less desirable categories follow. Both aesthetic and health considerations appear in the list, with risk increasing in going from categories 1 to 5. A few examples, not exhaustive, illustrate some of the situations involved. As shown in chapter 3, there has to be a corresponding increase in the security of protective measures used to counter the increasing risk.

3 Water Installations

Whilst aesthetic qualities and health aspects of the water supplied to buildings are fundamental, the requirements of the user are generally wider than those factors covered by the bylaws and regulations reviewed in chapter 2. A water service installation in a building should be capable of supplying the required numbers and types of sanitary appliances with flows of wholesome water adequate for the users, and at a suitable temperature. The water should be distributed quietly and at the point of use delivered in an acceptable fashion. The controls available should be simple and readily operated and adjust the flow effectively throughout their range. Contamination of the supply, e.g. through back-siphonage or by cross-connection to a source of lower quality, should be avoided, and the system should not give rise to any other form of hazard to the user, e.g. explosion. The potential for waste should be reduced as far as possible. Protection against freezing is necessary. The supply should be reliable both in the short and long term, and thus it should be possible readily to maintain the system in good order. Appearance is important in relation to fittings and fixtures — a matter of design — and in the system generally, a need met by the proper integration of pipework and structure.

Some of these requirements and ways of meeting them are dealt with in more detail elsewhere in this book. Thus chapter 1 reviews water usage in buildings and the estimation of a design load and storage requirements, chapter 8 sets out the background to the hydraulic formulae for pipe sizing, and chapter 10 deals with noise and its control. The present chapter is concerned with the broader description of types of systems, pros and cons of water storage in buildings, features of foreign practice that are increasingly of interest in the UK, the preservation of water quality and related questions of backflow protection and pressure relief. Notes on sizing are also given in appendix 2.

Résumé of conventional UK practice

The familiar practice in the UK, typified by figure 3.1, makes use of storage within the building as a general source for both hot and cold water services, whilst retaining one or more mains-fed supply points for drinking water. This method enables several of the major requirements to be met. It ensures that water will be available to the user, at least for a period, in the event of a failure in the mains or during repairs. It provides a source of water at a known and often moderate pressure both for general cold and hot supply and thus is a factor in noise prevention, flow control and safety. By separating the water supply points in general from the mains, and leaving only a restricted number of mains-fed outlets for drinking water, it greatly reduces the risk of contaminated water at the appliances being drawn back into the mains under conditions when back-siphonage may occur. In certain areas in the UK, the mains have sufficient capacity to permit all the cold water outlets in low-rise housing to be fed directly, and the storage cistern may then be smaller than when both hot and cold taps are cistern fed.

With multi-storey buildings it is often possible only with the lower floors to provide a drinking water supply direct from the mains, as in low-rise housing, and various methods have been used to ensure an equivalent quality of water at upper levels. The alternatives used reduce in broad terms to two:

(i)   supply direct from a rising main in which the pressure has been boosted,

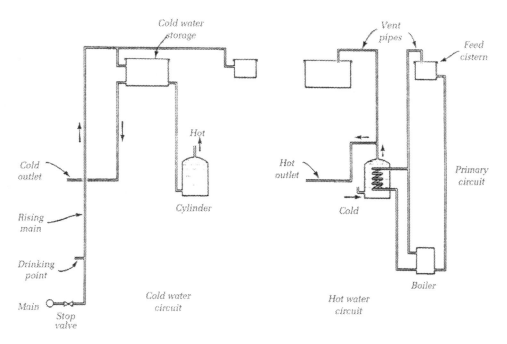

Figure 3.1 Conventional water supply installation for a dwelling in the UK — diagrammatic

or;

(ii) supply taken from a suitable storage vessel at roof level.

In the larger buildings additional storage may also be provided at ground or intermediate levels.

Examples of arrangements using the first of these alternatives is given in figures 3.2 and 3.3. In the former the upper nine floors are served directly with mains water by a riser in which the pressure has been boosted pneumatically and the lower five are served without pressure boosting. In the latter mains water is provided from the riser on each floor. Pneumatic boosting, involving a small pressure vessel topped up by an air compressor, provides a means of securing a ready response to any individual demand for water in the building. The vessel is partly filled with filtered air and partly with water and may be designed and controlled to provide a predetermined volume of water, meeting a short sequence of demands, without the main pumps having to operate. Another difference in these two examples lies in the provision for storage to supply the majority of the sanitary appliances. In the first storage is provided at roof level; in the second

local requirements have led to the provision of storage for each dwelling.

An example of an arrangement using the second alternative is given in figure 3.4. Here a limited amount of storage capacity is provided in the main *en route* to the primary storage cistern. The latter serves the majority of the sanitary appliances but the water intended for drinking in the upper floors is supplied by gravity flow from the smaller storage vessel. The storage at the upper level is fed through conventional centrifugal pumping sets.

In substantially higher buildings it is necessary to divide the building into zones, each perhaps 30 m high and served by its own water cisterns, with drinking water separate from the more general cold water supply. The drinking water cisterns have sealed covers, and warning and vent pipes fitted with filters. The main water reserve is stored at ground level. Table 3.1 exemplifies recommended storage volumes.

Enquiries some years ago suggested that perhaps half of all installations for multi-storey flats in England and Wales use a scheme with a pneumatically boosted riser, of which the riser serves all cold water outlets in some cases but serves only the drinking water outlets in the majority whilst also feeding a high-level

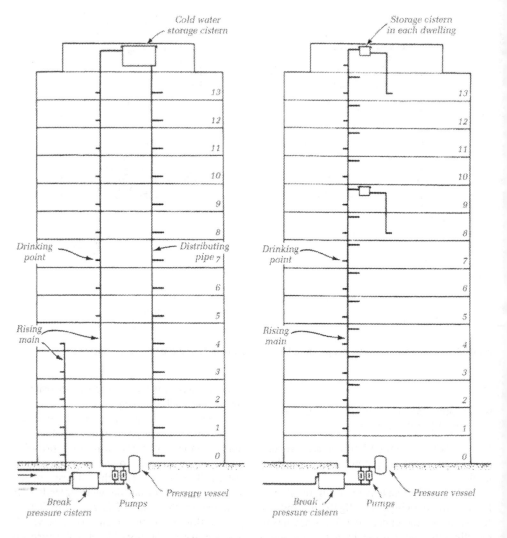

Figure 3.2 Drinking water direct from rising main, with boosting, and roof storage for general use — diagrammatic

Figure 3.3 Drinking water direct from rising main, with boosting, and storage in individual dwellings — diagrammatic

storage cistern for general supply. All the remaining installations follow the general pattern of figure 3.4 although with local variations. Storage is commonly provided in pressed-steel section or galvanized mild steel cisterns, with asbestos cement or fibre glass as an occasional alternative. Covers, overflows and venting arrangements vary considerably, from the loose lid to a well-sealed cover with trapped overflow and filtered-air inlet. Control of the level is effected by means of a ballvalve in most installations, sometimes linked to a float switch, probe or pressure switch.

Water storage within buildings has several disadvantages which might be summarized as follows. Contamination is one possibility, especially if the cistern provided is not well covered, and this is a matter for careful consideration since stored water may be used for drinking and might be drawn back into the mains through the ballvalve in the event of a suction occurring in the riser. Again stored water may become warm. Storage cisterns, moreover, occupy space, although not necessarily useful space, and may involve more, and larger-diameter, pipework and thus greater

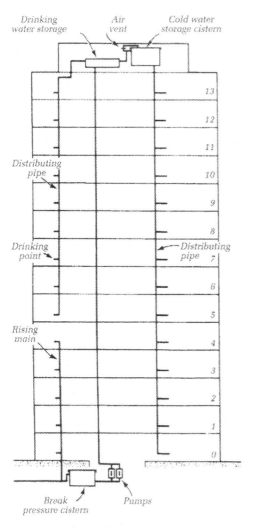

Figure 3.4 Drinking water from subsidiary storage vessel, and general supply from main storage vessel — diagrammatic

Table 3.1 Examples of recommended storage based on BS 6700 for cold water and *Plumbing Design Guide* (Institute of Plumbing) for hot water

Building	Storage (1)	
Smaller houses	100–150 for cold water only (115 minimum) 200–300 for cistern supplying hot and cold water (230 minimum)	
Larger houses	100 per bedroom	
	Cold	**Hot**
Hotel	200 per bed space	45 per bed space
Hostel	90 per bed space	32 per bed space
Offices		
with canteen	45 per employee	4.5 per employee
without canteen	40 per employee	4.0 per employee
Day school		
nursery/primary	15 per pupil	4.5 per pupil
secondary/ technical	20 per pupil	5.0 per pupil
Restaurant	7 per meal	3.5 per meal
	Minimum allowances are given	

Notes: 1. For houses, minima given are based on former bylaws. Cold water values are to allow for 24 hour interruption to supply
2. For houses minimum hot water storage in BS 6700 is given as 35–45 per occupant, but 100 for systems heated by solid fuel and 200 for systems heated by off-peak electricity

costs than where all cold water outlets are mains fed — although this depends on local requirements and the general design of the building. Storage at high levels in multi-storey buildings, in any case, has design and cost implications in view of the bulk and weight of water commonly to be stored.

The potential hazard from domestic stored water was investigated by a survey (DOE 1974) which gave some reassurance in this respect. In this work the type of property and detail relating to the water service installation were recorded for some 1300 properties throughout the country and samples of water were taken from storage cisterns and from taps fed from the mains. Chemical and bacteriological tests were carried out on the samples. Generally it was concluded that where a storage cistern is adequately covered and only receives water supplied by the water authority, stored water would not be harmful to health.

Whilst the storage cistern has merit from a back-siphonage standpoint in separating outlets from the mains, protection in conventional UK domestic installations also derives from the use of an air gap at individual appliances. A gap is provided, for example, with wash-basins, baths and sinks where taps generally are fixed with outlets 12 mm or more above the overflow rim of the appliance. In cisterns the ballvalve is intended to discharge at a distance above the overflow pipe not less than the diameter of the pipe, although this is not always achieved in flushing cisterns. The common use of silencing pipes with ballvalves eliminates the air gap, and the small anti-siphonage hole often provided is not an adequate substitute when suctions are more than nominal. More effective backflow preventers can be devised and this general subject is discussed later.

Hot water supply installations divide into central and local, with the former being particularly used where a compact installation is possible and where substantial hot water demands are to be met. Local water heaters, served by the cold water supply system, are popular where points of use are well separated and where demand is more intermittent. Choice of system is often difficult since a proper comparison needs to involve not only the costs of plant and pipework and related building costs but also the differing costs of fuel for the two types of installations and heat losses in the various pipe circuits and equipment. Enquiries suggested that in multi-storey flats in the UK about one-quarter use a central hot water supply, sometimes via calorifiers in the flats; perhaps one-third use individual hot water cylinders with immersion heaters, whilst combination tanks and instantaneous gas heaters each serve about one-fifth.

British Standards have in the past generally laid down dimensions for all classes of appliances, fittings and piping. The use of performance standards, in which the standard sets out the performance to be met by the component instead of or as well as giving a detailed dimensional specification for that component, offers scope for new developments in this general field. Some use is made of this method in Europe and a start was made in the UK with the drafting of a performance element into standards for taps, namely, BS 5412 and 5413. The gradual introduction of performance standards should help towards the harmonization of standards in EU countries.

Practice abroad

In contrast to the UK, cold water is not stored in buildings in many parts of the world, including the USA and continental Europe. Storage cisterns are commonly regarded as liable to contamination, wasteful of space, adding to the cost of construction and not justified as a reserve of water. In very tall buildings covered and vented cisterns may be provided as sources at intermediate levels in the building for the zone immediately below them — perhaps 15 or 20 storeys — but otherwise they are little used. The emphasis is on providing to a building used for human occupancy a supply of cold water that is fit for drinking and is also used for food processing, bathing, WC flushing and so on. Water for drinking is not distinguished from water for uses such as these. Thus it is usual to find all cold and hot supply points fed directly from the mains, boosted as necessary; and such requirements as backflow protection, safety and noise control are treated through design of the system and appliances and the use of devices such as vacuum breakers and pressure relief valves. The piping systems resulting from this approach are generally simpler than those found in conventional systems in the UK but, as against this, the necessary devices for pressure relief and vacuum control offset some of the saving in cost that may be possible by omitting storage, whilst storage vessels for hot water may have to withstand higher pressures. With both hot and cold outlets commonly fed by supplies at similar (and considerable) pressures it has been possible to devise and market various compact mixing and blending taps for supplying the sanitary appliances.

Practice varies from country to country but typically in Europe hot water supply systems on the indirect principle have the secondary circuits under mains pressure and a suitable arrangement for pressure relief. The primary circuit may be served by an expansion tank, sometimes hand filled, or by diaphragm-type expansion vessels. Single and multipoint gas or electric instantaneous water heaters, mains fed, are common.

With all-mains-fed installations, backflow protection is commonly ensured by providing an air gap at the point of use, i.e. between each water supply outlet and the flood rim of the sanitary appliance, but also by the judicious use of non-return valves and vacuum breakers, and by building in upstands at various locations in the pipework. Again practice varies but non-return valves may be required on the service main outside the dwelling or nearer the cold water inlets to calorifiers and hot water cylinders. Vacuum breakers may be required at the highest point of a rising main to admit air in the event of a fall in pressure

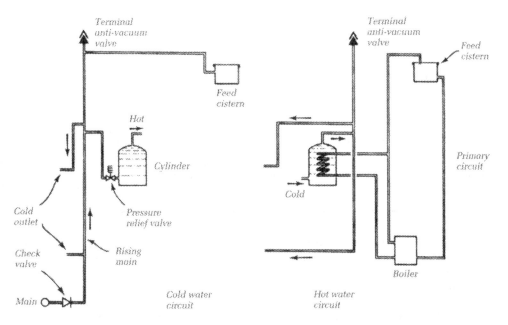

Figure 3.5 Principle of foreign mains-fed cold and hot water supply using hot water cylinder with secondary circuit

below atmospheric in the riser. Vacuum breakers may also be required at individual appliances, e.g. on the discharge side of a ballvalve serving a flushing tank, on the discharge side of a flushing valve, and with hoses and hand showers.

American practice is in many respects similar to that in continental Europe, with the avoidance of storage of cold water and with mains-fed cold and hot water in general use. Pressure boosting for taller buildings on the

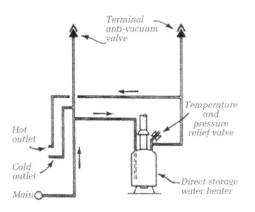

Figure 3.6 Principle of foreign mains-fed cold and hot water supply using direct automatic storage water heater

hydropneumatic system is common. Newer developments include the use of an on-line pump with variable speed drive or with constant speed drive and pressure-regulating valves, operating automatically by pressure signals received from the water circuit. Backflow precautions are based on the use of the air gap where possible, combined with suitable non-return valves and vacuum breakers.

As illustrations of some types of practice used abroad figures 3.5 and 3.6 have been prepared. Figure 3.5 shows the cold water supply mains fed with a hot water cylinder, also mains fed, and with the primary circuit topped up by a feed cistern. Figure 3.6 shows a direct automatic storage water heater mains fed on the lines of some American systems, also used to some extent in Europe.

Preservation of water quality

The main factors that can influence water quality in plumbing systems are summarized in table 3.2. Chapter 2 describes some of the principal forms of deterioration associated with such factors, both aesthetic and as regards potability. Decisions made in design and installation, in the specification and selection of materials, in commissioning and in the

Table 3.2   Factors which affect water quality in plumbing systems

Factor	Comment
Cross-connection	Connection between potable and non-potable water: for example, water not supplied by the undertaker or already drawn for use
Backflow	Flow in the opposite direction to that intended; owing to back pressure or back-siphonage — water sucked back into a potable supply pipe against the normal direction of flow
Materials of pipes, fittings and jointing	Soluble components of materials may be leached out by the water; particles may be picked up by the water; materials may support the growth of micro-organisms
Stagnation	Vessels, pipes and crevices in fittings or equipment where water is held for a substantial time and may be warmed: often significant for the leaching of materials and growth of micro-organisms

maintenance and operation of systems may all be of practical significance. The Water Supply (Water Fittings) Regulations 1999 and the Water Byelaws 2000, Scotland and the associated Guide form an essential reference, complemented by the Water Fittings and Materials Directory published by the Water Regulations Advisory Scheme. Relevant British Standards include BS 6700 and BS 6920, along with many other Standards on various aspects of pipework and fittings (and see Appendix 1).

The situations into which pipes and fittings are to be placed are important. Materials may deteriorate under certain conditions and water quality may be affected. It is, therefore, necessary to exclude water piping from such places as manholes, drains and sewers, and ashpits, and to avoid the possible contact of plastic pipes with oil and petrol and with soils contaminated by petroleum materials and phenols, e.g. at a gas works site. The National Joint Utilities Group provides guidance on the positioning of main services in relation to each other, e.g. NJUG (1986) and in effect recommends a distance of not less than 350 mm between a water service and gas piping. An example of the importance of correct commissioning is given in chapter 2.

The use of lead for pipes and cisterns has been prohibited (see chapter 2). Lead-based solder used in capillary fittings for copper pipework has also been found to cause lead concentrations above the level specified, and lead-free solder is now recommended. Guidance on connections to existing lead piping is given in BS 6700, which draws attention to the importance of avoiding corrosion by galvanic action. Lining pipes and cisterns with coal tar is also prohibited.

The occurrence of legionnaires' disease in a variety of buildings has focused attention on the possible growth of micro-organisms in plumbing systems. As a result, design, operation and maintenance have been re-examined in recent years and this has suggested various measures to combat the problem. The 1999 Regulations call for every system to be tested, flushed and, where necessary, disinfected before use. To summarize the main points of design and operation:

(i) Size cisterns and tanks to ensure adequate turnover of water; cover to avoid contamination; inspect and clean systems, storage vessels and filters regularly; disinfection should be in accordance with BS 6700.

(ii) Store and distribute cold water at below 20 °C as far as practicable.

(iii) Design to ensure uniform temperatures throughout the hot water storage vessel. Ensure sufficient heater capacity to heat water in the calorifier and system to 70 °C. Store hot water with a thermostat setting at not less than 60 °C, and distribute it at not less than 55 °C. It has been recommended that hot water coming out of taps should be between 55 and 60 °C at the furthest point from the calorifier where *Legionella pneumophila* has been identified and calorifiers have been cleansed. However, the temperature of water in domestic pipe systems should not exceed 60 °C, or about 43 °C in some hospital ward areas, to avoid hazard from scalding. At 50 °C, the risk is likely to be negligible for most people.

(iv) Design and construct to avoid local warming of cold water and dead ends of static water.

For the latest information on these matters, advice should be sought from the Department of Health and Health and Safety Executive, whilst design guides issued by the Chartered Institution of Building Services Engineers and Institute of Plumbing are also helpful. The former's advice on legionnaires' disease (CIBSE 1991) forms the basis for a credit award in the environmental assessment of a building (chapter 7).

Cross-connection and backflow

As more complicated and larger water installations came to be built in the early part of the century, new problems arose. Avoiding cross-connections between potable and non-potable water, and avoiding backflow of possibly contaminated water into potable supply, due either to 'back-siphonage' or 'back pressure', assumed greater importance. The risk of backflow received great publicity as a result of incidents during the Chicago World Fair of 1933. Two hotels were involved, and the outcome was over 1400 cases of dysentery and nearly 100 deaths (NIH 1936, DOE 1974). These and other less serious incidents in various countries led to the gradual evolution of protective measures, especially against back-siphonage, written into regulations, codes and standards. Attention was given to the hazardous situations that can arise in industrial, commercial and agricultural premises, in hospitals and laboratories, and in a variety of other locations including ship–shore connections, sewer-cleaning machines and so on. Domestic premises were also dealt with.

Back-siphonage is said to occur when liquid is sucked back into a supply line against the normal direction of flow. The nature of back-siphonage risk is such as to suggest probability analysis using techniques on the lines of chapter 1. The approach would be to calculate the probability of its occurrence on the basis of assumptions about the frequency of the various events necessary to take place simultaneously. For example, the likelihood of a suction arising in a supply main due to pumping by a fire engine in the event of a fire nearby; the likelihood of 'soft' pumping being used, in which the suction hose collapses once pressure falls below atmospheric; the likelihood of some appliance served by a riser being flooded with dirty water into which a hose connection dips — such are the kinds of events that would merit consideration in any risk analysis. Simple calculations of this kind suggest that the probabilities of back-siphonage occurring are small in typical circumstances. There is uncertainty also about the magnitude as well as the frequency of occurrence of suctions in supply installations. Whilst the likely major causes — fire pumping or mains fracture, for example — are known, quantitative data are sparse and hence the performance requirements are not readily specified. Nevertheless some cases have been reported (DOE 1974), particularly in industrial installations but also in domestic situations. The protective measures necessary thus form one of the key issues in the introduction of new types of water installation and in reviewing existing methods.

The classification of risk is considered in chapter 2. The differing classes of risk merit differing protection measures. Some measures are inherently more reliable than others. Thus an arrangement that provides a permanent and substantial air gap (figure 3.7(a)) between a water outlet and the upper rim of a receiving appliance is likely to be more reliable in the long term than a mechanical device incorporated in the supply line and intended to admit air in the event of a suction in the pipe. The latter may be affected by corrosion and scaling and become ineffective. On the other hand the more reliable arrangements may have certain disadvantages, e.g. a ballvalve discharging into a cistern without a silence pipe is noisy. Again, highly reliable measures applied indiscriminately may have substantial cost implications.

It is necessary to define and classify the various types of protective measures available as a basis for considering their application. In the study (DOE 1974) and subsequently in the UK, air gaps were distinguished (figure 3.7) as type 'a', with unobstructed gaps of height between about 1.5 and 2 times the supply pipe diameter (see table 3.3), and type 'b', in which the gap may be less and is not easily measurable. The former recommendation was based on studies of the distance which water could jump into fittings under substantial

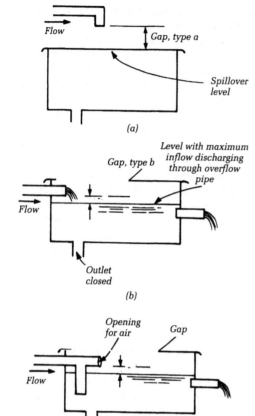

Figure 3.7   Types of air gap

Table 3.3   Relationship between pipe diameter and air gap

Bore of inlet or feed pipe (mm)	Minimum height of type 'a' air gap (mm)
≤14	20
15–21	25
22–41	70
>41	2 × bore

Note: Free flow from the water outlet to be not more than 15° from vertical

suction. In continental Europe there has been an attempt to specify more clearly what in the UK was described as type 'b'. This has given rise, for instance, to the categories exemplified in figure 3.7, i.e. type 'a' plus at least two further categories. This trend is also apparent with the various forms of mechanical device and fitting used for backflow protection. A greater number of such items have been defined in Europe and America, mainly as variants of certain basic devices. The latter include the pipe interrupter, check (non-return) valve, and vacuum breaker (anti-vacuum valve) both in-line and terminal. Definitions of these basic devices are given later.

Protective measures against backflow — the 1999 Regulations

Backflow — described as flow in a direction contrary to the intended normal direction of flow, within or from a water fitting — is dealt with comprehensively in the Regulations, Schedule 2, Section 6. Every water system must contain an adequate means for preventing backflow, with the exception, in paragraph 15.2 of:

(a)   a water heater where the expanded water is permitted to flow back into a supply pipe, and

(b)   a vented storage vessel supplied from a cistern where the temperature of the water in the supply pipe or cistern does not exceed 25 °C.

The basic principle for protection has been long established and is carried through into the Regulations. It is to apply an appropriate device at the point where water is to be used. This goes together with proper design and layout of the installation, along with any additional devices considered necessary. The main categories concerned are non-mechanical methods, mechanical methods, and the use of vertical separation through design of the system.

Non-mechanical devices

Air gaps are taken as a vital line of defence, to give the highest measure of protection against the worst fluid categories, table 2.2. Gaps are required, generally, to be not less than 20 mm or twice the internal diameter of the inlet pipe, whichever is the greater, and with water discharging at not more than 15° from the vertical c.f. table 3.3. for earlier requirements.

An air gap is between the lowest level of water discharge and a so-called 'critical level' in the receiving vessel, meaning the level not less than 2 seconds after closing the water inlet, starting from maximum water level. The latter means the highest level reached in any part of the receptacle when operated continuously under fault conditions.

The Regulations define seven types of gap illustrated in figure 3.8, which gives the worst fluid category appropriate to each, varying from 3 to 5. Types AA and AB assume unrestricted discharge from the cistern. The gap in type AC is between the lowest point of the air inlet pipe to the critical water level. Type AG is to satisfy the requirements

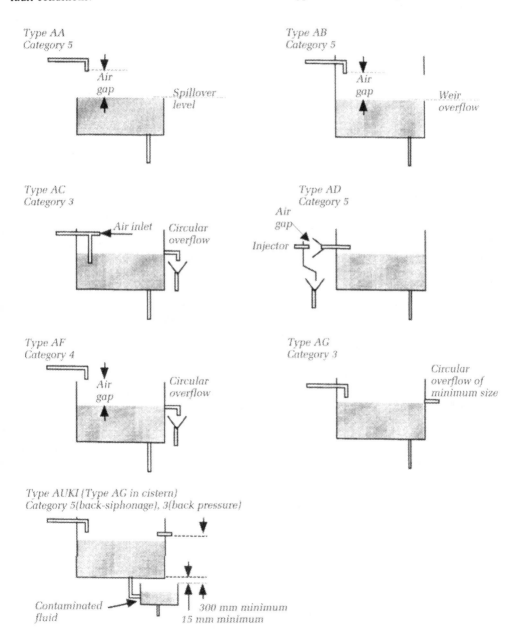

Figure 3.8 Diagrams to illustrate backflow protection through air gaps, based on the Water Supply (Water Fittings) Regulations, 1999. (Note: 'Category' refers to table 2.2.)

of BS 6281: Part 2: Specification for Type B air gaps. Type AUK 1 is exemplified by a WC suite. Overflows and warning pipes must discharge through or terminate with an air gap of Type AA. All these devices are suitable both for back pressure and back-siphonage protection.

Two further types of gap — tap gaps — are described. AUK 2 refers to gaps for taps or combination fittings discharging over domestic appliances such as washbasins, bidets, baths and shower trays with dimensions equivalent to those in table 3.3 but the maximum requirement being 70 mm. The worst fluid category is 3 with Type AUK 2 for back-siphonage protection; this type is not suitable for back pressure. Type AUK 3 relates to discharge over appliances where a category 4 or 5 fluid is present, for example, a domestic sink, some appliances in health care premises. The gap then should be not less than 20 mm or twice the diameter of the inlet pipe, whichever is the greater. Again, the device is suitable for back-siphonage but not back pressure. A pipe interrupter — type DC (figure 3.12) — which has no moving parts and contains a permanent opening through which air can enter, also comes under the category of non-mechanical devices. The water flow should be vertically downwards with no tap or valve downstream. It should be fitted not less than 300 mm above the spillover level of a WC pan or urinal and not less than 150 mm above the sparge pipe outlet of a urinal (and see figure 3.15, discussed under Application). It is suitable for back-siphonage protection against fluid category 5 but not suitable against back pressure.

Mechanical devices

Basic devices are defined in the Regulations. Used singly or in various combinations they provide fourteen different arrangements for backflow protection against fluid categories varying from 2 to 4. Table 3.4 provides a starting point for their consideration, describing the devices as simply as possible and giving their suitability against worst fluid categories. Some are illustrated in figures 3.9 to 3.12.

Type BA backflow preventers are commonly known as RPZ valves because they incorporate a 'reduced pressure zone'. They have a long history of use in America and for some 10 years in Europe but were not permitted under UK water byelaws. Trials and this wider experience have led to their inclusion in the 1999 Regulations, a significant advance for this country. They are intended for commercial and industrial applications where protection is required for fluid category 4, and not for normal domestic applications.

An RPZ valve (figures 3.9 and 3.10) consists of two check valves in series with an air space between them, this space having a relief valve that can open to atmosphere. The relief valve is normally closed, but can open under conditions of pressure reversal, and where a check valve fails to seal, in order to discharge potentially contaminated liquid. Installation should be in a horizontal pipeline with the relief valve outlet lower than the check valves, enabling the relief valve to discharge by gravity. Isolating valves should be provided upstream and downstream, together with an in-line strainer upstream to reduce the risk of particles fouling the check valves. The assembly should be protected against frost and be secure, either in a plant room or within a cage or cover. Further information including testing requirements is available in a note issued by the Water Regulations Advisory Scheme (2000) and in a paper by Grace (2000).

Details relating to use of the various devices include the following:

BA, CA : A type AA air gap should be provided between the relief outlet and the top of the allied tundish.

DA, DB, DUK1 : Such devices need to be able to drain down and open after use and hence there should be no control valve downstream.

DA, DB, DUK1 : To provide an additional measure of protection these should be installed at a vertical distance — not less than 300 mm — above spillover level of the sanitary appliance concerned. This important consideration is discussed in the following section.

Upstands

An 'upstand' in pipework, used abroad but not a general requirement of the UK water byelaws in the past, is a useful additional safeguard and features in the 1999 Regulations. It is illustrated in figure 3.13 and provides a vertical separation between the appliance and the connection to the supply pipe. If a backflow-preventing device at

Table 3.4 Mechanical devices for backflow protection

Description	Type	With back pressure	With back siphonage
An arrangement of water fittings with three pressure zones having differential obturators — means for closing off the waterway — that operate when possible backflow conditions arise; commonly termed RPZ valve assembly, e.g. figures 3.9 and 3.10.	BA	4	4
A device which provides disconnection by venting an intermediate zone between check valves to the atmosphere when the difference in pressure between that zone and upstream is 10% or less of the upstream pressure.	CA	3	3
Anti-vacuum valve (vacuum breaker) — a device through which water does not flow and with apertures for air admission that are closed when the device is under positive water pressure and open when there is a suction in the pipe. Sometimes termed 'terminal type', e.g. figure 3.11.	DA	X	3
As DA but with provision for water to flow through the device, outlet pressure then normally being greater than atmospheric. Termed 'pressurized air inlet valve' in the Regulations. Sometimes termed 'in-line type' of vacuum breaker.	LA	X	2
Check valve — a device that allows water to flow in one direction only, cf figure 3.10 showing check valves.			
Verifiable single valve	EA	2	2
Non-verifiable single valve	EB	2	2
Verifiable double valve (two valves in series)	EC	3	3
Non-verifiable double valve	ED	3	3
Verifiable implies test ports or some other test means.			
Pipe interrupter with moving parts, having an air inlet closed in normal use but which opens to admit air when there is a suction in the pipe; installed so that waterflow is vertically downwards, e.g. figure 3.12	DB	X	4
Hose union backflow preventer — a device fitted to a hose union tap, consisting of a single check valve with air inlets that open when the waterflow stops.	HA	2	3
Hose union tap with a double check valve incorporated in it.	HUK1	3	3
A diverter with automatic return — a device used in bath/shower tap assemblies which automatically leaves the bath outlet open to atmosphere if a suction occurs at the inlet to the device.	HC	X	3
Combinations			
Type LA vacuum breaker with check valve EA downstream	LB	2	3
Type DA vacuum breaker with check valve EA upstream	DUK1	2	3

the point of use fails, an upstand makes it less likely that contaminated water in an appliance will be drawn back as far as the supply pipe. In the Regulations, two types are defined:

Type A — a pipe, upward flowing, surmounted by a Type DA vacuum breaker or a Type DUK1 combination, any part of the outlet of which is not less than 300 mm above the spillover level of an appliance. cf figure 3.13(b).

Type B — a branch pipe serving an appliance, where the height of any part of the branch connection to the vented distributing pipe is not less than 300 mm above :

(a) the spillover level of the appliance; or
(b) the highest possible discharge point served by the distributing pipe whichever is the highest, cf figure 3.13 (a).

Figure 3.9   Reduced pressure zone (RPZ) valve (source: Arrow Valves Ltd., Tring)

The basis for the requirement of 300 mm minimum height difference may be approached as follows. If a suction at S (figure 3.14) is imposed suddenly, e.g. at the base of a riser, an air relief valve opening at A must have sufficient airway to prevent the level at B from rising to the weir W if the contaminated water is to be prevented from passing into the riser. Calculations based on the principles discussed in chapter 8 indicate that a difference in level of 300 mm is adequate for the airways in typical vacuum breakers. Further information is given there, and testing and proving requirements are also discussed. The application of a suction of 0.8 bar in such a way as to ensure that the suction does not fall below 0.5 bar for at least 5 seconds has been suggested (originally DOE 1974) for test purposes. The figure of 0.5 bar has some theoretical basis as outlined in chapter 8 — it is a requirement hardly likely to be exceeded under any practical conditions that can be imagined.

Tests have also been specified for the capacity, strength, tightness and durability of the various backflow prevention devices (see BS 6280–6282).

Application of measures for back-siphonage protection

Examples of the application of protective measures are given in table 3.5 for domestic and non-domestic situations. With WCs, the fluid in the pan is category 5 and measures have to correspond to that risk, both for cistern-fed and valve-fed pans, whether in domestic or non-domestic situations. Flushing valves for WCs and urinals, long used abroad, are increasingly applied in the UK for non-domestic situations. The Regulations distinguish between valves served by a dedicated cistern (with type AG airgap) and those served from a pipe system. With the latter a permanent air vent must be incorporated within the

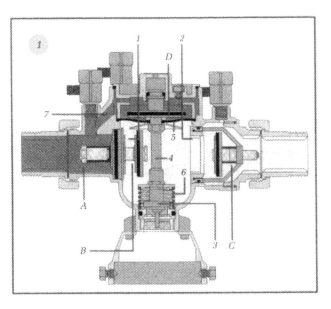

Figure 3.10 Operation of an RPZ valve (source: Arrow Valves Ltd., Tring)

Key

1, 2 Check valves	A. Upstream zone
3. Relief valve	B. Intermediate zone (reduced pressure zone)
4. Spindle	C. Downstream zone
5. Diaphragm	D. Washer plate
6. Spring	
7. Waterway	

Figure 3.10.1 Initial connection: Check valves 1 and 2 closed, relief valve 3 open. Pressure at A and above diaphragm 5 increases to 0.3 bar and closes valve 3. With pressure reaching 0.5 bar, valve 1 opens, water enters zone B. At 0.6 bar, valve 2 opens, normal flow occurs

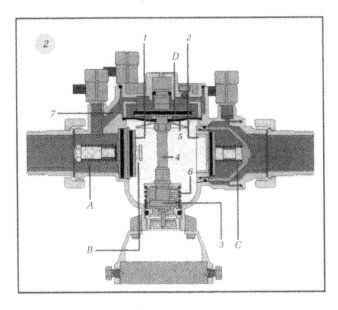

Figure 3.10.2 No flow: All valves closed, system under pressure. If pressure at A falls, relief valve 3 can open

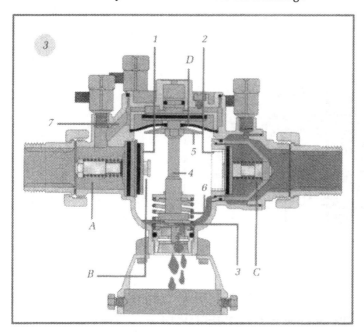

Figure 3.10.3 Failsafe: If pressure at C rises, valve 2 prevents backflow. BUT — if valve 2 fails to seal, relief valve 3 can open before pressure in B reaches pressure in A (with 0.14 bar margin)

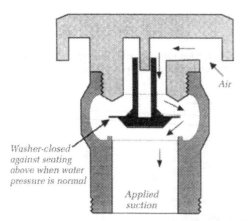

Figure 3.11 Diagram to illustrate the principle of anti-vacuum valve — shown in backflow position, with applied suction at valve base

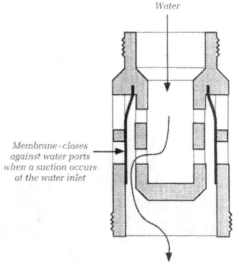

Figure 3.12 Diagram to illustrate the principle of pipe interrupter — shown in normal flow position

valve or provided by means of a pipe interrupter, type DC, immediately on the outlet side of the valve. Requirements for the vertical separation between a valve and the appliance with both types of installation are given in figure 3.15 The basis for a figure of 300 mm is discussed above under 'Upstands'. With basins and baths, it is assumed that a higher level of protection may be required in the non-domestic situation, for example, in hospital or other health care premises, hence the requirement for a type AUK3 device. Sinks

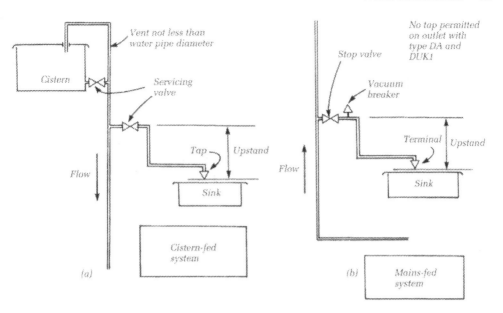

Figure 3.13   Examples of upstands in pipework

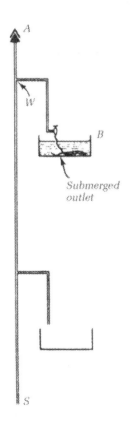

Figure 3.14   Definition sketch

require AUK3 in all situations. The terms over-rim and below rim mean above or below the spillover level of the appliance.

Hose connections, not covered in table 3.5, represent a hazard that is dealt with thoroughly in the Regulations. Amongst the situations to be considered are flexible hose and spray fittings for bidets, hoses fed from a supply pipe for general use, hoses that can be attached to outlets at washbasins, baths or showers, hoses for garden use, and for a variety of commercial and industrial purposes. The concept of a 'zone of risk' is introduced to deal with the possibility that the outlet end of a hose might be immersed in a fluid that represents a much higher risk than experienced in normal use. For example, the outlet end of a hose attached to a tap at a washbasin might be immersed in a nearby appliance such as a WC pan, bidet bowl or squatting type WC, when the protection required would be against a fluid category 5 risk instead of a category 3. To determine this need, a 'zone of risk' should be established as illustrated in figure 3.16, an area covered in a vertical and horizontal plane by a radius equal to the length of the hose. In the illustration, the fluid in the WC pan as well as that in the washbasin is within the zone of risk, that in

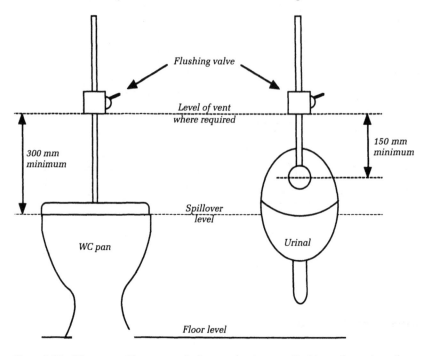

Figure 3.15   Diagram to illustrate vertical separation between flushing valve and appliance

Table 3.5   Examples of measures for back-siphonage protection

Source of risk	Fluid category	Protective measure
Domestic situations		
WC pan with flushing cistern	5 (in WC pan)	AUK1 — Flushing cistern discharged by gravity, with type AG air gap at the supply inlet
Bidet with over-rim water supply	3	AUK2 — Domestic tap gap
Taps at basins or baths		
(a) over-rim	3	AUK2 — Domestic tap gap
(b) below rim	3	EC — Double check valve on cold and hot supply pipes
Taps at sinks	4 or 5	AUK3 — Higher risk tap gap
Clothes and dish washing machines	3	Appropriate measure built in
Non-domestic situations		
WC pan with flushing valve	5	DC — Pipe interrupter with permanent vent
Taps at basins, baths or sinks, over-rim	4 or 5	AUK3 — Higher risk tap gap
Clothes and dish washing machines	4 or 5	Appropriate measure built-in, plus external protection if necessary

the bidet is outside it. The fluid in the WC pan would be the determining factor for the protective measure required in the hose supply pipes.

Further examples are given in table 3.6 of the fluid category that should determine the protection required on the hose supply pipes assuming the hose outlet is immersible in the appliances shown. If a hose outlet is immersible in two or three different appliances, the protection provided should correspond to the highest risk.

Whole site and zone protection is recommended for buildings in multiple occupancy, when additional precautions should be considered by means of 'secondary backflow protection'. This is to reduce the risk of backflow from, for example, one flat to another in a multi-storey block. The procedure incorporates appropriate devices and upstands

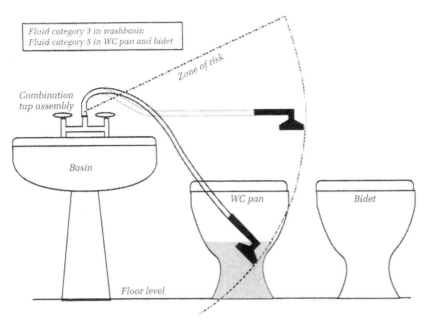

Fluid category 3 in washbasin
Fluid category 5 in WC pan and bidet

Zone of risk

Combination
tap assembly

Basin

WC pan

Bidet

Floor level

Figure 3.16 Diagram to illustrate the zone of risk of backflow via a hose connection

Table 3.6 Basis for the protection of water supplies to a flexible hose

Flexible hose outlet immersible in	Water supplies to the hose to be protected against fluid category
WC pan	5
Bidet	5
Urinal	5
Sink	5
Washbasin	3
Bath	3
Shower tray	3

in the system as a whole. Examples are given in figure 3.17 for both mains-fed and cistern-fed situations, showing the use of double check valve assemblies for this purpose and with upstands of 300 mm minimum. The additional (secondary) devices could be omitted at the lowest level in the block. With a cistern-fed installation, secondary protection might also be provided by means of a vent and with upstands of not less than 300 mm, on the lines of figure 3.13(a). The same principles are recommended for the potentially more hazardous cases such as industry, trade and hospitals. Such situations, of course, generally require the highest class of protective measures.

At first sight it may be supposed that the cost of such measures would be substantial but this does not appear to be the case (DOE 1974). Studies of domestic installations suggested that, for ordinary houses, protection at the point of use should not lead to any overall increase in cost. The cost of whole-installation protection in ordinary houses, if required, might amount to 10 per cent of the cost of the water installation but would represent a much smaller proportion of unit costs with multiple occupancy as in blocks of flats, hotels, institutional buildings and so on. An exercise on a hospital ward suggested that the recommendations might add 1 per cent to the cost of the pipe system.

Temperature and pressure relief in hot water services

Whereas conventional heating installations of the kind illustrated in figure 3.1 rely on vents open to atmosphere for pressure relief, mains-fed installations as, for instance, in figure 3.5 require other methods. The approach is to consider the

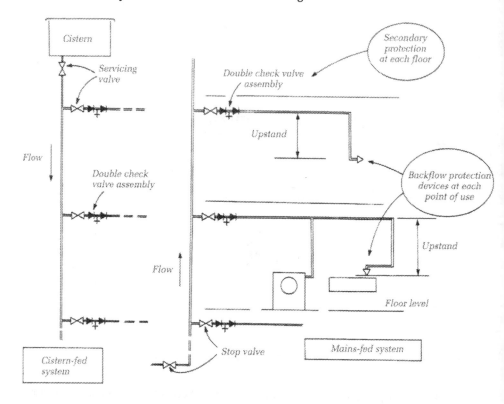

Figure 3.17   Examples of backflow protection in a multi-storey block of flats

control of energy supplied to the system, through thermostatic control, temperature relief and the safe dissipation of heat; and then the safe control of the expansion of the heated water with as little waste of water as possible. Pressure and temperature relief valves and closed expansion vessels are, of course, commonplace in some branches of engineering. They have been used, too, in domestic water installations in continental European and American practice in this field but little in the UK.

One consideration is the possible flow of hot water back into cold water supply mains as a result of a suction in the mains or due to expansion of water in a hot water storage vessel. This need not necessarily be regarded as a hazard, and is not to be regarded in the same light as the backflow of polluted water into supply lines. Whilst a check valve may be used, therefore, to prevent the backflow of hot water, it need not be. In the latter event measures must, nevertheless, be provided to prevent a storage vessel from draining back into the mains when these are empty for some reason, cf figure 3.18. These differences are significant as far as pressure relief is concerned. The demands are more onerous when a check valve is present than when it is not. German and American practices appear to differ in this regard, the latter often preferring not to use a check valve in the supply line to hot water apparatus in order to reduce the risk of explosion.

The foregoing suggests a simple classification for a review of measures for pressure relief:

(i) situations where backflow is permitted but means are provided to prevent a storage vessel from draining down;
(ii) situations where a check valve or similar device is installed to prevent backflow.

Figure 3.18 (a)–(c) illustrates the principles of temperature and pressure relief in these two cases by reference to three simple examples of heating installations, drawing on the results and recommendations of a review (DOE 1976).

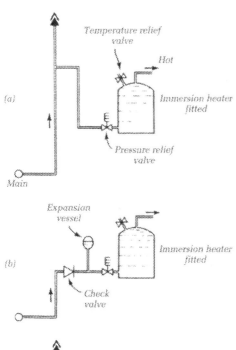

Temperature relief valve

Hot

Immersion heater fitted

(a)

Pressure relief valve

Main

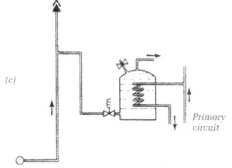

Expansion vessel

Immersion heater fitted

(b)

Check valve

(c)

Primary circuit

Figure 3.18 Diagrams to illustrate principles of temperature and pressure relief

The basic requirement behind these recommendations was to prevent the instantaneous production of steam in the event of a sudden reduction of pressure — due to bursting. Control is effected by devices that operate in sequence as the temperature rises.

In the first two examples, figure 3.18 (a) and (b), a storage cylinder fitted with an electric immersion heater is considered as the source of hot water. It is assumed that the heater itself is fitted with a means for controlling the heat energy or fuel input through a thermostat, in order to limit the hot water temperature to, say, 60 °C. A second line of defence is required,

a non-self-resetting type of temperature-actuated energy cut-off set to operate at some temperature a little below 100 °C, say 90 °C, in case the thermostat fails. The additional measures recommended as further lines of defence are shown. In figure 3.18 (a) no check valve is installed and water is thus free to expand back into the supply line. A simple calculation shows that the quantity involved is small — a matter of a litre or two with an ordinary domestic storage cylinder. In these circumstances a pressure relief valve in the supply line would be necessary, and a temperature-actuated relief valve fixed at the point of highest water temperature and set to prevent the water temperature exceeding, say, 95 °C. In figure 3.18 (b) a check valve is assumed to be installed and in this situation additional protection such as a closed expansion vessel, as shown, would take up the expansion of water on heating. Similar measures would be appropriate if, instead of a check valve, a pressure reducing valve that did not permit backflow was to be used.

The role of the pressure relief valve (expansion valve) fitted close to the storage vessel is important, but it is not to be seen as a first line of defence against expansion. The first line of defence is to provide space for the expanding water either within the supply system or by adding an expansion vessel. The relief valve provides a fail-safe mechanism whereby water can be discharged safely when the pressure in the system rises owing to a malfunction elsewhere. Typically, such a valve should operate when the water pressure rises to 0.5 bar above that to which the storage vessel is normally subject. The aim is to keep any loss of water to a minimum; such discharge can be a useful indicator of malfunction.

In figure 3.18 (c) the situation of figure 3.18 (a) is repeated but heating is by an indirect method, the primary assumed to be topped up by and vented to an open feed cistern. The temperature relief valve is again fitted, as shown, to the indirect cylinder. The primary circuit would have a pressure relief valve. The heat source, if oil or gas, would be provided with a thermostat and an energy cut-off, temperature actuated at just below 100 °C. A solid fuel boiler would require alternative measures, if permitted in these circumstances, and a thermostatically controlled damper with a pressure relief valve would be a possibility.

Many variants of the simple examples in figure 3.18 can, of course, be conceived. With the arrangement in figure 3.18 (a), cistern-fed instead of mains-fed, for example, similar precautions would be required. With the primary circuit of figure 3.18 (c) assumed to be sealed, on the other hand, a closed expansion vessel would be needed to supplement the pressure and temperature relief arrangements outlined above. Multiple installations may be built up from the basic principles outlined.

Hot water services — the Building Regulations and the Water Regulations

Changes in practice are possible in view of changes to Building Regulations in the UK in 1991 and the issue of Water Regulations in 1999. These have come about through experience in the UK and abroad and the kind of study reviewed above. Basic requirements may be summarized as follows.

1. *Building Regulations* specify requirements on water storage in Schedule G3. If hot water is stored and the storage system does not incorporate a vent to the atmosphere, precautions should be adequate to:
   (i) prevent the temperature of the stored water at any time exceeding 100 °C; and
   (ii) ensure that the hot water discharged from safety devices is safely conveyed to where it is visible but will cause no danger to persons in or about the building.
   It is required to have a non-self-resetting thermal cutout.
   These requirements do not apply to:
   (i) a system having a storage vessel of 15 litres or less;
   (ii) a space heating system;

(iii) a system which heats or stores water for the purposes of an industrial process.

Systems must be installed by a person competent to do so.

2. *Water Regulations* specify a range of requirements in Schedule 2, Section 8. Both vented and unvented installations are dealt with, and the Water Regulations Guide elaborates on a range of situations. Matters covered include:

   the provision and effective operation of temperature and pressure control devices to limit water temperatures to not more than 100 °C; and see table 3.7; means for accommodating the expansion of hot water; the safe discharge and disposal of hot water from control devices; the avoidance of cross-connection and back-flow hazards; other aspects of the distribution and storage of water, including temperature considerations, as discussed earlier in the chapter.

The foregoing outline of basic principles provides a starting point for considering the application of some of these requirements, especially as they relate to unvented systems.

3. *Applications.* Figures 3.19 and 3.20 illustrate features of a simple system and means for accommodating expansion. A capacity of 4 per cent of the volume of the storage system is normally required to deal with expansion, since water expands by about this amount as temperature rises from 4 °C to 100 °C. In figure 3.19 the term 'restriction' relates to the provision, or not, of a check valve, a type of pressure reducing valve, or stop valve with loose washer plate. Such restriction could prevent backflow as temperature rises. Point A may not be close to the cylinder; it could be some

Table 3.7 Specification for temperature and pressure control devices

BS 6283	*Specification for safety devices:*	Intended for use with mains-fed hot water
	1 Expansion valves, pressure actuated	systems; deal with the design and construction
	2 Temperature relief valves	of devices and their testing for strength,
	3 Combined temperature and pressure relief valves	endurance, pressure, flow, discharge capacity and easing gear operation as appropriate
	4 Drop-tight pressure-reducing valves	

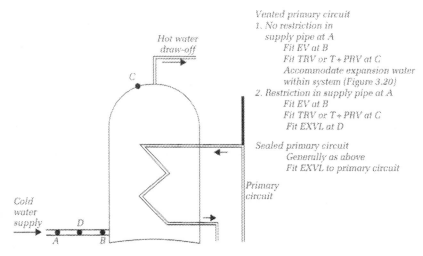

Vented primary circuit
1. No restriction in
   supply pipe at A
   Fit EV at B
   Fit TRV or T+PRV at C
   Accommodate expansion water
   within system (Figure 3.20)
2. Restriction in supply pipe at A
   Fit EV at B
   Fit TRV or T+PRV at C
   Fit EXVL at D

Sealed primary circuit
   Generally as above
   Fit EXVL to primary circuit

Figure 3.19 Diagram to illustrate features of unvented hot water storage system requirements. Key: EV expansion valve TRV temperature relief valve T+PRV temperature and pressure relief valve EXVL expansion vessel

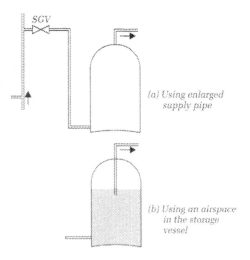

(a) Using enlarged supply pipe

(b) Using an airspace in the storage vessel

Figure 3.20 Diagram to illustrate means for accommodating expansion water
Key: SGV servicing valve (with fixed washer plate)

distance away. Point C is near the top of the vessel, within the top 20 per cent of volume of the water. The measures shown in figure 3.20 may be appropriate for simple installations. A reservation especially regarding figure 3.20 (a) is that a change upstream in the system at some future date, for example, by a later fitting of a valve offering a restriction, could prevent the reversal of flow during expansion. Similarly

there could be a problem for installations in older properties where the stopvalve at entry to the premises might have a loose washer plate. A more secure method is the provision of a separate expansion vessel that incorporates a bag membrane; water entering the bag expands against a cushion of air or inert gas. This method is preferred, in any case, where the cold water supply line incorporates a restriction (figure 3.19). As noted earlier, the expansion valve referred to in figure 3.19 is to be seen as a fail-safe device.

The Water Regulations Guide refers to a possible hazard with expansion vessels that have a single entry for water. This is that bacteria may grow in an area where there is no through-flow of water, as in a dead leg of pipe (cf table 3.2). Equipment in which the water is continually changed, as in a 'flow-through' type of vessel, is, therefore, to be preferred.

Recommendations in the Approved Document of June 1992, related to the Building Regulations, distinguish between systems of up to 500 litres capacity and 45 kW power input, and systems of greater capacity or power input. The latter (section 4) are likely to be individual designs and not appropriate for approval through the European Organization for Technical Approvals (EOTA), which has the British

Board of Agrément (BBA) as a member, or the National Accreditation Council for Certification Bodies (NACCB), as discussed below. They should, however, be designed to the same safety requirements by a qualified engineer.

Systems up to 500 litres and 45 kW are recommended to be factory and not site assembled, in order to reduce the risk of safety devices being omitted. They should be in the form of a proprietary unit or package (BS 7206) subject to approval by EOTA or NACCB procedures, or an equally authoritative independent procedure. With a unit, all safety devices described below and also all other devices such as backflow preventers and pressure controls are fitted by the manufacturer. A package includes the safety devices described below, fitted in the factory; other devices are supplied by the manufacturer in a kit to be used by the installer.

To meet requirements a unit or package should have a minimum of two temperature-activated devices operating in sequence, namely a non-self-resetting thermal cut-out (BS 3955 or BS 4201) and a temperature relief valve (BS 6283). These are additional to any thermostatic control fitted to maintain temp-

erature and means to control expansion. The temperature relief valve should be located on the storage vessel so as to limit the temperature of the water to just below 100 °C, and be sized in accordance with BS 6283. With indirect heating, the thermal cut-out should be wired up to a means for shutting off the flow to the primary heater approved as indicated above, e.g. a motorized valve. If a unit incorporates a boiler, the thermal cut-out may be on the boiler. A thermal cut-out would also be needed on any other direct source of heat if provided in an indirect unit or package.

Discharge from safety devices is, subject to specific requirements, via a tundish fitted vertically and in the same space as the storage system. The discharge outside the building should be visible and safely located, preferably to a gully below a fixed grating. The pipe leading from the tundish should be larger than the normal outlet size of the safety device by one or more pipe sizes according to length. A simple means of sizing is given, based on fluid mechanics principles (chapter 8). Finally the importance of regular checks and maintenance of the devices considered in this chapter is to be emphasized.

4 Principles of Soil and Waste Pipe Installations

Water has been supplied and waste carried away through pipes since ancient times (Staniforth 1994). Supply and drainage systems came into more general use early in the nineteenth century, but the precautions to be found in modern design were not taken. The interceptor and the gully had not been introduced into drainage, and waste and soil were usually discharged into one vertical pipe inside the building leading directly to a sewer or cesspool, often without traps at the appliances. This was the original 'one-pipe system' and because the principles of design with traps and vent pipes were not understood and, in addition, the quality of materials used and of construction were often poor, foul smells in buildings were common. An idea of conditions prevalent even towards the end of the 19th century is given in figure 4.1 reproduced from Teale (1881). Eventually it was required in the UK that washbasins and baths should be discharged outside the building into a vertical pipe leading to a gully trap which also served the sinks, and WCs were served by a separate stack also outside the building — the 'two-pipe system'. Besides the traps at the appliances and the gully trap, the interceptor was introduced to provide a third seal between building and sewer.

It also began to be realized that the water seals in the building could be sucked out by discharge through the pipework, and hence anti-siphonage piping was installed in all but the smallest buildings to ensure that a proper seal was retained. The London County Council bylaws made in 1900 required that where more than one WC was connected to a soil pipe each

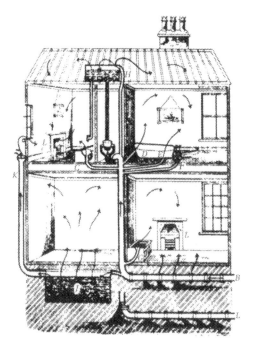

Figure 4.1 House with every sanitary arrangement faulty

WC should be ventilated by a pipe not less than 50 mm diameter fixed between 75 and 300 mm from the trap, and traps of waste appliances were required to be vented if there was a chance of seal loss. An example of what these requirements led to in terms of soil, waste and vent piping is shown in figure 4.2. As a result of these precautions and of improvements in the quality of pipes and methods of construction, unhygienic conditions in buildings became the exception rather than the rule.

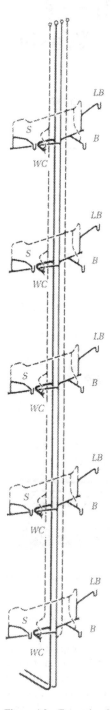

Figure 4.2   Example of a two-pipe system

The development of these installations followed similar lines elsewhere but an extra factor was often important, the climatic conditions. In many countries it is not practicable to have pipes outside because of the severe frosts that occur. Pipework was kept indoors and, to save space, pipes were reduced in number and size as much as possible. Hence the two-pipe system did not find favour and attention was given instead to the design of the one-pipe system, which remained the usual method.

American use of the one-pipe system interested designers in the UK in the early 1930s and it was seen to have several advantages over the two-pipe system. A saving is achieved because one stack is used instead of two while, with fewer pipes, there is an improvement in appearance. In addition, it becomes easier to use an internal system. Its hygienic advantages are elimination of the hopper head and gully method for basins, baths and sinks which, although it ensures a barrier to drain air, tends itself to lead to smells. A further advantage claimed was that carriage of hot as well as cold discharges in the soil stack reduce the chance of deposits accumulating. These considerations led to the reintroduction of the one-pipe system into the UK in 1933. The depth of seal for waste appliances was fixed at 75 mm compared with the traditional 37.5 mm of the two-pipe system. The WC pan could not easily be increased in depth and hence its seal remains at 50 mm. All traps were required to be individually ventilated between 75 and 300 mm from the seal as shown, for instance, in figure 4.3. The one-pipe system gradually came to be accepted for larger buildings, whilst the two-pipe system remained general practice in houses and small blocks of flats, either with hopper head and down pipe or with a complete waste stack, often with full trap ventilation, besides the separate stack for soil.

UK local authorities became interested in the possibilities of simplifying soil and waste pipe design and in 1949 the Building Research Station began an investigation of the problem with the object of providing data for a comprehensive code of practice. Laboratory research and studies in blocks of flats in use showed the possibilities of simplified methods and several publications making recommendations were issued in 1953 and 1954. It was shown that, by the observance of several simple rules in design and installation, vent pipes could be omitted from one-pipe installations and a so-called 'single-stack system' (figure 4.4)

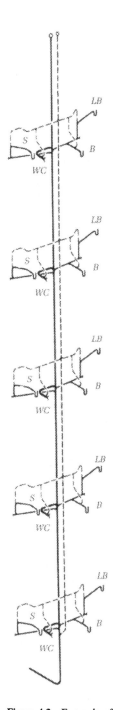

Figure 4.3 Example of a one-pipe system

achieved with a saving in cost up to 40 per cent or thereabouts; for example, see table 4.1 based on this early work and relating to a simple plan for a bathroom and kitchen such as shown in figure 5.4, example 2.

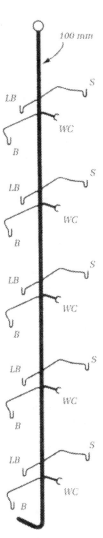

Figure 4.4 Example of a single-stack system for a block of flats

These early recommendations covered the use of single-stack installations up to five storeys and the simplification of one-pipe installations in taller buildings. Since then research has enabled the use of single-stack installations to be extended to application in 30 storey flats and more recently to office buildings. In general the venting systems required in all types of large building have been simplified compared with traditional practice, and rational methods have been developed for their design. In other European countries various systems have been developed, ranging from two-pipe methods to single-stack systems, to suit local requirements and forms of construction. The present chapter is concerned with general principles relating to the design of such installations. Detailed design recommendations follow in the next chapter, and the research background is given in chapter 8. The future may see measures introduced to recycle waste water within buildings and to reclaim heat from wastes, for reasons of economy. Some possibilities are outlined in chapter 11. Where normal gravity drainage is unsuitable, there is the option nowadays of using a vacuum drainage system inside buildings, covered by EN 12109. This method draws on the great experience of such systems in the marine industry, aircraft and railways and is useful for situations where gravity drainage cannot readily be installed and where special considerations apply.

Performance requirements

Schedule H1 of the 2000 Building Regulations requires that any system which carries foul water from appliances within the building to a foul water outfall shall be adequate. 'Foul water' means soil and waste water but not trade effluent; outfall implies a sewer, cesspool, septic or settlement tank. This requirement may be elaborated as follows.

Ensure adequate flow   Requirements as to the discharge rates from appliances should be a primary consideration. Rates of flow vary during discharges, with a maximum or peak flow in the early stage and much reduced rates towards the end of a discharge. To satisfy the user the sizes of outlets, traps and pipework should not be such that the rates of flow are much reduced.

Exclude smells and foul air   Soil and waste pipe installations rely on water-filled traps at the appliances for the exclusion of smells and foul air from buildings. The water seal depth should be chosen taking account of possible loss due to evaporation and pressure variations. With WCs there should be sufficient water for the containment of excreta.

Limit noise   Noise generated by discharges should be limited so as to maintain a satisfactory environment in buildings. The discharge

Table 4.1   Comparison of costs and weights of materials for one-pipe and single-stack installations for a five-storey block of flats

Item	One-pipe system (figure 4.3)	Single-stack system (figure 4.4)	Approximate saving
Cast iron	464	326	138
Brass	40	26	14
Copper	39	19	20
Lead	52	30	22
Labour			50%
Estimated total cost of one installation	100%	56%	44%

The table has a spanning header: "Comparative economy — Approximate weights (kg) and costs"

from appliances is one source of noise, pressure variations causing seal loss are another.

Be and remain leaktight   Leakage of contaminated water and foul air into a building is to be avoided, taking account of the conditions of service including the volumes and temperatures of discharges likely and the possible activities of the occupants.

Be and remain free flowing   The pipework should be designed to minimize the number of blockages so as to maintain the expected service and to prevent the overflow of contaminated water from the appliances into the building.

Be durable   The whole system including materials, joints, supports and fixings should remain serviceable, bearing in mind the expected life of the building.

Be traceable and accessible for maintenance   Pipework should be readily accessible and traceable, with proper access for cleaning and maintenance equipment, and to permit clearing of all parts of the system.

Be replaceable   The piping and fittings should be designed to facilitate replacement bearing in mind the type of building concerned.

Be able to be tested   It should be possible to check that the performance required is attained; for example, access should take account of the need to make water and air tests.

The requirements relating to seal depth and pressure variations in the pipework are now considered in more detail. Traps should be deep enough to resist pressure variations occurring during use and provide a margin for evaporation. Diffusion of foul air through seals has been shown to be unimportant.

(a) Seal loss by evaporation Evaporation from small-bore metal traps in ordinary circumstances in the UK takes place at a rate of about 2 or 3 mm a week, whilst the rate of evaporation from WC seals with a larger exposed water surface may be a little greater than this. In fixing a margin for evaporation it is not reasonable to allow for the highest temperatures and driest conditions possible and for a great length of time. Even seals 75 mm deep will disappear under certain conditions. Bearing in mind the various factors involved, a retention of 25 mm of seal is usually considered satisfactory.

(b) Seal loss due to pressure disturbance The effects of self-siphonage and induced siphonage of trap seals must be considered. The former is not important with baths, large sinks and WCs because any loss due to self-siphonage is replenished by the trailing flow at the end of the main discharge. In any case, self-siphonage does not normally occur with WCs because the soil pipe is too big for full-bore flow to develop (see figure 4.14). Washbasins, because of their shape and small-bore waste pipes, are more subject to self-siphonage and may lose all their seal in extreme circumstances. Controlling factors are the length and slope of the waste, and shape and size of the trap.

Loss due to induced siphonage depends on the magnitude of the suction and on the dimensions of the trap. With a full trap of uniform bore and curvature, the loss is about half of the suction when the drop in pressure is less than or equal to the trap depth, as shown in figure 4.5. With 500 N/m² or 50 mm (water gauge) suction, for instance, the loss is about 25 mm. With WCs, on the other hand, the appliance side of the trap is larger in bore than

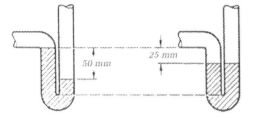

Figure 4.5 Seal loss is about half the suction with a uniform trap

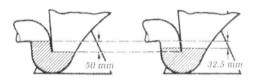

Figure 4.6 Seal loss is more than half the suction with this non-uniform trap

the stack side and the loss is greater than half the suction, as shown in figure 4.6. With a typical washdown pan the loss is approximately 32.5 mm for a 500 N/m² or 50 mm suction. A fall in pressure greater than the trap depth generally gives a much bigger loss, whatever the trap shape, because air is pulled through the seal and pumps out the water.

Two values for permissible pressure variations must be specified. These can be expressed as: the amount of seal to be retained in traps after self-siphonage, and the allowable pressure variations due to induced siphonage and back pressure. They should be chosen on the assumption that a minimum of 25 mm of seal is to be retained to allow for evaporation, as already discussed. The permissible induced suction is limited by the minimum depth of trap in use. The WC trap is normally 50 mm deep and can be subjected to a steady suction of 375 N/m² or 37.5 mm wg and still retain the required minimum seal of 25 mm. With a 75 mm deep, uniform trap subjected to this suction, the seal retention is about 56 mm. The permissible back pressure has also to be considered, and it is reasonable to limit this also to 375 N/m² or 37.5 mm wg. It remains to decide upon the minimum retention of seal required after self-siphonage has occurred. The smallest possible value is required so as to make the scope of the designer as wide as possible. As before, this should be 25 mm of seal to allow for evaporation.

Using the above arguments it is, therefore, possible to state standards of performance relating to air pressure variations as follows:

25 mm minimum seal retention after discharge of individual appliances

± 375 N/m² maximum pressure variation in the system in general.

It may be seen that these standards provide a factor of safety. For example, in a uniform trap with a seal retention of only 25 mm a suction of 500 N/m² (50 mm wg) can still occur without air bubbling through the seal. The above values are used in BS 5572 (see Appendix 1). Some codes may adopt a simple guideline of not less than 25 mm seal retention for both self- and induced siphonage.

Variable discharge

In fixing the seal retention required after self-siphonage, it has been assumed that the same result is obtained after each discharge. The suction, in fact, varies slightly with successive discharges — for example, in 10 test runs with a particular basin–trap–waste pipe arrangement the seal retentions might be 27, 28, 25, 27, 27, 30, 30, 25, 23 and 28 mm. The average is 27 mm and the minimum 23 mm. Should the suggested standard of 25 mm minimum retention refer to an average or to a minimum in 10 runs or to a minimum in 100 or even 1000 tests?

It is helpful to know the probable distribution of values about an average of 25 mm over a period of time, and it can be shown that the following is likely:

two-thirds between 23 and 27 mm
95 per cent between 21 and 29 mm
99 per cent between 19 and 31 mm

Basing design on an average retention of 25 mm from 10 tests means, therefore, that the actual retention frequently will be 23 mm and occasionally only 19 mm. A figure of 25 mm as a minimum in a set of 10 test runs, when only a small percentage of results in practice will be less than 25 mm, is a better standard to adopt and was used as a basis for the methods of design described later.

Simultaneous discharge

The pressure range for the seal having been chosen, it remains to decide what discharge conditions should be selected for determining the maximum pressure or suction that can develop, i.e. whether it should be when all the appliances discharge simultaneously or when some fraction of all the appliances does so. Experiment shows that simultaneous discharge of all the appliances in a system does not necessarily give the worst result and, in any case, the probability of this happening is negligible, as the following example shows. Consider an installation with five WCs connected to one stack. A WC flush takes about 5 seconds, and a study of usage in flats has shown that a WC is likely to be used on average once every 1140 seconds during the morning peak period. Under these conditions the chance of finding any one WC flushing at some particular instant is 5 in 1140, a probability of 0.0044, and the probability of finding all five discharging at any particular instant is 0.0044⁵ or 1.6×10^{-12}. If the peak period of demand lasts for 1 hour each day it can be shown that this event will probably occur once about every 500 000 years! It follows that design on the basis of simultaneous discharge of all appliances would mean allowing for effects that in practice would never occur. A less extravagant method of design is reached by choosing some more likely event to couple with the permissible pressure range. The following standard used in developing simplified systems has been found satisfactory in practice — that the pressure range of ±375 N/m² should not be exceeded for more than 1 per cent of the time. From a knowledge of duration of discharge and frequency of use of appliances the numbers of appliances that should be discharged together in carrying out tests may be calculated. An example of what is required is given in table 4.2. More comprehensive information is given in chapter 1.

Table 4.2   Discharge together to test for morning peak use

Number of each installed	WC	Basin	Sink
1–9	1	1	1
10–24	1	1	2
25–35	1	2	3
36–50	2	2	3

Characteristics of individual discharges: branch flows

The sanitary appliances supply the flow load to the soil and waste pipe systems: knowledge of the characteristics of the individual discharges, types of branch flow, and the ways in which surges of water from the separate appliances change as they flow through the pipework is fundamental to design. In practice, simplifying assumptions are made for design purposes, but some understanding of the basic hydraulics is useful in giving insight into the design data.

Consider first the waste appliances such as baths, basins and sinks. When a tap is discharged continuously into such an appliance, the rate of outflow to the waste pipe equals the rate of inflow, probably between 0.1 and 0.5 l/s, but less than this when a spray tap is used on a washbasin. When the appliances are partly or completely filled and the plug removed, a higher rate of outflow normally occurs. A basin may discharge at between 0.5 and 1 l/s for perhaps 5 to 10 seconds; a bath at 1 to 2 l/s for perhaps 60 seconds or more; a sink at about 1 l/s for perhaps 15 to 25 seconds. These are the ranges of flow rate common in the UK and which are generally regarded as acceptable by users. With each type of appliance the flow trails off to a low rate which lasts for a few seconds although usually a good deal longer than this with a bath.

The size of the waste fitting and its grille, of course, affects the rate of outflow, but the downstream pipework has much less effect on the flow rate than might be expected from conventional hydraulics. The vortex that is commonly formed during the discharge limits the rate of flow by entraining air. With some appliances the overflow connects to the tail pipe of the appliance and this, when open, acts as an air supply channel, maintaining the pressure in the tail pipe at about atmospheric pressure and setting a limit to the rate of discharge. The flow in a small branch from a washbasin illustrated in figure 4.7 is typical of such conditions. In contrast, blocking the overflow eliminates most of the air and

Figure 4.7   Example of flow in a trap and waste pipe from a washbasin (Courtesy of Heriot-Watt University)

discharge is more rapid. The discharge rate then depends on the design of the trap and downstream piping as well as on the size of the waste fitting and grille.

Larger waste pipes — 40 or 50 mm diameter — are less likely to run full and this also has the effect of limiting the rate of discharge by limiting the head that can be developed. The flow from a water closet also is generally independent of the pipework into which it discharges because at the usual rate of discharge the receiving pipe — typically 75 or 100 mm diameter — can only flow about one-quarter to half full. The head causing flow is therefore limited to that which can occur within the appliance itself. With siphonic pans, a suction is developed locally to assist the discharge. In UK practice the total discharge takes between about 5 and 10 seconds. Past American practice used greater quantities of water, sometimes twice that in the UK, and the discharge then takes correspondingly longer. The peak rate of flow also varies with the design of suite and may reach 2 to 3 l/s.

So far we have considered the flow at the outlet of an appliance. Downstream the initial surge is constrained by friction and the layout of the piping, and the flow thus spreads out as it passes through the pipework. For immediate purposes it will suffice to note that the characteristic curve (e.g. figure 8.18) does not change much in the first 5 or 6 m from an appliance. Thus with branches up to several metres long, the flow entering a stack can be taken for practical purposes as having the same characteristic as that leaving the appliance.

For design it is necessary, as already stated, to assign representative values of flow rate and discharge time to the various types of sanitary appliances and some currently accepted values have been listed in chapter 1. They form the basis for subsequent design recommendations.

Self-siphonage of trap seals

If at the end of the discharge of an individual appliance its trap is left empty by self-siphonage, foul air may find a way into the building. Furthermore, a conventional water closet trap should retain water to perform its function of receiving excreta. The design of the system should therefore ensure that the traps remain adequately filled with water on com-

pletion of flow through them. Research has provided an understanding of the self-siphonage of individual appliances and the factors that determine the amount of water remaining as a trap seal after use (Wise 1957b).

The occurrence of a suction and hence of seal loss at the end of a discharge depends on the branch running full somewhere in its length. No suction can occur in a soil branch when there is only a shallow flow and the downstream side of the WC trap is in communication with air substantially at atmospheric pressure in the stack throughout the discharge. Self-siphonage, in fact, occurs with WCs only when the soil pipe is of considerable length and then only when a large wad of material such as newspaper is flushed, momentarily filling the branch.

With full-bore discharge in smaller pipes that serve waste appliances, a form of siphon is set up in which the outlet leg of the trap forms the short leg of the siphon and the branch pipe provides the long leg. A stage in the process of siphonage is shown in figure 4.8. Some water remains in the trap but, owing to inertia, the long plug of water in the branch continues its motion. The resulting suction — also a cause of noise — removes some of the water left in the trap and that still trickling in from the appliance. As this happens the plug slows down and comes to rest in the branch and is then drawn back towards the trap by the suction still existing in the pipe. If the trap is a P-type and the plug has come to rest sufficiently close to the trap, the return flow partly or completely refills it. Refill from the branch pipe is not possible with S-type traps and hence the seal loss is often greater than with a P-type.

With larger waste pipes the flow may not be full bore and self-siphonage does not then occur. Nevertheless, at certain locations, e.g. at a sharp change of direction, or upstream of a junction, the pipe may run full. This is also possible with drains. What is known as a hydraulic jump (figure 4.9) can occur, and, in a waste pipe leading from a trap, the resulting full-bore flow may result in a siphonage action and seal loss.

Self-siphonage may occur with any waste appliance, but the effect is masked with fitments of a very large internal surface area by the water trickling into the trap from the appliance itself and refilling the seal. With

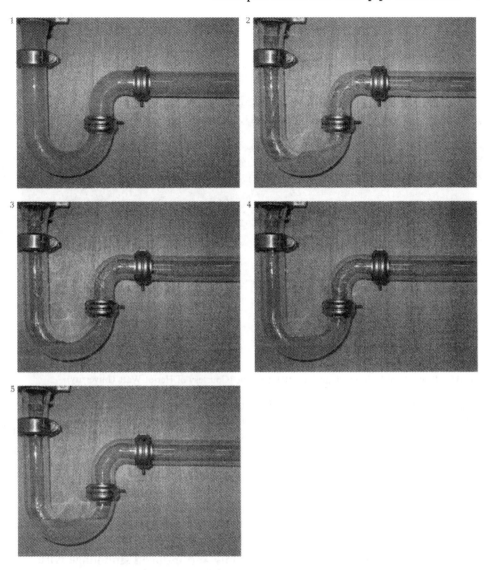

Figure 4.8 Example of self siphonage in a trap and waste pipe leading from a basin. Sequence illustrates trap seal loss due to the momentum of the discharge (Courtesy of Heriot-Watt University)

large sinks, for instance, self-siphonage is important only with very long wastes, perhaps 6 or 7 m long, when the plug movement and hence the suction may continue long enough for a large proportion of the water entering the trap from the appliance to be siphoned out. With washbasins and small sinks the amount of refill from the appliance is small and the possibility of suction is more important.

The research which provided these insights resulted also in methods for predicting seal losses due to self-siphonage in terms of the discharge rate from the appliance, the gradients and dimensions of the pipework and the design of the trap. Chapter 8 summarizes these findings, but for present purposes it is sufficient to note that the length and slope of the branch are of prime importance in determining the effect of self-siphonage with a pipe running full. The plug of water causing siphonage travels further from the trap the longer the pipe and the steeper its slope. This both increases the duration of the suction, and also the noise, and reduces the amount of water

Figure 4.9   Example of hydraulic jump in a 100 mm pipe (Courtesy of Heriot-Watt University)

reaching the trap in the reverse flow. Typical results showing the increase in seal loss with length are given in figure 4.10. One approach

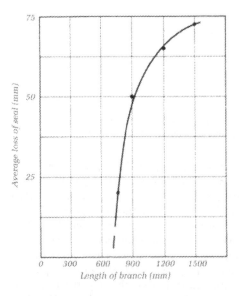

Figure 4.10   Example of the effect of pipe length on seal loss by self-siphonage — 32 mm trap and pipe serving a washbasin

to retaining a proper seal and controlling the noise is, therefore, to limit the length of pipe between the trap and vent or the trap and stack; hence compact planning of the sanitary accommodation tends to help in reducing the need for individual vent piping, e.g. figure 5.4, example 1. The gradients of pipes and some other factors should also sometimes be limited as specified in chapter 5.

Flow and induced siphonage with combined branches

Induced siphonage is the suction effect of the flow from one or more appliances on the traps of others not discharging. For combined wastes the principles can best be illustrated by reference to a simple installation where the flow is likely to be full: a branch serving both a basin and bath as in figure 4.11; the vertical pipe is known as a wet vent. Consider first what happens when the basin is discharging. The bath is most likely to lose seal when the entry of the wet vent into the bath branch is swept as in the top figure. The discharge from the basin entrains air from the portion A of the bath branch between the wet vent and the trap.

This causes a partial vacuum in A and results in some seal loss. Water builds up backwards in A towards the trap, however, and if this pipe is short enough and its slope flat enough, the backflow reaches the trap and tends to maintain a full seal.

The type of entry of the basin branch into the wet vent — whether straight or swept — also has some influence on the seal loss. When the basin branch runs straight into the vent, water tends to build up as shown in figure 4.11(b). This prevents any air being entrained by flow down the wet vent and the rather larger negative pressure in A continues until the build-up falls away. In this case a single discharge may give a bath seal loss up to about 25 mm and subsequent consecutive discharges increase the loss only slightly. When the basin branch is swept into the vent as in the upper figure, flow down the wet vent can entrain air from the vent almost straight away. The partial vacuum in A is, therefore, quickly relieved and seal loss is smaller. Consecutive discharges give rise to cumulative losses, however, due to the short-lived negative pressure at the beginning of each run, and the

maximum loss is about the same as when the basin waste runs straight into the vent.

The situation is different when the entry of the wet vent into the bath branch is practically straight as in figure 4.11(c), as with some compression and capillary tee fittings. Here the initial surge creates an immediate positive pressure which forces the bath seal up into the waste tail. The water fills up in A unless the trap is some distance from the vent or the slope is steep. When the wet vent is fairly close to the trap the bath seal and some of the basin waste may be forced out into the bath.

The addition of the bath discharge to that of the basin normally tends to reduce any losses of seal from the bath trap, because:

(i) a considerable quantity of the bath discharge usually remains in the bath branch at the end of the run which encourages trap refill;

(ii) there is a trail from the bath itself which tops up the seal.

When the bath discharges on its own, the basin trap is likely to lose seal if the vent is closed but not if the vent is open.

Figure 4.12 illustrates conditions where a waste connects into a much larger branch or drain. With the flow in the latter quite shallow, as shown, the air pressure in the larger pipe remains unaffected, and so there is no siphonage effect on a trap at A.

Water flow and air pressure distribution in stacks

Discharge stacks are most nearly filled with water at the points where branches are discharging into them, before the water has begun to accelerate downwards. With a small branch discharging into a large stack, a cylinder of water approximately equal in diameter to the branch lies across the stack and partly blocks it temporarily (figure 4.13). When the branch and stack diameters are equal, as may occur with a WC discharging into a 100 mm stack, rather more of the cross-section of the stack is occupied by the discharging water especially when the entry is at nearly 90° to the stack (figure 4.14).

Below an inlet the water flow accelerates to a terminal velocity, normally within a fall of a few metres, and takes up an annular form

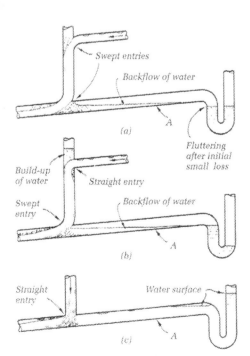

Figure 4.11 Examples of conditions in combined waste pipes

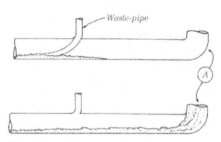

Figure 4.12 Examples of conditions with a large branch or drain

around the inner wall of the pipe. A central core of air is drawn downwards by the water and, within this, solids and paper fall down the stack. At the base of the stack the water falls to the bottom of the bend and then flows along the invert of the drain carrying solids and paper with it, whilst the air moves along above the water. The flow discharges into the main drain or sewer and the air exhausts through the adjacent vents or stacks. The annular flow in the stack may reach speeds of 3 m/s or more and induce an air flow in the region of 10–200 l/s.

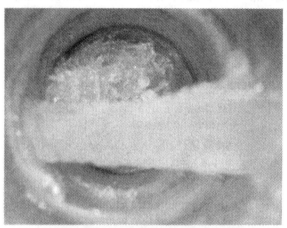

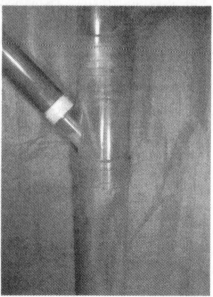

Figure 4.13 Example of flow from a waste branch into a vertical stack. Internal view illustrates the reduction in air path cross section at and below the discharging branch (Courtesy of Heriot-Watt University)

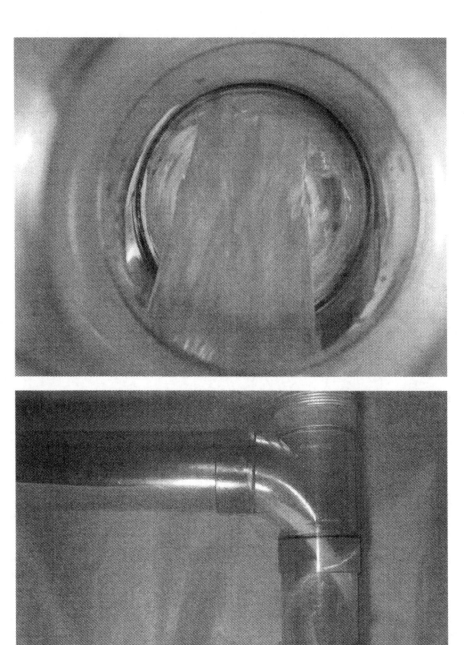

Figure 4.14 Discharge from a WC branch entering a vertical stack, illustrating the formation of annular flow. The internal view illustrates the reduction in air path cross section below the active junction (Courtesy of Heriot-Watt Universtity)

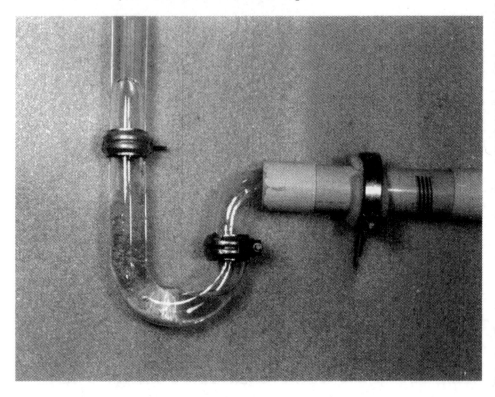

Figure 4.15 Trap displacement due to back pressure (Courtesy of Heriot-Watt University)

Under these conditions the air pressure within the stack departs from atmospheric, and the pressure distribution has certain characteristics. Figure 4.16 illustrates in diagrammatic form the main features of the hydraulic and pneumatic conditions in a 100 mm stack and drain with three appliances discharging as an example. The static pressure falls below atmospheric immediately below the top of the stack, i.e. there is a suction. The suction increases downwards due to friction loss as the air flows along. A marked drop in pressure occurs across each stream of water entering the stack. Below the lowest branch that is discharging the air pressure gradually increases down the wet part of the stack to above atmospheric at the base, the actual pressure depending on the friction in the bend and drain and any restriction downstream. An offset in the stack modifies the pressure curve, its effect depending on its dimensions and its position in the stack.

The distribution and magnitude of pressures that occur are, of course, important in relation to trap seals. Pressures below atmospheric tend to remove water from the traps and destroy the seals by induced siphonage; pressures above atmospheric tend to push the water back into the appliances (figure 4.15) and may force foul air into the building. As well as the qualitative picture given above research has provided a means for predicting the maximum suctions and pressures that occur in installations in practice in terms of the numbers and types of appliances installed, diameters and heights of stacks proposed, shapes and sizes of branches and bends, and so on. The method, which is outlined in chapter 8, is based on a theoretical analysis of the fluid mechanics of such installations supported by a variety of information obtained in the laboratory. It is sufficiently general and well enough supported by full-scale trials of installations in blocks of flats and office buildings to have formed the basis for the broad design recommendations that have appeared in various publications.

A separate problem is that when the water flows down a stack past a branch that is

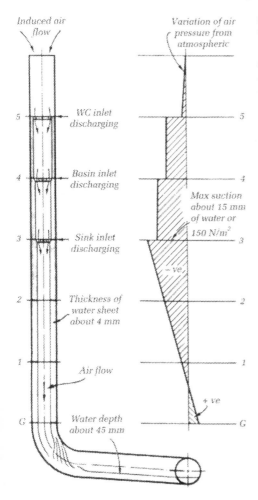

Figure 4.16 Diagram to illustrate hydraulic and pneumatic conditions in a discharge stack

attention in very large heavily loaded installations (Wyly and Eaton 1961).

Finally it should be noted that, under the conditions of turbulence and aeration that develop in a vertical pipe down which water is flowing, quantities of foam are formed when the water contains detergent and has, therefore, a lower surface tension than usual. Experimental investigations of this matter have been conducted which showed that under some conditions of heavy loading, and with detergents as formulated in the 1950s, backflow of foam into lower-floor rooms could occur under the local conditions of excess pressure. The design rules in chapter 5 take account of this possibility and, as a result, in recent years in the UK little trouble in this respect has arisen.

Flow and siphonage in 'horizontal' installations

Figure 4.17 illustrates the type of installation now being used in some large buildings such as hospitals built into a deep plan. Even peak usage of the appliances during the day is unlikely to produce a continuous flow. Water normally passes along the main drain in surges which move solid materials in stages along the pipeline to the vertical stacks. With designs used for hospitals, for example, wads of material flushed away through a WC may travel in smooth, straight, 100 mm pipes for distances up to 10 m or so on one flush and are then left stranded, being moved along further by the next discharge. Junctions and bends, especially of short radius, retard such movement, and also impede the transport of the range of articles that experience shows can find their way into drainage pipework — mop heads, syringes, spatulas and rubber gloves, for example. Such considerations, confirmed by experience, suggest that blockages are particularly likely to occur in this type of installation. Special care is therefore needed in design and construction.

The concept that solid materials may require several flushes to be able to clear a 'horizontal' drain and reach a stack is particularly significant in relation to the use of disposable items such as bedpans made from papier mâché. Bedpan disposal units discharge a pulp which may be thick when several bedpans are dealt

discharging, the outflow from the branch has to deflect the vertical stream in order to enter the stack and, as a result, a back pressure is created. This has the effect of retarding the branch flow and, if the back pressure greatly exceeds the head causing flow, water may re-enter the appliance. This problem is distinct from the pneumatic effects already considered. Discharge tests with the baths of a four-storey experimental system showed the extent to which this is a problem in buildings of several storeys. It was found that flow rates on the lower floors were reduced by some 10 to 15 per cent when a substantial flow was passing down the stack. Whilst this is not a serious problem in most buildings in the UK it may require

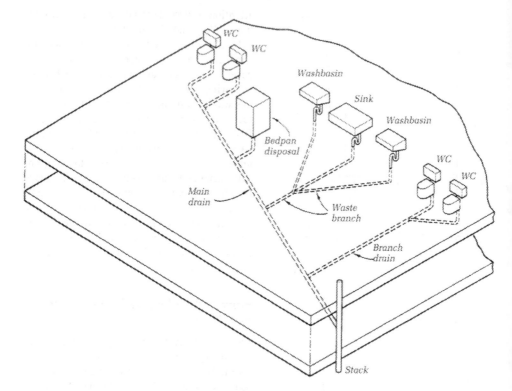

Figure 4.17   Diagram to illustrate the drainage arrangement in a deep-plan building

with at the same time. The pulp tends to lie in the drain and cause a blockage. Flushes from WCs upstream of the connection from the bedpan unit to the main drain help to keep such a drain clear (c.f. figure 4.17).

The question arises as to whether flow in the main drain may induce siphonage of traps connected to it. From this standpoint, conditions in the main drain are much less severe than in vertical stacks. Drains of 100 mm diameter normally run only partly full and water speeds do not exceed 1 m/s. Air induction is much less than with flow in vertical stacks and the air pressure variations are also much less. In an installation of the type shown in figure 4.17, for example, the seals of WCs, about 50 mm deep, could not usually be

significantly affected even with ancillary venting completely absent. With such installations self-siphonage is a more important factor. Architectural considerations encourage the use of S-traps in these circumstances which tend to accentuate the self-siphonage problem as already discussed. Since architectural considerations and cost also inhibit the use of individual vent pipes close to each trap, which might be used to cure the problem, one solution is to fit special resealing traps on the waste appliances in such installations in hospitals. Air admittance valves are also sometimes used. The lessons from these findings have been incorporated into subsequent design recommendations. The flow in 'horizontal' installations is further discussed in chapters 8, 9 and 11.

5 Design of Soil and Waste Pipe Installations

Planning considerations usually result in the soil and waste pipe system in modern buildings taking one of two forms. By far the most common stems from the more or less close grouping of the sanitary appliances in columns up the building. The system consists of several 'horizontal' branches, usually fairly short, serving the sanitary appliances on each floor and leading to a vertical discharge stack that conveys the soil and waste water by the most direct route to the underground drains. Such installations are to be found in many types of building. They are often on the single-stack principle outlined in the previous chapter, and an example where such a system might be installed in a hospital (figure 5.1) and a system for a block of flats (figure 4.4) are typical. In contrast, hospitals, schools and offices are sometimes planned nowadays with a considerable horizontal spread. The sanitary accommodation is arranged as 'islands' on each floor and it may then be difficult to link the appliances by short branches to a vertical stack. Extensive 'horizontal' piping is used instead to connect the appliances to a vertical stack some distance away, e.g. figure 4.17.

The move towards European standardization has focused attention on the various approaches to the design of such systems found in different countries. Methods to determine water flows on the discharge unit principle for the main stacks and drains vary, as outlined in chapter 1. A good deal is known about the hydraulics of stacks and drains — see chapters 8 and 9 — but the methods used to establish discharge capacity also vary. A simple relationship given in chapter 8 has formed the basis for the calculation of stack capacity in BS 5572 and in some other countries and was

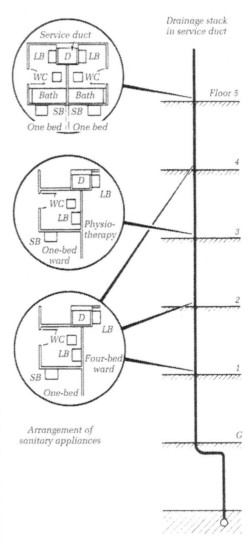

Figure 5.1 Examples where single-stack drainage might be installed in a hospital

proposed as a basis for a CEN standard (De Cuyper 1993). Further tables in BS 5572 are based on the approach to the determination of vent sizing outlined in chapter 8, but the CEN proposals included some rules of thumb. Now BSEN 12056, with National Annex based on BS 5572, brings together the relevant information (see Appendix 1) and offers a choice of design approach. In the long term, the new methods set out in chapter 9 and the advanced procedures discussed in chapter 1 may lead to a new approach to design calculation.

The design of individual and combined branches also varies in European countries and three main categories may be identified:

1. Traps from waste appliances such as basins, baths and sinks connected into larger pipes which do not normally run full, thus avoiding siphonage problems; for example, a 30 mm trap connected to a 50 mm branch or to a 100 mm soil branch (exemplified by figure 4.12).
2. Traps with waste appliances connected to pipes of similar or slightly larger diameter. Full-bore flow may occur and the resulting suction effects are limited through limitations on such factors as lengths and gradients of pipes, numbers of bends and vertical drops, thus ensuring adequate trap seal retention.
3. What is, in fact, a variant of category 2, smaller trap and pipe sizes being used with appliances having lower discharge rates associated with water-saving devices.

Practice in the UK on the whole has followed category 2, research having provided a sound and economic design procedure set out in publications of the Building Research Establishment, in BS 5572 and in Approved Document H associated with the 2000 Building Regulations. Awareness of the possibilities with category 1 has led to the inclusion of certain layouts on this basis in UK procedures, research having provided the necessary information. European standardization provides opportunities for the use of each of the three broad categories. Their application will depend upon the forms of construction and sanitary layout found in the countries concerned.

The present chapter draws on all this information to provide data and procedures for design, taking account of the direction of European standardization. The performance requirements given in chapter 4 form the general basis for design whilst research background is provided in chapters 8 and 9.

Traps and valves

With WCs the trap is an integral part of the appliance but with other appliances it is fixed to the outgo and a wide variety of shapes, materials and depths of seal are available. Appearance may be an important feature, also ease of access for cleaning, and flexibility of outgo position. Figure 5.2 shows basic types including bottle traps, in one of which the body of the trap is a glass container which indicates the need for cleaning. There are three-jointed traps which can give a wide range of outgo positions, and the bend of which can be removed in the case of blockage. Table 5.1 gives the commonly accepted minimum diameters of traps for various appliances.

In addition to the features described above the depth of seal must be specified. In the past traps were used with seals ranging from a nominal depth of a few millimetres to a depth of 75 mm. The latter figure was specified for waste appliances on the original one-pipe system in the UK and has subsequently been used on simplified one-pipe and single-stack systems. This depth of trap creates an installation problem under baths and showers, where a 50 mm depth causes less difficulty. The latter can be used for these two appliances, and also for floor drains, bearing in mind that WC traps commonly have 50 mm seals. The performance requirements for seal retention in use have been specified in the previous chapter; this retention is normally achieved by suitable design of the pipe system into which the traps discharge.

Special traps in which the seal is retained through a particular feature of the trap itself are also on the market. They allow air to pass through while retaining sufficient water to re-establish the seal. Figure 5.2 shows several traps of this type. Whilst it is possible by their use to make substantial pipework economies, deposits inside such traps over a period can render their operation less effective and regular cleaning may be required. Some can be noisy.

Air admittance valves, described in chapter 9 may be useful in some circumstances — see the

Outline of trap	Features
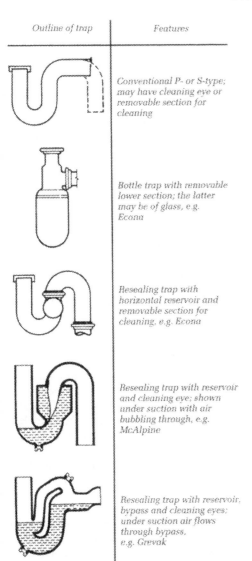	Conventional P- or S-type; may have cleaning eye or removable section for cleaning
	Bottle trap with removable lower section; the latter may be of glass, e.g. Econa
	Resealing trap with horizontal reservoir and removable section for cleaning, e.g. Econa
	Resealing trap with reservoir and cleaning eye; shown under suction with air bubbling through, e.g. McAlpine
	Resealing trap with reservoir, bypass and cleaning eyes; under suction air flows through bypass, e.g. Grevak

Figure 5.2 Examples of traps — diagrammatic

Table 5.1 Minimum internal diameters of traps

Type of appliance	Dia. (mm)	Type of appliance	Dia. (mm)
Washbasin	30	Drinking fountain	20
Bidet	30	Bar well	30
Sink	40	Hotel or canteen sink	40
Bath	40	Urinal (bowl)	40
Shower tray	40	Urinal (stall, 1 or 2)	50
Washing machine	40–50	Urinal (stall, 3 or 6)	65
Kitchen waste disposal unit	40	Waste disposal unit (commercial)	50

Note: Where there are more than six urinal stalls in one range, more than one outlet should be provided

surfaces. The main requirements have been set out in chapter 4. Cast iron has been the main material for soil stacks, for combined soil and waste stacks, and for main vent stacks. It is still used but PVC-U pipes and fittings are also much used. This pipework in the early 1960s sometimes failed as a result of hot water discharges causing local softening. Greater wall thicknesses and improved properties and designs have greatly reduced the numbers of failures with the result that PVC-U is now highly competitive. Copper, stainless steel, galvanized mild steel in a prefabricated form, and some other plastics (see chapter 14) are also used for stacks. Waste pipes are commonly in copper, with fittings in brass. Plastics and steel are also used, with glass and plastics for applications such as laboratory wastes where there are special requirements.

Design of individual branches

In order to reduce the risk of stoppages, pipework should not decrease in diameter in the direction of flow. With this requirement and with the sizes of traps and branches in common use, e.g. 30 mm for basins, 40 mm for sinks, 40 mm for baths (see table 5.1), a rate of discharge to satisfy the user is readily achieved. Design, then, is not primarily a matter of achieving a required flow rate but of limiting noise and the seal loss associated with self-siphonage. The same kind of measure fortunately tends to limit both effects, and experience shows that adopting the following recommendations for controlling self-siphonage so as to retain 25 mm depth of seal also gives

final sections of this chapter — provided that such valves, as with other special devices, remain effective over long periods and are accessible for maintenance. A waterless trap is also described in chapter 11.

Pipes and fittings

Only a few types of pipe are in common use for soil and waste pipe installations, chosen on the basis of cost and such factors as durability, ease of working and jointing, and their

generally a satisfactory result as regards noise. Where appropriate, the avoidance of backflow and blockage is also considered in this section.

It was customary to control self-siphonage by installing a vent pipe close to the trap, but research has shown that a special vent is usually unnecessary. It is now common for a trap to be connected directly to a stack without venting and to limit self-siphonage by suitable design of the waste, with particular reference to its length and slope. A gradient of 2 per cent is a reasonable practical minimum.

(i) Washbasins

Figure 5.3 gives a relationship between slope and length of waste from trap to stack to be used to control self-siphonage with ordinary basins having the normal outlet and grating and 75 mm seal traps. Originally established for 32 mm pipes, it also forms guidance for the 30 mm size. For lengths of about 1.7 m, and above, the slopes required are hardly suitable for practical purposes and a vent pipe near the trap may then be necessary. An alternative to a vent pipe with the longer wastes is to use this diameter P-trap discharging into a larger waste pipe — 40 or 50 mm diameter — with not more than two bends in its length, of large radius, not less than 75 mm. The pipe length should be limited to 3 m and the slope not more than 4.4 per cent. This method prevents the waste running full and, therefore, avoids suction on the trap.

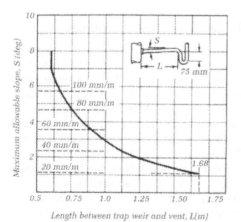

Figure 5.3 Waste pipes from washbasins — maximum slope for various lengths of 32 mm pipe

The use of S-traps is on the whole less common in UK practice, although they may sometimes be convenient from an installation standpoint. The conditions at the end of discharge without a vent tend to be more variable than with P-traps and large seal losses and noise are more likely. Successful unvented installations may be achieved by careful choice of pipe sizes, using the larger pipes, and layout, but tests may be best to establish the performance to be expected. S-trap installations are more common in other European countries, also in the USA where the trap outlet is normally vented to control self-siphonage and noise.

Special resealing traps may also be used successfully with waste pipes many metres long but, as already mentioned, such installations are often noisy, whilst the trap is likely to require regular cleaning to maintain performance.

(ii) Sinks, baths and showers

Self-siphonage is not normally a problem — it may be with small sinks — because the trailing flow at the end of the discharge refills the seal if any suction occurs, but noise can be a problem with long, unvented bath and sink wastes. There is, therefore, an advantage working to some broad limits to the length of unvented wastes from these appliances. With P-traps and 40 mm pipes, a limit of 3 m is a reasonable practical rule, with pipe slopes up to 9 per cent. With 50 mm pipes the length might be up to 4 m. Any vertical drops in the pipe should not exceed 1.5 m. The waste pipes from S-traps used on sinks should also be limited in length, say to 3 m. Wastes from showers on their own are commonly 40 mm diameter but there seems no reason why they should not be smaller, say 32 mm.

(iii) WCs

Traditionally the WC branch was required to be not less in diameter than the outgo from the pan. This was to ensure that a stoppage occurred in the pan rather than in the pipework and could, therefore, be more readily cleared. Washdown pans thus, commonly, have an outgo of somewhat less than 100 mm and discharge into 100 mm branches. Siphonic pans with outgoes perhaps as small as 55–60 mm may discharge into 75 mm branches. In these circumstances, moreover, there is no danger of full-bore flow in the branch and hence of self-siphonage with the WC in any normal position in relation to the soil stack. Thus with P-trap WCs, there should be no need to limit the length of branch; a practical limit of 6 m is suggested in Approved Document H, Building Regulations 1991. Branches from S-trap WCs contained within the floor are normally quite short, the pan being sited close to the stack so that the branch at the normal angle does not appear in the room below. Self-siphonage does not normally occur under these conditions, nor with much longer branches. Experience has not shown any special precautions to be necessary in this respect with siphonic pans.

Application of information on individual branches

Figure 5.4 shows a few examples of compact sanitary layouts against which the data on branches may be considered. The appliances are so positioned in relation to the service shaft that each branch to the stack is normally well within the limits for P-traps given in the previous section. Pipe connections of the type shown in figure 5.4 may, therefore, be used. With figure 5.5, the basin or sink is some way from the main stack and would, therefore, probably require a separate stack or venting — the arrangement shown in figure 5.6.

As regards the tolerance in fixing the basin branch (for which the slope of pipe is an important factor, figure 5.3), the plans in figure 5.4 may be grouped as follows.

Plans 1–4: the basin waste typically would be some 600 mm long and the slope would not be critical. Plans 5–7: the basin waste typically would be about 1000 mm long. The slope is important; a working rule is to ensure that the difference in level between the trap outlet and stack connection does not exceed 50 mm.

An alternative with some constructions and layouts may be to connect waste appliances into a larger 'horizontal' branch at or below floor level — see next section.

Figure 5.9 gives recommendations, elaborated later, regarding the relative positions of WC and bath connections to the stack, with the object of reducing the risk of cross-flow from one branch into another.

Design of combined branches

(i) Bath and basin wastes combined

From a siphonage standpoint some arrangements of combined bath and basin wastes perform satisfactorily unvented, but general design recommendations are not readily formulated. Tests are needed to check the performance of particular installations. Installing a vent as shown, for example, in figure 5.6 avoids the need for testing. Care is also necessary to avoid water from the basin flowing into the bath, as can occur unless precautions are taken. From this standpoint, the bath and basin wastes should join at an angle of 45° on plan and the length of the waste between the bath trap and tee should be sloping, preferably using a bath trap with 9 per cent rake of outlet. The waste between the tee and the stack should, if possible, be straight. If a horizontal bend is necessary into the stack it should be of large radius. The whole length of waste from basin to stack should have a continuous fall, 4 per cent being reasonable.

(ii) Bath wastes combined

With the arrangement in figure 5.7 water from one bath may flow into the other unless precautions are taken. The waste pipes should, therefore, join as far from the traps as possible and have a good fall. It is often convenient to use a Y-junction connecting directly into a 50 mm boss in the stack wall with the angle of the Y nominally 60°.

(iii) Combined soil and waste branches

Bathrooms in which the WC is 2 or 3 m from the stack may be conveniently arranged with the WC branch, 100 mm diameter, containing connections for the basin and bath or shower. A typical arrangement is with an S-trap WC, with the WC branch beneath the floor, if construction permits, where the waste connections may readily be made in the upper half of the WC branch. The connections should be selected to ensure that the waste flows are

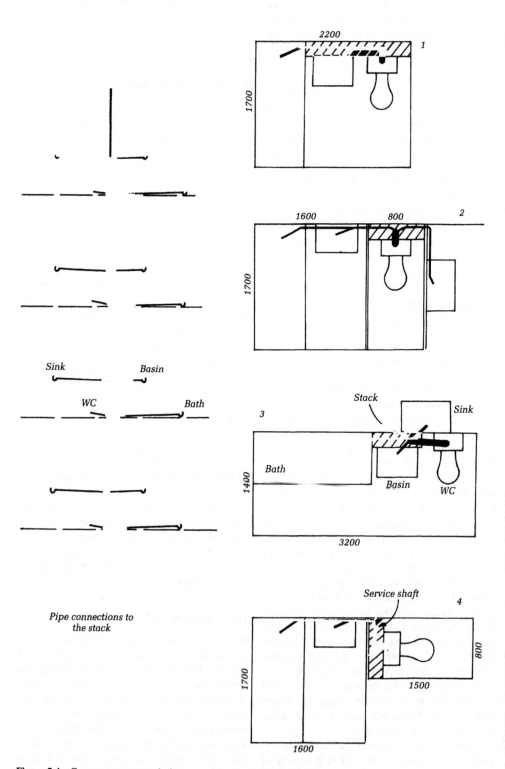

Figure 5.4 Compact accommodation served by single stack system

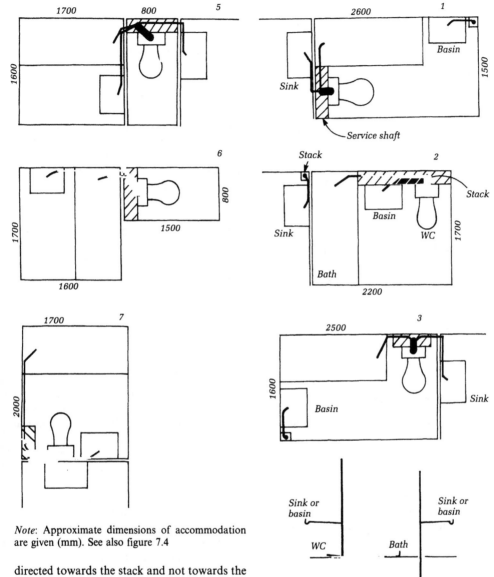

Note: Approximate dimensions of accommodation are given (mm). See also figure 7.4

Note: Approximate dimensions of accommodation are given (mm). See also figure 7.4.

Figure 5.5 Compact accommodation with separate waste stack

directed towards the stack and not towards the WC. Since the WC branch does not run full, noise and siphonage problems do not normally arise, except in so far as self-siphonage may occur in the basin waste between its S-trap and the WC branch. With a 30 mm S-trap discharging into a short, vertical 40 or 50 mm pipe the problem is not likely to be serious.

A variation on this arrangement, common in Scandinavia, is to use a floor gully trap in the bathroom, to receive flows from the washbasin and bath. Pipes from these two appliances discharge into the gully which can also take any spillage from bath or shower. The gully is connected into the WC branch located within the floor. There is sometimes a practical difficulty in accommodating the gully trap within the floor, without it appearing in the room below, especially with modern prefabricated concrete construction.

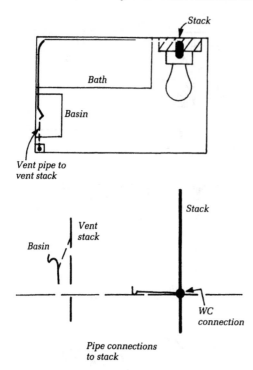

Figure 5.6   Bath and basin waste pipes combined

gives a safety margin, since the following recommendations are based on filled basins. The recommendations in figure 5.8 relate to traps with 75 mm seal used on ordinary basins, and are intended to ensure not less than 25 mm seal retention.

Wastes from spray taps do not normally run full even when the branch is only 30 mm diameter, and trials have shown that up to eight basins may be connected to this size of pipe without venting. Such waste pipes are, however, likely to become blocked by the build-up of deposit, and regular cleaning is usually necessary. Pipes should be kept short.

Resealing traps are sometimes used for basin ranges, as noted earlier, but may be noisy and require periodic cleaning to perform well. The above venting considerations do not apply when a running trap is used, serving several basins. The latter is probably a less satisfactory installation than the fully trapped type which should be used where possible.

(v) Ranges of WCs

Ranges of WCs are common in public and commercial buildings. Branch pipes serving such ranges normally are 100 mm diameter and do not run full, whatever the slope. There is, therefore, usually no need for branch venting, although, as a precaution, it is recommended for more than eight WCs. Where there are bends in the pipe it may be necessary to fit a vent pipe to the appliance furthest from the stack. Hydraulic considerations indicate that WC connections to the common branch should be swept in the direction of flow.

(iv) Ranges of washbasins

Ranges of basins are common in public and commercial buildings. Design recommendations (figure 5.8) have been formulated with the aim of limiting siphonage in the event of full-bore flow anywhere in the branch. Full-bore flow depends upon such factors as discharge rate, pipe size, the length and slope of the branch and the shape of tee connections. It is unlikely if washing is done under a running tap and, as the latter is common, this

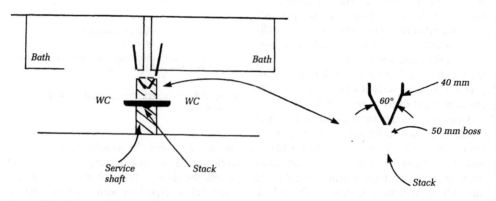

Figure 5.7   Bath waste pipes combined

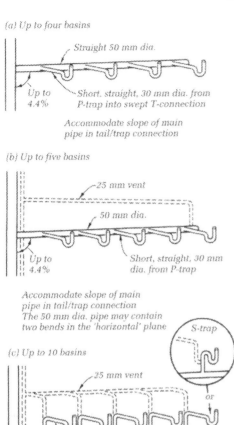

(a) Up to four basins

Straight 50 mm dia.

Up to 4.4%   Short, straight, 30 mm dia. from P-trap into swept T-connection

Accommodate slope of main pipe in tail/trap connection

(b) Up to five basins

25 mm vent

50 mm dia.

Up to 4.4%   Short, straight, 30 mm dia. from P-trap

Accommodate slope of main pipe in tail/trap connection
The 50 mm dia. pipe may contain two bends in the 'horizontal' plane

S-trap

(c) Up to 10 basins

25 mm vent

or

50 mm dia.

Short, straight, 30 mm dia. from P-trap, alternatively S-trap

Figure 5.8 Pipework for ranges of washbasins

(vi) Ranges of urinals

On flow considerations alone, venting is not normally necessary for a branch pipe, 50 mm minimum, to a range of urinals. As with wastes for spray tap installations, however, regular cleaning of urinal branches may be necessary, especially in hard water areas where substantial deposit may build up. Pipes should be as short as possible.

Design of stacks and ancillary venting

(i) Some general considerations

Both hydraulic capacity and the control of air pressures have to be considered. Stacks generally should be open at the top to admit or remove air, and terminated outside the build-

ing at least 900 mm above the top of any opening within 3 m. The details of the larger branch fittings that admit water to a stack, the dimensions of offsets and the curvature of bends influence performance and hence the need for venting. Research at the Building Research Establishment has shown that suctions in stacks tend to be greater when WC branch entries are straight and at an angle not far from 90° to the vertical. Thus the trend in the UK has been for the larger sizes of branch fitting at angles of between 90° and 112.5° to be made with entries curved into the stack with a 50 mm radius, or similar curve. At steeper angles there is less need for curvature since the branch itself feeds water into the stack in the direction of flow. The advantage of curved entries has not been realized in some European countries which have relied on straight (normal) entries without a significant curve. The CEN Standard now takes account of both types. The shape of the smaller inlets, i.e. waste branches, is not critical as far as stack performance is concerned. In the main tables of data, stacks are assumed to be straight; separate comments cover offsets. A large-radius bend is assumed at the base of all stacks, minimum radius twice the internal diameter, e.g. 200 mm for 100 mm pipes.

Stacks serving urinals only or sinks only block up readily as deposits accumulate. If possible, therefore, urinals and sinks should be connected into stacks that are frequently flushed out by flows from WCs or baths. If separate stacks for urinals and sinks are essential, the need for regular clearing must be reckoned with and adequate means for access provided.

(ii) Domestic installations

Stack diameters range typically from 100 to 150 mm. The smallest size is normally sufficient for single-family dwellings of one, two or three storeys and may be used with systems on the single-stack principle, some recommendations for which are assembled in figure 5.9. These are intended to help avoid cross-flow from one branch to another, and to limit pressure variations at the base of stacks.

The branches generally should be designed in accordance with the recommendations given earlier in the chapter. Typical arrangements are

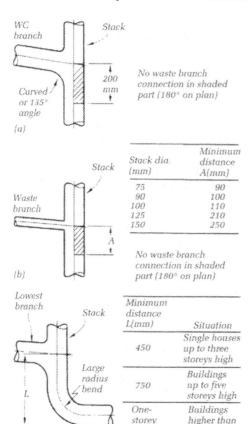

Figure 5.9 Recommendations for stacks and branches

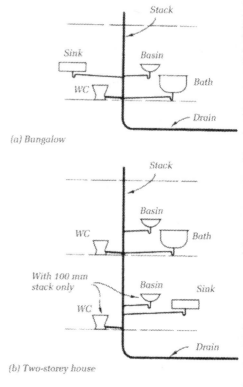

Figure 5.10 Examples of single-stack installations for single-family dwellings

shown diagrammatically in figure 5.10. In circumstances typified by figure 5.10(a) and with a 100 mm stack, it may be an advantage to use a 'stub stack' in which the top is closed. In this case, the stub stack should connect into a ventilated stack or drain nearby. No branch on the stub stack should be more than 2 m above the invert of that connection or drain. The drop from the crown of a WC trap to the invert should not exceed 1.5 m. A discharge stack may be terminated inside a building with an air admittance valve subject to a British Board of Agrément Certificate, provided this does not adversely affect the ventilation of the underground drains.

With blocks of flats or maisonettes, stacks should be 100, 125 or 150 mm in diameter. Again the single-stack principle may be used in many situations. Table 5.2 provides informa-

tion in a simplified form, not requiring calculations, based on BS 5572. It was originally established for WCs with 9 l flush, and is likely to be on the safe side with smaller flushing quantities. Indeed the general procedure given later may well give smaller stack sizes than table 5.2. To prevent the passage of effluent into a cross-vent from discharge stack to a vent stack, that vent should slope upwards from the discharge stack at an angle of not less than 135°. The recommendations of figure 5.9 should be applied. The practice of connecting ground floor appliances directly to the drain or manhole has been found to help to prevent backflow of detergent foam from stacks into buildings and in combating the effects of stoppages at the base of stacks.

In developing single-stack systems in the UK it was recognized that stacks do not work as independent units but may be affected by discharges from adjacent installations. From the early research it was recommended that a 50 mm relief vent should be fitted to the lowest branch of a single-stack installation if it was

Table 5.2   Stack sizes and vents for various domestic loadings

Flats with one group of appliances on each floor:
 100 mm stack for buildings up to 10 storeys high
 125 mm stack for buildings up to 15 storeys high   }   Single-stack system
 150 mm stack for buildings up to 30 storeys high

Alternatives:
 100 mm stack for buildings 11–15 storeys high — add 50 mm vent stack

Flats with two groups of appliances on each floor:
 100 mm stack for buildings up to 10 storeys high   }   Single-stack system
 150 mm stack for buildings up to 30 storeys high

Alternatives:
 100 mm stack for buildings 11–15 storeys high — add 50 mm vent stack

Maisonettes with one or two groups of appliances on alternate floors:
 100 mm stack for buildings up to 10 storeys high — single-stack system

Notes: Each appliance group includes WC, bath and/or shower, basin and sink, and may include a washing machine. WC branches should have entries swept or at an angle of 135° into the stack. Bend at base of stack should be large radius. There should be no offsets in the 'wet' part of the stack. One connection of vent stack to discharge stack in each storey should suffice

known that the drainage system downstream was likely to run sufficiently full to prevent the escape of air travelling with the water down the stack. Experience since the 1950s has shown that back-pressure effects of this nature are rare in practice. Only a few cases have come to notice: situations with 25 storey flats discharging into heavily loaded drains running surcharged; relief vents of 75 mm diameter minimum were necessary. Where surcharging is expected a relief vent should be fitted.

An offset in the stack above the topmost connection to the stack has little effect on the performance of the system. Offsets below the topmost connection do not have a large effect in simple installations with light loading, e.g. in dwellings of three storeys, but greater pressure fluctuations may occur with heavier loadings and in taller buildings. In general such offsets should have large-radius bends and, in the taller buildings and with heavier loadings, may require venting to reduce pressure fluctuations in the stack.

Discharge stacks may take small flows of rainwater. The highest likely rainwater flow from the roof of a small house or flat (with an area of approximately 40 m²) is equivalent to the maximum discharge from a lavatory basin. For this size of roof the rainwater can normally be ignored in the calculation of hydraulic loading on the stack. Where easy access to the roof exists, the roof gully linking the roof to the stack should be trapped.

(iii) Office buildings

Ranges of appliances are commonly used in office buildings. Stacks of 100 or 150 mm diameter are usual, and the latter is likely to be large enough for most buildings in the UK. Figure 5.11 illustrates a traditional installation for an office building in which all the traps are vented. Considerable economy is now possible compared with traditional practice and recommendations are given in table 5.3. As with table 5.2, the sizes are likely to be on the safe side when WCs flush less than 9 litres, and the general procedure given next may be more economical still. With 100 mm discharge stacks used up to 12 floors and 150 mm up to 24 floors, table 5.3 gives the minimum vent sizes recommended for use with such stacks serving equal ranges of WCs and basins. For example, table 5.3(a) shows that with a 100 mm discharge stack serving four WCs and four basins on each of eight floors, a 40 mm diameter vent stack is needed. The effect of the flow from washbasins on induced siphonage is small and hence the recommendations can apply when the number of basins is less than the number of WCs. Stacks serving basins only are usually less than 100 mm in diameter and table 5.3 does not therefore apply.

The effect of urinals may be allowed for as in table 5.4 to be used with table 5.3 for the design of stacks serving WCs, washbasins and urinals. It is not comprehensive but illustrates

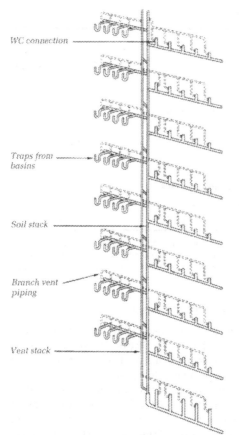

WC connection

Traps from basins

Soil stack

Branch vent piping

Vent stack

Note: The complete venting shown is not usually necessary — see text

Figure 5.11　Example of a traditional soil and waste pipe system for an office building

WC/urinal/basin combinations that may be taken as equivalent from a hydraulic standpoint to the WC/basin combinations in table 5.3.

In table 5.3, the 10 min interval, 'Commercial' use, section is likely to be sufficient for most purposes; the 'Congested' use (5 min) table is included to cover special situations where a concentrated peak use may be expected.

The table does not cover offsets in the 'wet' part of the stack, nor a series of changes of direction between the lowest connection to the stack and the sewer, which may increase back pressure above that likely in a simple situation comparable with figure 5.11. As with domestic installations where above-normal back pressures are likely to arise as a result of downstream conditions, a relief vent connected to the stack near the lowest branch connection is recommended, especially where table 5.3 shows no vent stack needed. Table 5.3 assumes the use of a large-radius bend at the foot of the stack, whilst a bend and drain of 150 mm diameter is recommended for the type of situation shown in figure 5.11. This figure also shows one way in which a vent stack may be connected to a discharge stack at each floor; alternatively the vent stack may simply be connected directly to the WC branch on each floor. The vent stack should join the discharge stack just above the bend at its base to help relieve back pressure. Both stacks should end away from parapets and the corners of roofs

Table 5.3　Stack sizes and vents for office buildings

Diameter of discharge stack		100 mm			150 mm		
Number of floors		1–4	5–8	9–12	1–8	9–16	17–24
(a) 10 min interval 'Commercial' use	WCs and basins						
	1 + 1	0	0	32	0	0	0
	2 + 2	0	0	32	0	0	0
	3 + 3	0	32	40	0	0	0
	4 + 4	0	40	40	0	75	75
	5 + 5	0	40	////	0	75	75
(b) 5 min interval 'Congested' use	1 + 1	0	0	32	0	0	0
	2 + 2	0	50	50	0	0	0
	3 + 3	0	50	////	0	0	75
	4 + 4	32	////	////	0	75	75
	5 + 5	32	////	////	0	75	75

Notes: 0 means no vent stack needed; /////// means overloaded on 'one-quarter-full' basis. WC branches should have entries swept or at an angle of 135° into the stack. Bend at base of stack should be large radius. There should be no offsets in the 'wet' part of the stack because the substantial suctions due to wind in these areas may cause considerable loss of seal.

Application of the recommendations will commonly involve the use of table 5.3 with figure 5.8. Suppose, for example, that in a 'commercial' building there are four floors with four WCs on each. The individual ranges may be served by 100 mm straight branches without venting; from table 5.3, the four branches may discharge into a 100 mm discharge stack without the addition of a vent stack. Suppose that four washbasins on each floor are also connected to the discharge stack. With the arrangement shown in figure 5.8(a) the individual branches do not need venting and table 5.3 shows that the addition of these basins does not necessitate a vent stack. With the basins fitted with S-traps according to figure 5.8(c), however, the individual ranges require venting as shown and a vent stack would be required to link this vent piping together and with the outside air. The information available suggests that a vent stack of 30 mm diameter should be sufficient. A vent stack remains unnecessary as far as table 5.3 is concerned; limitation of induced siphonage and back pressure associated with flow in the discharge stack itself is virtually independent of the way in which the basins are installed.

As a further example, consider 12 'commercial' floors with four WCs and four basins on each, all discharging into the same stack. The individual WC ranges do not need venting, nor do the basins if served by P-traps as shown in figure 5.8(a). The installation does, however, require a 40 mm vent stack according to table 5.3, and this should be cross-connected to the discharge stack at each floor and near its base. If the basins were to be fitted with S-traps, venting would be required as in figure 5.8(c) and this could be linked by the 40 mm vent stack derived from table 5.3.

(iv) General sizing method

The basis is set out under the section on water flow capacity in chapter 8. Based on equation (8.11), which is not dependent in any way on shape of water entry to the stack, two relationships yield tables of capacity against cast iron stack diameter. The values depend upon the assumption made about the proportion of the

Table 5.4 Conversions to be used in conjunction with table 5.3

WC	Urinal	Basin		WC	Basin
2	+ 1	+ 2	equivalent to 2	+ 2
2	+ 2	+ 3		3	+ 3
3	+ 3	+ 4		4	+ 4
4	+ 4	+ 5		5	+ 5

stack cross-section to be occupied by water under idealized conditions of annular flow. The 'one-quarter-full' criterion has been assumed in most UK procedures (BS 5572), including the data in tables 5.2 and 5.3, along with swept WC entries. Proposals in CEN (De Cuyper 1993) took a greater safety margin and assumed around 'one-sixth full', along with 'normal' WC entries, which gives smaller capacities. Figure 5.12 gives the calculated information from the equations in chapter 8. The capacities that appear in the CEN standard are rounded to satisfy the wide range of views in different countries. An attempt to allow for the greater capacity permissible with swept WC branch entries can be made through the recommendation of an arbitrary increase on the 'one-sixth-full' values.

In a sizing exercise the diameter of the discharge stack should be determined from figure 5.12 after establishing the likely flow using one of the discharge unit procedures of chapter 1. The diameter of vent stack required, if any, is then determined from table 5.5 (as in BS 5572) or other rule.

Design of 'horizontal' installations

In this section we consider 'horizontal' installations of the kind illustrated in chapter 4, with long drain lines or branches taking a considerable number of inflows (figure 5.13). Where applicable the information given in the foregoing sections may be used. The following notes supplement this from the experience available.

Estimation of the peak flow load in the main drain may be done on the basis of the methods set out in chapter 1. As indicated, information on the frequency of use of appliances in hospitals is scanty and the best course is probably to assume that the frequency of use

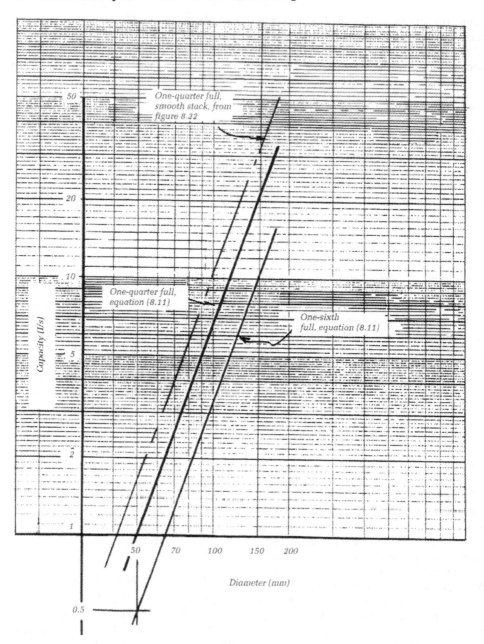

One-quarter full,
smooth stack, from
figure 8.22

One-quarter full,
equation (8.11)

One-sixth
full, equation (8.11)

Capacity (l/s)

Diameter (mm)

Figure 5.12   Relationships between stack diameter and capacity

Table 5.5   Diameters of ventilating stacks

Diameter of discharge stack, D	Diameter of ventilating stack
Less than 75 mm	Two-thirds D
75 mm and over	Half D

corresponds to not less than that assumed for peak hours in public buildings.

The capacity of the drain to carry the peak load has to be considered. For underground drains the assumption is often made that pipes should be designed for flow three-quarters full, thus providing some spare capacity. Some

Figure 5.13 Example of 'horizontal' drainage installation in a hospital (Nottingham Teaching Hospital; courtesy of Schott UK Ltd)

assumption of this kind must be made for 'horizontal' systems inside buildings. It might reasonably be assumed that the 'horizontal' drains in the building should not flow more than half full, thus providing a greater safety margin against surcharge in the branches from the appliances, and permitting ready circulation of air. Such considerations apply particularly to the main drains such as that illustrated in figure 4.17. As to gradients of pipes, limited experience with main drains laid at 1 in 80 (1.25 per cent) has been successful and it is suggested that 1 in 80 should be regarded as a reasonable practical minimum for pipes of 100 and 150 mm diameters within buildings. At this gradient these two pipe sizes will take around 5 l/s and 13 l/s flowing half full (compare figure 12.4). The equivalent discharge units may be obtained from chapter 1. Figure 1.7, for instance, gives around 300 and 2300 units respectively with mixed appliances, as in BS 5572 (1994).

For the reasons given in chapter 4, blockages are particularly likely to occur in this type of installation, and the design and construction should take account of this point. Offsets and short-radius bends should be avoided as far as possible as present evidence suggests they are a main source of blockage. When a change of direction is necessary it should be made by using either a large-radius bend or by combining two 135° bends. All such bends should be fitted with access doors. Junctions should be made obliquely in the direction of flow using

120° or 135° fittings. They should have access doors fitted to the side opposite the branch drain. Double-branch junctions on plan should not be used. Access door openings should be rectangular in shape, large enough for cleaning equipment to be easily inserted into the pipe, and should be positioned where the removal of the door and the insertion of the equipment can be easily carried out. They should not be obstructed, for example, by other pipes. Access points should be provided at each end of every main horizontal drain, at the top end possibly by a screwed cap and socket ferrule and at the lower end by a rectangular door on the crown of the pipe.

Waste pipes should as far as possible run in straight lines at even gradients of not less than about 2 per cent, and with any bends of not less than 75 mm radius on plan. Such piping should be well provided with cleaning eyes and with means for readily dismantling sections for the clearing of stoppages. Wastes should enter the main drain in the direction of flow using 120° or 135° unequal branch junctions. Where a boss is used it should not discharge vertically downwards into a horizontal drain.

It is usual in such installations to fit S-traps to waste appliances. Following the principles set out earlier, the trap either should be vented to prevent self-siphonage or should be of a special resealing type. As a guide, the length of unvented waste pipe in the latter case should be limited to 6 m in order to limit noise. In general the guidance given earlier for limiting self- and induced siphonage in combined branches should be followed where applicable. For example, where P-traps with gradually sloping wastes can be incorporated in the scheme, 75 mm seal traps should be used and the piping installed in accordance with the recommendations for avoiding self-siphonage.

Traditional practice with an installation such as that shown in figure 4.17 would require individual venting of all traps to avoid problems of induced siphonage due to flow in the main drain. Whilst it is clear that such precautions would be excessive, it is difficult with present knowledge to make precise recommendations. Experience suggests, however, that a vent pipe should be provided at the top end of all such horizontal drains more than about 10 m in length, and at intervals of 10–15 m in longer drains. A pipe of 50 mm diameter is likely to be adequate. The use of air admittance valves may be considered in such circumstances.

Appliances special to hospitals often require special treatment. Disposal units that grind papier mâché bedpans and can discharge a porridge-like effluent are an example. To assist the flushing of such effluent through the pipe system, the design should be arranged so as to ensure that the piping receives the discharges from regularly used WCs upstream of the disposal unit. New special purpose appliances may pose further problems in hospital installations, and the designer should obtain all the information possible about the appliances to be installed and their use before planning the installation. There is also some evidence that highly infectious areas in hospitals should be served by separate pipe systems. Experience has shown, for example, the importance of keeping separate the drainage from clean and infected parts of an animal house.

Again in hospitals, and particularly in installations in some laboratories, there is a risk that some reagents and solutions may be particularly corrosive. A further risk, although uncommon, is that some solutions may contain components that can react with metals and form compounds that readily detonate and explode, e.g. when some mechanical means is used to clear a blockage and a mechanical shock occurs. It is important, therefore, particularly in the more complex buildings such as hospitals and laboratories, for the designer to consider in detail not only the appliances to be used but also the nature of the effluent likely to result from processes going on within the building. Information of this kind is basic to good design, the specification of materials, and the preparation of maintenance schedules.

Note on pipe sizes

In preparing the foregoing design information the small variations between pipes of different materials, although nominally similar sizes, were considered. Between such pipes in cast iron, plastics, copper, steel and glass there can be differences in bore of several millimetres. In general such differences are not critical for performance in the fields considered in this chapter and thus a range of pipe specifications may be regarded as equivalent in relation to the recommendations made. The actual bore and nominal size of pipes should, nevertheless, be compared because appreciable differences in bore may have an effect to be considered on flow capacity or on flow regime — for example, from partial-bore to full-bore flow. Information on nominal and bore sizes is given in Standards.

6 Solid Waste Storage, Handling and Recovery

Solid waste collection and disposal fall within the domain of municipal and civil engineering and traditionally have not been a prime concern of the architect or services engineer. Increasingly, however, the designers of buildings and their services are required to take an interest in the storage and handling of waste within the curtilage of a site. This is resulting partly from the increasing amounts of waste and partly from the increasing use of engineering hardware to be accommodated in and around buildings. The reclamation and reuse of materials are of increasing importance too. The hardware involved is used partly for reasons of hygiene and aesthetics and partly to ease the handling problems and achieve economies. A saving in manpower for refuse collection may result, for example, if money is spent on a machine for reducing the volume of waste on site. With the more complex mechanical installations such as those involving pipelines, costs-in-use may well be higher than with conventional collection but may be justified on environmental grounds. Waste handling in and around buildings has, therefore, to be treated increasingly as an engineering service requiring proper planning, design, specification and installation, with reference to the needs of the particular buildings, the environment of the surrounding area, and recycling.

A range of methods is involved in this field. The traditional dustbin with manual collection has offered a cheap and generally acceptable solution. This is often combined with the provision by local authorities of sacks, usually in plastics or sometimes paper, to line dustbins. Some authorities provide sacks for the separate collection of paper and card and of metals and plastics, for recovery and reuse, also bins for holding organic waste. Wheeled bins are increasingly common and storage chambers may be provided either for an individual dwelling or for communal use to facilitate handling and separation. A simple method in the larger buildings consists of chutes, leading to storage chambers, to reduce the distance over which the waste has to be carried by the occupants and to reduce the number of points from which it has to be collected. Mechanical devices for reducing the volume of waste collected in storage chambers are used. The incineration of waste in storage chambers in buildings is practised abroad, although discouraged in many countries including the UK for reasons of air pollution. Within dwellings, waste grinders installed in sinks afford a means for rapidly disposing of putrescible waste into the sewers and thus supplement the normal collection system. The Garchey system within buildings provides a pipeline installation operating under gravity that is intended to cope with a greater proportion of waste than dealt with by sink waste grinders, including tins and bottles. Disposal at a central point within the site is by incineration or into a tanker vehicle. The Centralsug system provides a pipeline installation underground operating pneumatically; it is connected to those waste chutes within the area served and is intended to transport waste passing through chutes to a central disposal point where it can be incinerated, or collected by vehicle. Methods of collecting pulverized waste by pipeline have also been tried.

Building Regulations 1991 state the basic requirements: adequate means of storing solid waste; adequate means of access for the users of a building to a place of storage; and adequate means of access from the place of storage to a street. Government initiatives relating to recycling and composting are likely to lead to new requirements in the future, and changes in methods and systems.

Types and quantities of wastes

'Controlled waste', regulated under the Control of Pollution Act 1974, includes household and commercial wastes, the prime concern here. They amount to some 20 and 15 million tonnes respectively, each year, a little under 10 per cent of total waste arisings in the UK. Surveys of local authorities in England and Wales have, in the past, provided published information on which the left hand columns in table 6.1 are based to indicate historical trends. Such data in recent years have been much less comprehensive. The estimates in the right hand columns for the 1990s are based on information from several sources put together in a report on the UK environment (DOE 1992).

Table 6.1 illustrates some of the changes in waste arisings. The volume has increased with a growing use of disposable articles and more elaborate packaging. At the same time the density has decreased with a decline in solid-fuel heating and a consequent reduction in dust and cinder. Plastics wastes have increased, both film and containers. At an assumed density overall of 130 kg/m^3, the weekly volume per household in the 1990s was about 0.09 m^3 on average. For comparison, the weekly output from an American household in recent years

approaches twice that from a similar British one. Household waste is likely to continue to change for a variety of reasons, demographic, commercial and environmental. Evidence suggests that by 2000 compared to table 6.1 and in broad terms:

the proportion of 'fine dust' has further reduced;
plastics have increased to over 10% by weight and to some 20% by volume;
metal and glass arisings have fallen a little;
wheeled bins used more, tend to generate appreciably more waste for collection than the use of plastic sacks.

The foregoing data are, of course, broad averages. Arisings in detail depend on such factors as household size, occupancy, type of dwelling, socio-economic status and so on, and are likely to vary substantially. A single-person household is quoted, for example, as producing around 4.5 kg each week, a four-person household nearly 13 kg (Institute of Wastes Management 1984). The standard deviation in some individual properties tested was 5.5 kg/week against a mean value of 10.5 kg. It is thus significant that the size of UK household has fallen from the average of 2.5 in 1991, and is expected to reach 1.4 by 2016 (Meikle 1996).

Studies of buildings in use

Very few detailed studies within buildings have been reported in the literature. Amongst these are several which, although of some years ago, are outlined below as useful pointers to the factors that must influence the design and use of waste facilities. Data for seven blocks of one- and two-bedroom flats, centrally heated, in Watford, Brighton, Lambeth and Westminster

Table 6.1 Composition and yield of domestic waste

Historical	1935	1963	1974	1990s	
Fine dust or cinder	57	40	20	10	Dust etc.
Vegetable, bone, etc.	14	14	23	20	Kitchen
Paper and rag	16	25	32	37	Paper (33) Textiles (4)
Metal and glass	7	16	18	18	Ferrous metal (7) Aluminium (1) Glass (10)
Miscellaneous incl. plastics	6	5	7	15	Plastic film and containers (7) Miscellaneous (8)
Density (kg/m^3)	291	200	164	—	
Weight per dwelling per week (kg)	17	14	13	11.5	Weight per household per week (kg)
Volume per dwelling per week (m^3)	0.06	0.07	0.08	—	

Note: Figures in the upper part of the table are percentages by weight

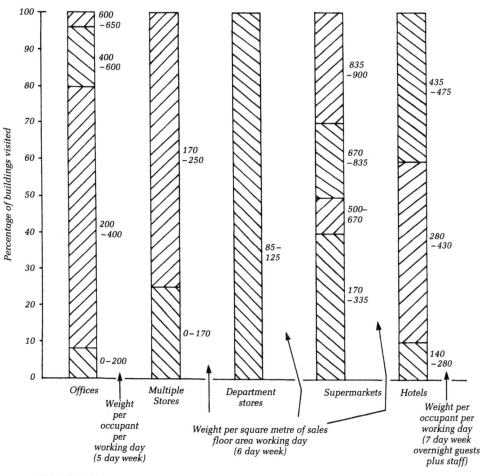

Figure 6.1   Weights of waste arising in several types of building

were obtained by the Building Research Establishment (Sexton and Smith 1972) as part of a study of waste compaction and incineration. The loads passing down the waste chutes in these buildings were determined by weighing and were found to range from 7 to 9 kg/dwelling/week. This amounted to about 80 per cent of the total waste output of the buildings, the remainder consisting of larger items such as furniture and so on. The density of the chute loads ranged from 75 to 80 kg/m³ and the volume of this waste was, therefore, about 0.1 m³/ dwelling/week, the figures reflecting the absence of dust and cinders from these centrally heated flats.

A detailed investigation of the output of non-domestic waste was also conducted by the Building Research Establishment (Anon 1971).

Based on this work the overall weights of refuse are given in figure 6.1 whilst table 6.2 combines data on its observed composition by weight and bulk densities. The data may be used to make an assessment of storage space needed, although this information does not necessarily obviate the need for a more detailed study for specific projects and the consideration of possible future trends, e.g. in business practice. The data were used in BS 5906 and in advice by the Institute of Wastes Management (1984).

The offices studied, 24 in all, were in the south of England and included examples in central and local government, administration, commerce, banking and insurance, some in single occupancy, others in multi-occupancy, some with canteens, others without. Figure 6.1

Table 6.2   Commerical waste

	Typical bulk density (kg/m³)	Multiple stores	Depart- mental stores	Super- markets	Hotels	Offices
Folded newspaper: cardboard packed or baled	500 ⎫	81	65	50	8	80
Loosely crumpled paper:	⎬					
office stationery	50					
Wastepaper (loose in sacks)	20 ⎭					
Mixed general refuse, similar to domestic (no solid-fuel residues)	150	13	31	40	55	16
Separated food wastes:						
uncompacted vegetable waste	200 ⎫	4	2	—	33	4
well-compacted, moist pig swill	650 ⎭					
Salvaged bones and fat	600	2	2	10	—	—
Empty bottles	300	—	—	—	4	—

Note: Apart from densities, figures are percentages

suggests that a figure of 400 g/occupant/working day might reasonably be taken as a guide to specifying the total storage space needed in a typical office building. As table 6.2 shows, the bulk of this waste is paper; indeed, waste paper from offices and also shops is one of the principal sources of stocks for recycling. The information in figure 6.1 taken in conjunction with that in table 6.2 provides a rough basis for estimating the make-up of the storage require-ments for the different categories of refuse likely to come from office buildings.

Waste output from the six large department stores visited did not vary greatly in terms of weight per square metre of sales floor. An upper figure of 125 g/m²/working day was obtained, and with the make-up and densities shown in table 6.2. Whereas such stores tend to provide plenty of space for display pur-poses, multiple stores (eight studied) and supermarkets (10 studied) tend to make greater direct use of the available sales area and waste outputs per square metre of sales floor were higher as shown in figure 6.1. With super-markets, in particular, turnover also is an important factor and this is reflected in the shape of the histogram. The block up to 40 per cent applied mainly to smaller supermarkets, whereas higher outputs, especially the maxima, applied to large supermarkets. Upper figures in the two categories ranged from some 300 to 900 g/m²/working day.

For hotels, of which 10 were studied, the data are presented in terms of occupants taken as the sum of the maximum number of guests that can be accommodated overnight plus the total number of staff. The histogram suggests a figure of 450 g/occupant/day of a 7 day week as a guide to the specification of storage space.

Well established practice

(i) Storage containers; sacks

The traditional dustbin remains a cheap method of storing and disposing of waste from buildings. For a satisfactory installation adequate capacity must be allowed, the bin must be placed satisfactorily both for the user and the collector, protection from the sun and adequate ventilation ensured, and an imperme-able and easily swept standing place provided. The local authority may provide a sack to go in the dustbin each week, a cleaner alternative. The cost of sacks may be offset by a saving in walking time when bins do not have to be returned empty. The noise associated with galvanized steel bins is largely overcome by the use of rubber or plastic bins. Small blocks of flats, for example, as well as houses can be served acceptably by dustbins placed conv-eniently at ground level, in groups of up to eight. Widespread use is being made of wheeled dustbins in sizes of up to 0.24 cubic metre. As a guide, location may be determined on the basis that a householder should not have to carry refuse more than 30 m, and that collectors should not have to carry bins more than 25 m.

Approved Document H6 associated with the 2000 Building Regulations recommends that space should be provided for storage of containers for separated waste (i.e. waste recyclable stored separately from waste not to be recycled) having a combined capacity of $0.25\,m^3$ per dwelling, or as agreed with the waste authority.

To improve storage, simplify collection and make mechanical handling possible, larger communal containers placed in chambers are often used in blocks of flats and in commercial areas. Special collecting vehicles are required to handle and empty the containers which are much larger than dustbins. About $0.08\,m^3$ of storage per dwelling has sometimes been required for a weekly collection; table 6.1 gives food for thought, but also see above. Containers up to $1\,m^3$ in volume are common and require vehicles that lift and tilt. Bulk containers of $10\,m^3$ or more may be used and require more sophisticated transport arrangements. Suitable cleaning arrangements for containers are to be recommended.

(ii) Chutes

Chutes for the disposal of waste have application in various types of building including, for example, specialized requirements in hospitals, but they are by far the most common in housing. In residential buildings of more than four storeys it is not practicable to require that all rubbish is carried by the occupants to containers at ground level. One way of overcoming this problem is to provide a chute running down the building which will take refuse from the upper floors and deliver it to a container at ground level. Occupants should have no more than 30 m to walk to a point of access to the chute, and for reasons of hygiene and maintenance not more than six dwellings should share one hopper. Waste, preferably wrapped, is introduced by means of close-fitting access hoppers at each floor containing a dwelling, and falls into a container which is housed in an enclosure at ground level. The chute is carried up to roof level and vented.

Chutes may be free standing, e.g. figure 6.2, or incorporated in builders' work ducts. They are required to have smooth non-absorbent inner surfaces, be non-combustible, and typically are constructed of glazed fire-clay or concrete pipes with specially designed hopper units for receipt of refuse. A pipe diameter of

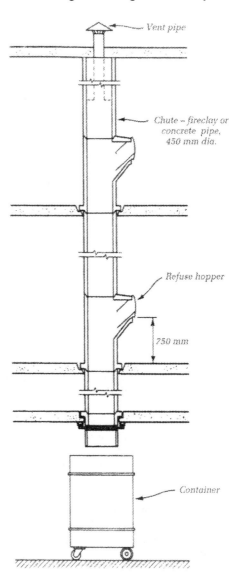

Figure 6.2 Diagram of a free-standing chute for a block of flats

380 mm has been used for some installations but experience generally is that 450 mm is to be preferred, reducing the risk of blockage. The top vent pipe should be not less than $350\,cm^2$ in area. In field surveys of compactors to be described later it was found that even the 450 mm diameter chutes to which the machines were connected sometimes blocked. The blockages were, however, generally related to the dimensions and angle of the make-up piece between the chute and the machine. Figure 6.3

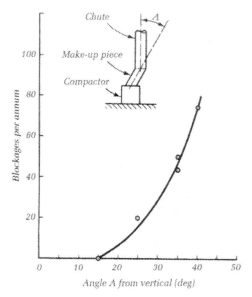

Figure 6.3 Relation between chute blockages and angle

has been prepared from the results of this work and suggests that connection pieces should as far as possible be vertical, as with the chutes themselves.

Waste chutes have an influence on planning. They must rise vertically through the building, and dwellings arranged in handed pairs can make maximum use of chutes. Nuisance and smell from waste, noise and the possibility of a fire in a chute are amongst the factors to be considered in planning and location. Building Regulations 1991 set out rules as to the minimum weight of walls to separate chutes from dwellings, to reduce noise. Fire requirements are also specified. Disposal points are often arranged on communal landings but this can result in nuisance from careless handling of refuse. Disposal points within the area of the dwelling, on the other hand, encourage careful use and maintenance. The kitchen, although convenient, is not acceptable from the point of view of hygiene, and disposal from a private balcony is often a reasonable solution.

Planning must also take account of the importance of vehicular access to ground level containers covering an area larger than the chute itself. Large items that will not go into the disposal hopper have to be carried down separately to a special communal container. BS 5906 suggests that about 10 per cent

of household waste consists of articles too large for a hopper of a chute-fed system. In the study reported in a previous paragraph it was found that these may amount to some 20 per cent of the total waste output from a block of flats.

(iii) Compactors

With increasing quantities of waste arising, interest has developed in compaction as a means of reducing the volume of waste on site, thus reducing the storage space required and easing collection problems. The machines available cover a wide range. An early example is shown in figure 6.4, a unit capable of being installed at the base of a chute in a small block of flats with the purpose of compressing the refuse received from the chute by paper sacks erected on a carousel. At the other end of the scale, compactors may serve large town centre developments as in an example in north London which includes 140 shops, a market with some 50 stalls, two large residential blocks and a few offices. Waste from the area is taken by electrically propelled hand trucks of some $3\,m^3$ capacity to a $7.5\,m^3$ container and a compactor of $128\,m^3/h$ maximum throughput. Some 15 containers are removed each week, each with an average load of some $2000\,kg$ and a compacted density of some $250\,kg/m^3$.

Detailed studies of small machines fairly recently introduced to the market were undertaken by the Building Research Establishment to provide information on their performance in use and running costs (Sexton and Smith 1972). The investigations were carried out in five blocks of modern flats. One of the machines studied is illustrated in figure 6.4. It has three moving parts operating pneumatically: a plate which moves across to isolate the chute from the machine before the compression takes place, a ram which compresses the waste into sacks, and a carousel which carries 10 sacks and rotates to bring an empty sack into the filling position when the previous sack is full. Variants of this machine exist, some incorporating a receiving hopper and feeder mechanism which provide some prepacking of the waste before it is fed to a sack for compression. The mechanism is set in motion by a trip arm activated by incoming waste. A machine cycle typically takes some 30 to 60 seconds depending on whether some means for

Figure 6.4 Example of a waste compactor (DEVA) in a block of flats (Building Research Establishment; Crown Copyright)

prepacking is incorporated. Continuous operation depends on manual removal of the sacks when full and their replacement with empty ones.

On-site measurements showed that the compactors produced neatly packed sacks weighing about 25 kg, with a variation of about ± 3 kg, in which the waste was reduced to about 25 per cent of its original volume. This weight of sack is convenient for manual collection in conjunction with a standard vehicle. The machines

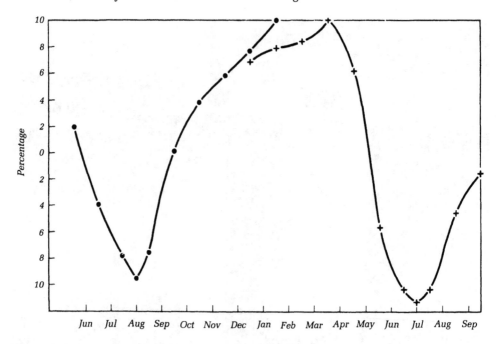

Figure 6.5   Examples of seasonal variation in waste output in a block of flats

were reasonably quiet in operation, and reliable. Investigation of the usage patterns of the machines was also carried out, both over short periods and over some months. It was found that maximum usage tended to be between the hours of 08.00 to 11.00 and 18.00 to 20.00, and in the winter months, although seasonal variation was small as figure 6.5 shows. Cost data from the study of flats in Watford are given in table 6.3 and it must be borne in mind that these relate to 1970. Little criticism of the machines studied was made, and this type and other machines have been increasingly used since 1970.

In the 1990s there are four main categories of compactor available, together with some individual types, according to Walsh (1993). Static ram compactors remain common, although superseded to some extent by more sophisticated machines that take up less space. Portable machines giving 3 or 4 to 1 compaction ratio; rotary compactors, either bag or bin types, giving a 7 or 8 to 1 ratio; semi-automatic balers giving a 9 or 10 to 1 ratio — such machines offer a wide choice for a range of circumstances. Walsh includes four case studies showing the potential for cost saving by introducing portable or rotary machines for a hotel and in three industrial applications. He notes that sites are kept cleaner and tidier through the use of compaction. He also notes their potential for use in recycling schemes.

Table 6.3   Summary of cost data for compactors (Sexton and Smith 1972)

	Deva 1	Deva 2	Deva 3	Deva 4
Dwellings served	52	48	52	48
Annual load kg (estimated)	21 800	17 400	19 400	20 600
Load/dwelling/week kg (estimated)	8.1	7.0	7.2	8.3
Annual cost (£)	354	298	323	315
Cost/head/annum (£)	3.22	2.86	2.94	3.03
Cost/dwelling/week (P)	13	12	12	12
Running and maintenance costs/dwelling/week (P)	5.0	3.1	3.8	3.8

Notes: Annual cost is for owning the machines and excludes the cost of chutes and hoppers. Running and maintenance costs exclude the cost of annual service

The successful introduction of a portable compactor for a block of flats in Hackney, London, is described elsewhere (Institute of Wastes Management 1984). The original system of cylindrical storage containers of 1 cubic metre capacity required very frequent emptying. A 6 cubic metre portable compactor was installed in the communal waste storage chamber. Residents did not have direct access to it but could insert refuse through an opening in the existing shutter door. Time clocks activated the machine to clear and compact the waste every 15 minutes. This safe and environmentally more attractive system, with its cost effectiveness, has led to a wider use of this method for blocks of flats. Walsh (1993) notes that the additional transport involved when emptying the container of its compacted refuse can be a drawback, but some contractors regularly replace a full compactor with an empty one in some applications.

In considering the selection of a compactor for a particular duty, the following requirements should be borne in mind as a checklist:

1. *Effectiveness.* Machines should produce significant and consistent reductions in volume.
2. *Reliability.* The equipment should be adequately engineered. It should have a reasonable life expectancy, need only nominal day-to-day attention and a minimum of servicing.
3. *Ease of use.* The machine should be simple to operate and not require any special skills.
4. *Safety.* Machines should be safe in use and should not cause excessive vibrations or a fire hazard. They should be adequately foolproofed.
5. *Noise.* Quiet operation is essential.
6. *Cleanliness.* The need to handle raw refuse should be reduced to a minimum. Machines should be easy to keep clean and designed to prevent liquids that exude during compaction from accumulating in inaccessible places. Filled containers should not leak whilst awaiting collection.

Some engineering developments

(i) Incineration on site

Incineration on site, as with compaction, has the objective of reducing the volume of waste to be removed from a building. Successful incineration can give a reduction to as low as one-tenth of the initial volume, leaving an inoffensive ash and allowing less frequent collection. The method thus has some attractions, particularly for larger buildings such as flats and hospitals. A few trials of a chutefed incinerator in a block of flats have been made (Sexton and Smith 1972). A combustion chamber stored and dried refuse prior to ignition by the gas burner. Larger objects could be fed directly into the chamber which had appropriate safety devices. Waste gases were cleaned by a secondary burner and a water spray unit before exhausting up the chimney. The incinerator operated automatically once the frequency of operation and duration of firing had been set to give a balance between the degree of burn-out of the waste and the gas consumption of the burners. Typically for such a machine there might be 10 main burner firings a day, each lasting 7 or 8 minutes. Generally the equipment was found to perform satisfactorily although some unburnt residue was found in certain of the tests.

The main problem is that such equipment adds to air pollution through the emission of smoke and smell, dust and grit; toxic gases may also be emitted. Continually efficient maintenance and operation is necessary for good standards to be maintained and it is doubtful whether this could be achieved in most situations where a caretaker for a block of flats has the responsibility for the plant. Smaller incinerators for individual dwellings can suffer even more from this problem and, in addition, with incinerators in low buildings it is usually difficult to ensure that gases discharge at a height sufficient for efficient dispersion. Larger-scale municipal incinerators, on the other hand, can be of a size to justify suitable effluent treatment plant and tall chimneys; and efficient maintenance is more likely to be achieved in hospitals with engineering maintenance staff. Higher emission standards, however, raise questions about such uses too, as does the disposal of the toxic ash residues.

Incineration on site has been much more widely used in the USA than in the UK but in the 1960s and 1970s gave rise to concern in the major cities for reasons of air pollution. In New York, for example, where some 15 000 buildings had such equipment installed, legislation was introduced to require existing plant to be fitted with smoke control devices and to prohibit incineration in new dwellings. Considerable research and development has been conducted by manufacturers and others in the USA in an attempt to overcome such problems. Experience shows, however, that incinerators are unsuitable for dwellings. Applications in such special situations as hospitals, for certain trade premises, and in larger municipal installations, and the problems that can arise, are discussed in the report of the Royal Commission on Environmental Pollution (1993).

(ii) Disposal into drainage systems

(a) Garchey system The Garchey system was developed in France but UK rights are held by a British firm. Its essential feature is a special receiver fitted beneath the kitchen sink as in figure 6.6, enabling refuse to be discharged into a type of drainage system. Refuse is introduced into the receiver through a large plug in the sink. A smaller plug inset in the larger one deals with ordinary waste water discharge. When sufficient refuse and water have accumulated in a receiver a plunger is raised and the contents of the receiver are discharged into 150 mm or 180 mm diameter cast iron stacks which may also receive the flow from other appliances but not appliances for soil. Some residues such as paper may have to be washed out of the receiver and the system may, therefore, increase water consumption.

In the original version of the system the waste was collected in pits near the foot of each stack and then drawn through pipes by suction to a central plant for dehydration and burning in a furnace intended to produce heat for laundry use. Such a central plant was expensive, and hence the system was limited to large installations. Combustion of the wet waste was difficult and the utilization of the heat from burning not always very effective. The British licensees, therefore, developed a collecting vehicle which replaces the central plant and itself draws the waste from the collection pits. The vehicle is very much cheaper than the plant and can serve a number of buildings. Early installations with central plant in the UK were in Sheffield and Leeds; more recently installations using collecting vehicles have been installed in several places.

The Garchey method can handle most types of refuse, including bottles and tins, but the size it can deal with is limited. Overall it may cope with some 50 per cent of the total output of a dwelling and, therefore, quantities of refuse besides large items must be carried down separately and deposited in a communal container. Difficulties that have arisen with the Garchey system include blockages in the stacks and drains. After long service, some corrosion of pipework and erosion of bends receiving the continual impact of refuse may occur. It is, therefore, essential to provide ample access into the pipework and space for the use of clearing equipment. A further problem in some installations has been smells from the receivers and from the collection pits, the latter necessitating careful location of the tops of stacks that act as vents.

(b) Grinders Grinders installed under sinks and using perhaps 4 to 8 l of water per person per day have been increasingly popular in recent years, particularly in flats as a means of improving on the service provided by dustbins. They can deal with food wastes and other small items, perhaps 15 per cent of ordinary domestic refuse, but not tins and bottles. Care must be taken when siting the grinder and when mounting and connecting it to minimize the risk of trouble from noise. Larger machines are available for specialist applications in catering establishments and hospitals, and for particular industrial uses.

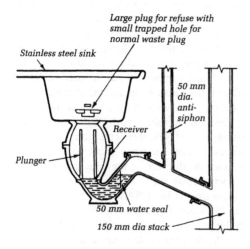

Figure 6.6   Refuse receiver for a Garchey system

(iii) Handling by pipeline; pneumatic methods

With pipelines used widely for conveying many different materials it was natural that attention should be directed towards their use for handling solid waste. Chutes and the Garchey system, as already described, make use of pipe stacks operating under gravity, but a greater sophistication is possible once the idea of using mechanical power is accepted.

One approach that has been considered both in Europe and in the USA is to provide equipment in dwellings for grinding domestic refuse and discharging it into the sewers, aiming to deal with a much larger proportion of the refuse than sink waste grinders. Costs militate against such a proposal and treatment works would have to cope with solids of a different character and several times larger in volume compared with those in sewage. In addition, the mixing of refuse with sewage militates against the reclamation of useful materials from solid wastes. As an alternative to the use of sewers, a separate pipe network for the district-wide collection of pulverized refuse suspended in a slurry has been considered. Costs again are substantial and the method introduces a new series of problems, in disposing of the wet material and treating the contaminated water that is reclaimed.

The pneumatic handling of dry waste shows greater promise than wet systems, and installations of a particular type originating in Sweden and first used in 1966 have been constructed in different parts of the world. This method, known as Centralsug, is simple in concept. In a situation where several blocks of flats are to be constructed on a large site, underground pipes are laid between the blocks and linked to a central plant house containing an exhauster and a storage silo. In each block normal chutes are provided and connected to the underground pipe by means of special valves. When a valve is open and the exhauster is working, waste from the chute is sucked to the central storage point where it may be stored and incinerated or removed by vehicle. The pipe network thus replaces vehicle collection within the site itself.

An installation of this type was built under licence at Lisson Green in London and began operating in 1972. It served some 40 buildings, mainly five- and seven-storey blocks of flats

with some terraces of town houses, in all about 1600 dwellings and with 56 chutes. The system was investigated by the Building Research Establishment (Smith 1976) in a period when occupation rose from 300 to 1000 dwellings. Generally satisfactory operation and performance were reported, all the normal types and quantities of waste and some dense, heavy and awkward articles being conveyed successfully. There was some evidence of the pipelines becoming dirty and increasing in roughness, suggesting for the longer term the likelihood of an increased power requirement. Blockages were not a problem and the main trouble on the site was associated with the central silo and its mechanism for discharging refuse to the incinerator. A more recent study (Carter 1979) has confirmed this problem and led to the replacement of the silo and incinerator by a large compactor producing compacted refuse for removal by vehicle. Blockages and their clearance have been found more of a problem than suggested by the earlier study.

Costs for such an installation are, of course, substantial and amounted in 1972 to £368 capital per dwelling. Operating costs include such items as wages for site staff, gas and electricity and were offset to a small degree by heat reclamation. The best estimate for a full-load annual operating cost was £15.5 per dwelling (Smith 1976).

Pneumatic handling has thus been shown to be feasible, at a cost. Further development of such schemes might involve the waste, apart from very bulky and difficult items, being pulverized before entry to the pipe system, thus permitting the use of smaller pipes (Courtney and Sexton 1973). One significant advantage of pneumatic systems over wet, pumped systems is that the waste remains in a state where useful materials can be most readily reclaimed by appropriate separation techniques.

(iv) Further notes on equipment

Shredding and crushing of refuse give special advantages in certain applications. Small machines for breaking bottles used in hotels and restaurants to reduce the space required for storing non-returnable bottles give volume reduction ratios of around 10:1. Similar reductions are obtained with machines which crush or cut tin cans into small pieces.

Shredding hardly reduces the bulk volume of paper and board products but, if used in conjunction with baling, a more dense bale is produced which makes the process economic. The type of feedstock determines whether shredding before baling is necessary; where the input is tightly bundled paper, books or similar dense paper products, shredding, or riffling as it is sometimes known, is justified. In contrast miscellaneous waste paper can be baled easily in its crude form without shredding. Grinding is similar to shredding in that the method does not generally produce worthwhile volume reductions when handling paper or refuse containing a high proportion of paper. Dry grinding is widely used on a municipal scale as a predisposal process — the ground product compacts steadily in a landfill and is unattractive to flies and rodents — but is not often used for on-site treatment.

Waste management — recovery

Whilst landfill has been the least expensive of proven methods of waste disposal, there is an increasing emphasis on the reclamation and reuse of materials, partly to aid disposal and for environmental reasons but more especially for reasons of resources, both materials and energy. The EU Directive 1991 on waste disposal stipulates that the competent authorities in a member state must produce waste disposal plans and that states are to encourage recycling. Recyling goes beyond the basic reclamation and reuse of materials from waste and implies processing to produce marketable products. Governments in various countries have set targets for recycling: in the UK the target is to recycle 50 per cent of the recyclable element of household waste by the year 2000 (DOE 1992), in effect some 25 per cent of all household waste. Subsequently in the White Paper Making Waste Work (DOE 1995) this target was retained as a broad aim and a new emphasis was given to the strategy for waste management as a whole. A hierarchy was identified involving prevention, reduction, reuse, recovery and disposal. Recovery could include recycling, composting and energy generation. Within this strategy several targets were set down including the recovery of 40% of municipal waste by 2005 (where municipal waste includes household, street cleaning and

some commercial and trade waste). Household waste minimization was also emphasized, in parallel with policies towards the responsibility of producers, as in the Producer Responsibility (Packaging Waste) Regulations, 1997. The scene has been reviewed by Shouler (1998).

Starting from the currently low level of recycling in the UK — some 10% only of household waste by the end of 2000 — various initiatives have continued and new schemes begun to meet the targets of 'Making Waste Work'. The potential is there. With reference to the right-hand headings in table 6.1, over 50% of paper and plastics and a much larger proportion of glass and metals are estimated to be recyclable. Some paper is compostible, as are kitchen waste and some dust. Considerable scope thus exists, with new schemes being developed to aid recycling (DOE 1991). Contributing to the development of this potential will be changes in some of the well-established practices of waste collection and handling already described, together with the development of new approaches and methods. Some possibilities are outlined in the following notes on collecting recyclables.

Published waste collection data show few opportunities for collecting recyclables provided by local authorities in the mid-1980s; some 28% provided such opportunities about 10 years later, and this trend is expected to increase to meet targets. Current options for collecting recyclables include the use of coloured boxes of some 40–50 litres capacity, plastic sacks of different colours, and a separate wheeled bin which may be subdivided to hold different materials. These are collected from kerbside; glass bottles and jars are normally excluded from such schemes. The householder is expected to undertake the separation in kitchen or utility area, perhaps using small trays or crates to assist in this operation. Separate facilities for this purpose can be foreseen in the future, a commercial need beginning to be recognized as material in planning policy for housing and commercial and industrial developments.

In multi-storey housing options based on those mentioned above may be introduced, with residents carrying recyclables to bulk containers at ground level or to a local facility. Many housing blocks include waste chutes as described earlier and attempts have been made to adapt this facility to accommodate recycling. An American system introduces an electronically operated turntable holding up to six containers below a chute. A resident using a chute at any floor level can line up the container at the base of the chute appropriate to the material being deposited, waste or recyclable. Effectiveness in practice is unclear; capital cost is considerable and it is untried in local authority housing. Another option to adapt chutes might be to introduce a 'recycling hour' — a set of containers for recyclables put in position below a chute at a designated time for an hour a day, perhaps using different times for different materials. This seems to be impractical for many reasons.

In developing schemes for handling, collecting and sorting waste, the potential hazards require attention. These include health and safety considerations both for the householder and operatives, perhaps physical, chemical or biological. In addition, fire can be a hazard, especially associated with the storage for longer periods of paper and card. The results of a 3 year study of recycling in Sheffield supported by government, industry and amenity bodies have been reported (Coggins 1993). The report deals with such matters as kerbside collection for recyclables and drop-off programmes, comments on both household and commercial waste recycling, and gives limited information on costs. It is noted that unstable markets and fluctuating prices for products lead to an irregular income, a problem when set against capital and operating costs. The experience has shown, not unexpectedly, that recycling must be planned within the framework of waste management as a whole and include both household and commercial waste. Central materials recovery facilities are likely to be important for the future.

Already in the UK there are several sophisticated central complexes for the receipt, handling and treatment of wastes, incorporating advanced techniques for their separation and reclamation, and for the production of fuels and other products. The design of products to facilitate recycling is likely to receive greater attention.

Environmental assessment

Greater attention to the role of buildings as regards environmental impact is being given through the UK Building Research Establishment's Environmental Assessment Method (BREEAM). Through this voluntary scheme developers, owners or occupiers can obtain an assessment for new or existing buildings based on criteria covering a wide range of issues. The aim is to cover qualities over and above those required by law. The list of 'credits' so obtained, together with an overall rating for a building, form an environmental statement, a result that is proving of increasing interest in assessing environmental impact. The scheme has been established through a series of reports (Building Research Establishment 1991–1993), and several topics within the present publication fall within its scope. Among these is recycling, given credit in environmental assessment on the following basis:

Credit for the incorporation of separate dedicated storage space for recyclable materials;

Superstores: 4 m² for up to 2000 m² floor area
4 m² for every extra 1000 m²
New and existing offices: 2 m² for each 1000 m² floor area with a maximum of 10 m² Very large existing offices may require more — at least 10 m²
New industrial and warehousing units: Hard standing area for at least two recycling skips, in addition to those for normal refuse.

Credit for a recycling facility for individual use:

Superstores: At least three separate storage containers for customer use, signposted and with easy access
New homes: Four containers, total capacity at least 0.24 m³ per household on hardstanding within 3 m of an external door of a single-family house, or 10 m from doors of other dwellings. Space at least 2 m², protected and not more than 20 m from a collecting vehicle stop.

7 Rationalization of Services

Traditional practice for many years was for buildings to be designed with little regard to the engineering service installations required. During construction, specialists arrived on the site and fitted in the services — plumbing, lighting and heating — as best they could. There was competition for space and problems of fitting everything together. Eventually the building was completed and the services on the whole worked satisfactorily. The eventual problems that might arise when repairs and maintenance were needed, or services needed to be extended or replaced, were not usually given close attention.

Whilst such methods were adequate when buildings and their services were simple, and are not unknown today, different approaches have gradually been developed since the 1920s and 1930s. New methods became necessary as buildings became larger and more complex, new constructional techniques were evolved, and the service installations themselves became more extensive and were designed to meet more onerous requirements. In many circumstances it was no longer sufficient for the architect simply to sketch in the positions of the heating unit and sanitary appliances and perhaps allow a small space on the plan for a service duct without considering the need for ventilation ducting and its accommodation, the extent of horizontal and vertical pipe-runs, and so on. The integration and coordination of services in relation to structures as a whole began to be studied, especially for buildings such as blocks of flats, offices and hospitals in which the engineering services could account for 25–50 per cent of the overall capital cost. Architects began to pay more attention to engineering requirements, and services engineers were consulted at earlier stages in the design and construction of larger buildings. For larger buildings comprehensive planning of the services was undertaken and techniques such as pipework models sometimes proved economic. Coordinating engineers were used to oversee the work and had responsibility to continue their co-ordinating role on site. The sanitary engineering services, with which this book is concerned, have not been neglected in this context. Even in the early days of the introduction of sanitation into buildings, it was obviously cheaper and simpler in terms of pipe runs to group sanitary appliances reasonably close together. In much low-cost housing and also in many multi-storey buildings with accommodation of a repetitive kind, sanitary and kitchen accommodation were grouped together. Studies were carried out to simplify and standardize pipe runs and to facilitate some degree of prefabrication where sufficient numbers of similar units were to be constructed. Units of varying complexity known as service or installation cores were devised, especially abroad, incorporating a range of components and intended for prefabrication either in a workshop on site or in a separate factory.

In the UK impetus was given during the 1950s and 1960s to the study of such methods as part of the development of industrialized or factory building. The urge for such study was economic. Large programmes of prefabricated building, particularly of blocks of flats, were known to be in progress in various countries and their application in the UK seemed overdue. Many firms, including those in the building industry and some with general engineering interests, investigated the possibilities and arranged either to offer foreign systems under licence or to develop new systems. For a period, several hundred different industrialized building systems were available in the UK and they incorporated various forms of service core, some with extensive prefabrication into a major service assembly, others relying on much simpler arrangements.

The boom in industrialized buildings in the UK did not last for several reasons, among the most important being cost, the lack of sufficient production runs to justify expenditure on factories producing a large number of more or less similar buildings, and the restrictions on design and planning which factory methods implied. By contrast, such methods were then more appropriate in a country such as the Soviet Union which had a massive building programme requiring up to 2 million dwellings a year and where substantial limitations on planning were accepted in the interest of housing the people to at least a minimum standard.

Since the 1960s many fewer industrialized building systems have been on offer in the UK. It has become apparent, for reasons of planning, design and cost, that the more sophisticated types of service core construction have severe limitations for British application. Attention has been given instead to simpler concepts, albeit with improved coordination and integration of services into buildings and with careful attention to dimensional coordination. Many different assemblies and methods have been used in different classes of building. The greatest emphasis has probably been given to the rationalization of services in housing and it is appropriate, therefore, to discuss the subject with reference to dwellings and with an eye to the three main approaches that have been used:

(i) prefabrication of services,
(ii) the use of service walls, and
(iii) the use of heart units in the form of rooms or part-rooms.

Prefabrication of services

There is a range of options including:

1. Precutting and prebending of pipe lengths, with fittings attached to the prepared pipes. Packs of such components may be made up for particular houses or flats and a diagram prepared for the plumber to use.
2. Preforming of assemblies in jigs, either in a site workshop or in a factory. Several manufacturers operate in this way and supply substantial numbers of assemblies for dwellings and for some other types of building.
3. Factory construction of packages consisting of a frame containing a range of services and with or without cladding. The package may include heating components and water storage vessels as well as soil and waste pipes and ventilation ducting as, for example, in figure 7.1 (small piping not shown).

For low-cost dwellings it has been reported by Purdew (1969) that by proper planning, including standardization of plumbing layouts, precutting and bending of pipes and the preparation of component packs, the time normally spent on a dwelling by a plumber and mate can be halved. The two fixings, i.e. installing tanks and the main pipe runs and later connecting the sanitary appliances to the pipes and taps, might each be done in half a day or thereabouts. The aim should be to organize the plumbing work to enable pairs of plumbers to be kept fully occupied on prefabrication and fixing in sequence with other trades.

In a review of the installation benefits of pre-assembled space and water heating systems, Hayhurst (2001) emphasizes the trend to carry out skilled work off-site where practicable. The benefits resulting from using a pre-plumbed and pre-wired system of unvented domestic water heating include saving in on-site installation time (more than 50 per cent), improved completion times and aids to maintenance; whilst the packages can be factory tested and greater overall reliability results from their use.

What is good practice in this field depends to a large degree on the size and extent of the building programme, whether it involves a large services contractor working on several sites at a time or is a much more limited operation. With the former it is often practicable to prefabricate assemblies of pipework on or off the site and to move teams of installers about as necessary, with the advantage of better productivity, quality control and less wastage of materials, and hence the likelihood of cost savings for the client. With a smaller operation it would be good practice to seek for standardization of plumbing layouts as far as possible, prepare the structure to receive

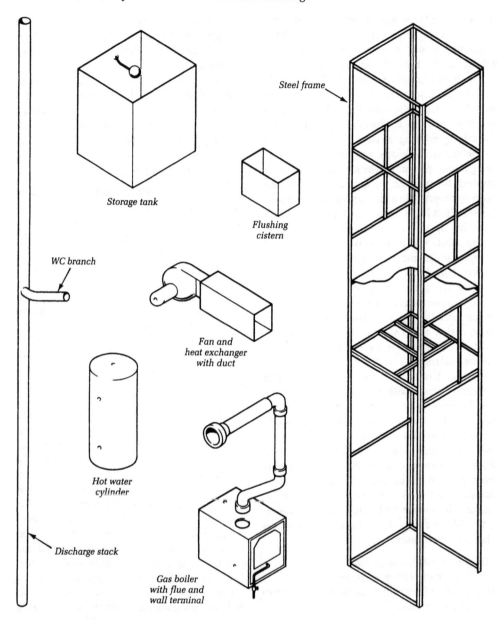

Figure 7.1 Diagram to illustrate a package service core — main components to be fitted in steel frame, excluding small piping

piping, and arrange for the assembly of small packs of components. The plumber may use a jig to prepare assemblies of these components for each dwelling.

Service walls

From the early days of the prefabrication of buildings and services the concept of a 'service

wall' has appeared attractive, i.e. a wall unit prefabricated to contain services that might include, as well as plumbing pipework, ventilation ducting, a flue and a refuse chute. Various methods have been tried, some incorporating the service wall as a structural component, others involving a lightweight unit not itself a structural wall. Figure 7.2(a)–(d) illustrates the principles of several methods

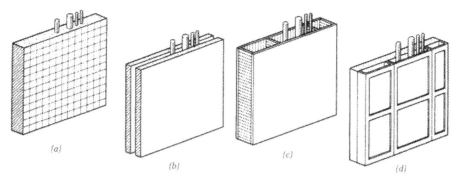

(a) Piping is embedded directly into a solid wall of concrete or possibly brickwork. The method foregoes easy access to the pipework and assumes that the building is sufficiently accurate to permit effective and easy floor-to-floor connections. Whilst the unit and pipework can be moved as a whole, its weight (its thickness must be sufficient adequately to accommodate the services) makes for difficulties in handling and lifting, and restricts transport

(b) Pipework is prefabricated off site, brought to the site, and built in between two massive walls. As with (a) the method foregoes easy access as the piping is bricked in between the structural walls during the construction of the building skeleton

(c) Pipework is prefabricated off site, brought to the site, and built in between two light walls, leaving the spaces between the walls unfilled. Access is improved compared with (a) and (b)

(d) Pipework is prefabricated off site, brought to the site, and incorporated into a light-framed partition in front of a structural wall. Access to the pipework is improved and the partition may incorporate cupboard space. Sound insulating materials may be incorporated and pipes suitably insulated. Good access is feasible

Figure 7.2   Diagrams to illustrate service walls

that have been tried. The captions refer to some of the factors to be taken into account in developing and using any form of complex service unit, e.g. weight, shape and size, transport, handling, lifting, construction tolerances, and ease of access to piping once fixed. Additionally the design of the pipework and the choice of piping materials and fittings are basic, as are such aspects as noise control, thermal insulation and fire precautions.

For a practical example of one such method that has been employed in a building system used extensively in Scandinavia, refer to figure 7.3. A concrete service wall is prefabricated as one of the basic building blocks. The service passages are formed using steel tubular inserts that are withdrawn after the concrete is cured. The wall incorporates a main vertical ventilation duct, vertical ventilation branches, a space to take soil, water and heating pipes after erection of the panel, and spare passages to save material and, incidentally, to permit additional ventilation if needed. The vertical ventilation branches are made as long as possible to help limit noise transmission; local fire regulations permitted this method which is the reverse of the shunt used in the UK. The

arrangement is chosen so as to limit joints to one per storey as an aid to leaktightness (which experience with concrete extract ventilation ducts in the UK has shown to be a substantial problem unless adequate jointing methods are used). In this particular system jointing is achieved in the following way. During erection, steel sleeves are inserted into the tops of the passages in the wall and concrete is poured in almost to a level with the top of the floor. The sleeves are withdrawn after the concrete is cured and a thin layer of fine mortar is spread to act as a bed for the next panel. As to the branches, each inlet contains a galvanized socket cast in the concrete to receive a control grille. Apart from the question of the ventilation tightness achieved in this system, the accessibility of pipework is extremely limited. Methods that provide better accessibility are to be preferred.

In the UK several of the industrialized building systems that came on the market from the 1950s onwards used standard service walls of one form or another. Such methods are not now common for the reasons advanced already against industrialized systems — costs, difficulties in ensuring sufficient scale and

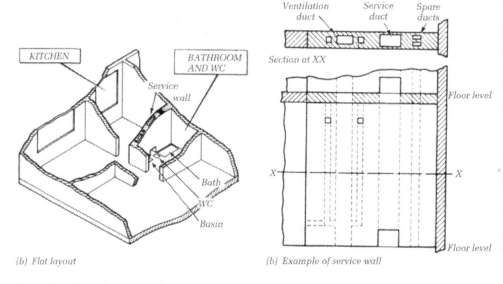

(b) Flat layout   (b) Example of service wall

Figure 7.3   Flat with service wall

continuity of production, lack of flexibility and restrictions on design and planning; in general the difficulty of competing with the simpler and more traditional methods of installation and construction.

Heart units

As an extreme in the sophistication of service cores, several firms in the UK and abroad have developed so-called 'heart units', i.e. complete bathroom/kitchen assemblies, factory produced, containing all services, fittings and fitments and ready for coupling on to the main drain, and to water and fuel supplies. This idea was pioneered in the USA in the 1920s and 1930s by Buckminster Fuller. Again for the reasons set out above, such units have not proved marketable except for limited use.

One early example was the Scandinavian unit for good-quality bungalows. A factory in southern Sweden produced several different layouts of unit including refrigerator, cooker, oil-fired boiler, sanitary appliances and pipes and ducts, for shipment both inside Sweden and across Europe. A typical unit might weigh 8000 kg. Construction on site was organized so that the plumbers were there early to connect the heart unit to mains water and drainage and to floor heating coils, after which they were not

needed. The installed heart unit represented some 15–20 per cent of the cost of the bungalow.

Considerably less sophisticated forms of packaged bathroom and/or core unit were developed for use by UK local authorities in the rehabilitation of older dwellings. The requirement was for a unit to be added to an existing dwelling, often 'clipped on' externally, with the minimum disturbance to occupants and as quickly as possible. Some bathroom units including the necessary roof, floor and external walls have been supplied, as have core units for building into older buildings or into bathrooms added during rehabilitation. Typically a core unit would include a cold water storage tank and hot water cylinder assembled in a steel frame with connections ready to be made to the sanitary appliances. Water heating may be by electric immersion heater or a gas heater, either attached to or separate from the core unit, or through connection to a solid-fuel boiler. Such a core unit might weigh 100 kg and be readily handled by two men, being small enough to pass through doorways and up most flights of stairs.

The production of sophisticated bathroom/ kitchen units by factory methods, such as the Swedish example described, has many advantages. Extensive preplanning, quality control, and the use of power tools and automatic machinery are all possible, whereas such methods are much less readily used on ordinary construction sites. With factory conditions, therefore, the standard of construction of the bathroom and kitchen and the associated engineering services can be higher than is commonly achieved on site, even though unskilled workers may be employed. It is unfortunate, therefore, that these advantages may be overridden by the overall considerations of planning, construction and costs.

Rationalization of sanitary layout

Developments in dimensional coordination associated with the change to the metric system in the UK in the 1960s and 1970s tended to assist the rationalization of design and services, especially in the then large local authority sector. In Imperial measure with inch increments several hundred different shell sizes for two-storey houses, for example, could be used and the recommended 300 mm planning grid substantially reduced this number. The National Building Agency studied the possibility of working to a smaller number of standard sizes and suggested the use of a range of 22 shells for two-storey houses, covering the requirement of four-, five- and six-person households. These shells ranged in frontage from 9.9 m to 4.2 m and in depth from 4.8 m to 9.9 m. They were put forward in the interests of cost savings and standardization, and encouraged a rationalization of sanitary layouts and of services.

At the same time there have been ergonomic and user studies (DOE 1972) concerned with the definition of an 'activity space' appropriate for each bathroom and kitchen appliance, and studies of the combination of spaces to provide layouts which give satisfactory solutions in terms of space and of accommodating the services. This information was used in the preparation of Part 2 of BS 6465 (1995) dealing with space requirements for sanitary appliances. Figure 7.4, based on these studies (and figures 5.4 and 5.5), suggests dimensions for a few typical layouts of appliances and of service shafts associated with them, which lend themselves to incorporation in house designs based on a small range of house shells. The depth of 1500 mm for a WC cubicle including a service shaft can be seen as a minimum; a further

200 mm would be preferable for the door to clear the WC pan, unless it opens outwards. Note the increased size of shaft necessary with multi-storey dwellings where ventilation ducting has to be accommodated in addition to the soil stack and a few small water supply or heating pipes. Figure 7.5 gives an example of the layout of ducts and piping used in the service shaft of a 17 storey block of flats, in this case representing the arrangement on an upper floor and serving a column of two-bedroom flats. Figure 7.6 shows a modified one-pipe drainage and vent system incorporated in a service shaft in a hospital.

Requirements for disabled people are dealt with in BS 5810 and in Building Regulations 2000, Part M. Fire protection of buildings is, of course, an important aspect of building control. The 2000 issue of the Regulations states requirements as regards fire spread; adequate resistance to the spread of flame over the surfaces of walls and ceilings is necessary — this includes service shafts and spaces. Concealed spaces in the structure or fabric of the building, or the building as extended, must be sealed and subdivided where this is necessary to inhibit the unseen spread of fire and smoke. This requirement would apply, for instance, to the service shafts described above.

Requirements in use

The importance of ready access to pipe systems for inspection and maintenance purposes once buildings are in use has been emphasized in the previous paragraphs. Lack of this feature has been one of the failures in some industrialized and other types of buildings. It has been taken up as a key feature in the 1999 Water Supply Regulations and Water Byelaws 2000, Scotland.

The methods shown in figure 7.2(a) and 7.2(b) contravene the Regulations which state that no water fitting should be embedded in any wall or solid floor. The associated guidance paragraphs describe several examples of how satisfactory access may be achieved in, for instance, situations such as those shown in figure 7.2(c) and figure 7.2(d), and also in floors. The principles involved are straightforward. In brief, the long-established practice of placing pipework in ducts or chases has merit on the basis that leaks become apparent

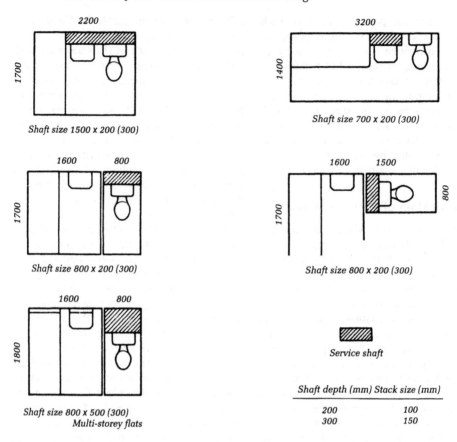

Figure 7.4   Examples of sanitary layouts with service shafts

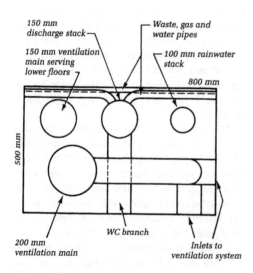

and access can be made easy via cover plates or doors or the removal of lightweight finishes, including plaster or screed. As a result, maintenance can be done on the spot or with the piping readily removed for repair or replacement. Warning is given that access openings with removable covers should be provided in continuous flooring if there are services beneath. Where inspection inevitably will be difficult or impracticable, the system locally should be tested before it is concealed.

Discussion

Variations in regulations and bylaws and their interpretation affect the development and use of service core installations. Such variations, although often a matter of detail, can have a considerable influence on the market for

Figure 7.5   Example of service shaft for multi-storey flats

Figure 7.6 Modified one-pipe drainage and vent system in a service shaft at the Queen's Medical Centre, Nottingham (Courtesy of Schott UK Ltd)

standardized units. Differences which can be important in this context include, for example:

The question of accessibility of piping for repairs and maintenance; the interpretation of what is 'reasonable' access can vary widely in different parts of the country (SLASH 1979).

Water supply from the mains; cold water taps may in some areas be fed direct from the main, but in others a storage tank is required.

Local requirements for the testing of systems and fittings.

Experience has shown the importance of undertaking a thorough study of such variations as part of the market investigation in developing units for industrialized buildings.

Each building type merits separate consideration and enough has been said to illustrate the complexity of the subject. The rationalization of services cannot be considered separately from the broad considerations of planning, design and construction of a building as a whole and indeed of the scale and financing of the whole building operation.

8 Fluid Flow Principles and Studies

Flow conditions in building services systems, whether ventilation or entrained airflow or free surface drainage flows, obey the rules identified in the wider study of fluid mechanics. In general, the flows encountered are incompressible and may be classified as steady (no change at a point with time), unsteady (conditions varying with time), uniform (no change in conditions with change of location in the flow direction) or non-uniform (conditions change in the flow direction). In the analysis of steady flows it is normal to assume that any system may be constructed of sections where uniform flow conditions apply as this allows the application of frictional resistance formulae developed historically.

In general, flow parameters are considered uniform across a flow cross-section — full bore pipe flow is assumed to move at the same velocity at all points across the section. Thus flow velocity is normally the mean velocity across the section. This is convenient but neglects the single most important point in determining frictional resistance, namely that at the pipe or channel wall the flow is stationary — the condition of no slip. A corollary to this is that two fluids in contact will share the same interface velocity, the effective rationale for entrained airflow in contact with a moving water film.

The estimation of flow conditions within pipes, ducts and channels therefore depends upon these criteria and upon the relationship between the kinetic and potential energies of the fluid, any energy input to the system, for example by pump or fan action, and the changes in internal fluid energy due to frictional resistance and the disruption caused by flow through the various fittings included in the network. The flow disruption at fittings generally leads to the formation of eddies in the flow that absorb available energy and often detach the flow from the duct walls. These effects are perceived as separation pressure or head losses but are more accurately energy transfers. An overall description of these balances and a determination of the subsequent flow conditions may be determined via the Steady Flow Energy Equation developed later. Considering an ideal, frictionless and incompressible fluid, Bernoulli in 1738 derived the theorem that provides the basis for the calculation of steady flows. This states that along the streamtube* the sum of the energies due to pressure (p) and height (z) above some datum, together with the kinetic energy, are constant, i.e.:

$$p/\rho + gz + \frac{1}{2}v^2 = \text{constant}$$

where ρ is the density of the fluid and v is its velocity. If we divide each term in this equation by g the result is a relationship between heights or heads with the sum equal to a constant total head, i.e.:

$$h + z + v^2/2g = \text{constant}$$

where h represents the pressure head (static head), z is the potential (position) head and $v^2/2g$ is the velocity (dynamic) head. The equation for continuity of flow along a streamtube gives the quantity of flow Q per unit time as vA, where A is the area of flow at mean velocity v.

* A streamline is imagined as a continuous line, drawn in a fluid, having the direction of the fluid velocity at every point. A streamtube is a tube enclosed by streamlines which form its surface. There is no flow through the walls of such a tube.

These relationships form the basis for considering flows of the kind normally encountered in building services. They relate to imcompressible fluids and thus may be used for water flowing in pipes or in open channels and also for ventilating air, since air may be regarded as incompressible for the conditions normally encountered in ventilation practice.

Pipe friction

Bernoulli's equation relates to an ideal fluid. Experiment shows, however, that the 'total' head at points along a straight pipe is not constant, as Bernoulli predicts, but reduces in the direction of flow.

Considering two stations along the horizontal pipe in figure 8.1:

$$H_1 > H_2$$

Bernoulli's equation is modified to:

$$H_1 = H_2 + h_f$$

where h_f is the 'loss' of head in the pipe run due to fluid friction. When a bend is added to the length of pipe a further loss of head is encountered owing to the turbulent disturbances caused by the bend. Features such as changes of section and fittings and valves also cause additional losses, which are variously known as fitting, secondary, minor or dynamic losses and are of importance in many building services installations. The total pressure drop in a circuit (total resistance) is obtained by summing the individual losses of fittings and those due to piping or ductwork.

Resistance to the flow of a fluid along a pipe is caused by viscosity. Viscosity is due to cohesion and interaction between fluid molecules and causes friction forces to be set up between layers of fluid travelling at different

velocities. Experimental observations show that when a fluid flows along a pipe the velocity increases from zero at the pipe wall to a maximum at the pipe axis. Reynolds in the 1880s demonstrated two types of flow in glass tubes. At low velocities a jet of dye introduced into the flow remains in a thin thread indicating that particles of fluid are moving in straight, parallel paths. This type of flow is known as laminar flow. At higher flows the thread breaks up and mingles with the fluid; the flow is then said to be turbulent.

Reynolds, investigating the loss of energy in a length of pipe, showed that the loss h_f was dependent on the type of flow existing within the pipe. At low velocities and with laminar flow, h_f is proportional to the velocity. At higher velocities in the turbulent regime, h_f is proportional to v^n (nearly proportional to v^2).

Reynolds reasoned that the character of the flow was determined by the dynamic viscosity μ, the density ρ, the fluid velocity v and the pipe diameter D. He grouped these together to give the dimensionless number now known as Reynolds number, $\rho D v / \mu$; ρ and μ are both properties of the fluid and are often combined into the kinematic viscosity $\nu = \mu / \rho$. Thus:

Reynolds number, $R_e = D v / \nu$

In general:

for laminar flow $R_e < 2000$
for turbulent flow $R_e > 4000$

Between these two limits is a critical region where flow conditions are variable. In engineering applications, including building services, R_e is usually greater than 4000 and flow is usually in the turbulent regime. This is the assumption in the following paragraphs which are intended to provide background to sections elsewhere in the book.

Turbulent flow; pipe roughness

Away from the entrance to a pipe a 'boundary layer' spreads across the complete diameter and a velocity profile is established which does not vary with distance along the straight pipe — the flow is then said to be fully developed. In turbulent flow, the profile is almost flat across most of the diameter. At the wall of the pipe the velocity is zero; very near the wall

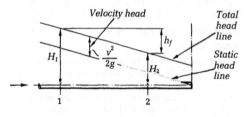

Figure 8.1   Head loss along a straight pipe

the velocity is small and a laminar type of flow occurs in a laminar boundary layer.

In turbulent flow the friction loss depends on pipe roughness and on Reynolds number. Research was done on the effect of artificial roughness and the roughness in commercially produced pipes. It was found that a value could be given to the mean height of projection of the roughness from the pipe wall — termed k — in commercial pipes which could be used to determine the friction loss in such pipes. The term k is sometimes referred to as the absolute roughness and is usually combined with the pipe diameter in the ratio known as relative roughness k/d. The type of turbulent flow — and the friction loss — depends on the ratio of the absolute roughness k to the thickness of the laminar boundary layer. When the roughness projections are small compared with the thickness of the laminar boundary layer, and remain submerged within it, the pipe behaves as though it were smooth and the friction loss is then independent of the relative roughness. The flow condition is said to be smooth turbulent. When the roughness projections are much larger than the thickness of the laminar boundary layer, this layer is virtually non-existent and flow is said to be rough turbulent. Between these two conditions, the roughness projections cause eddying which increases the loss but the laminar boundary layer is not completely disrupted. Flow is then in the transitional turbulent region.

Formulae for friction loss

Pipe friction loss thus depends on the velocity v, the pipe diameter D, Reynolds number R_e, and on the absolute roughness k. It is necessary to relate these factors to the friction loss to enable the loss to be worked out in practice.

(i) Friction factor; logarithmic laws

A rational formula for the friction loss in pipes was suggested by d'Arcy and Weisbach:

$$\frac{h_f}{L} = \frac{\lambda}{D} \frac{v^2}{2g}$$

This is known as the d'Arcy or the d'Arcy – Weisbach formula and λ is the d'Arcy friction coefficient or factor. An alternative approach

(taken by Chézy and Fanning) leads to the relationship:

$$\frac{h_f}{L} = \frac{4f}{D} \frac{v^2}{2g}$$

where $\lambda = 4f$.

Chézy also proposed a formula, written in pipe flow terms:

$$v = C \sqrt{\frac{D}{4} \frac{h_f}{L}}$$

C, the dimensional Chézy coefficient, equals $\sqrt{8g/\lambda}$. The term h_f/L is often replaced, especially in hydraulics textbooks, by S, the hydraulic or friction gradient.

The terms λ and f are not constant but vary with R_e and k/D. Prandtl and von Karman developed a theory for the turbulent flow in pipes and devised the Karman–Prandtl smooth and rough laws relating friction to Reynolds number and relative roughness

$$\frac{1}{\sqrt{\lambda}} = 2 \log \left(\frac{R_e \sqrt{\lambda}}{2.51} \right) \tag{8.1}$$

for smooth turbulent flow and

$$\frac{1}{\sqrt{\lambda}} = 2 \log \left(\frac{3.7D}{k} \right) \tag{8.2}$$

for rough turbulent flow. Colebrook and White combined these laws to give the transition law known as the Colebrook–White equation:

$$\frac{1}{\sqrt{\lambda}} = -2 \log \left(\frac{k}{3.7D} + \frac{2.51}{R_e \sqrt{\lambda}} \right) \tag{8.3}$$

In order to assess when flow is rough or smooth a further dimensionless number is used, the Reynolds roughness number:

$$R_* = R_e \sqrt{\frac{\lambda}{8} \frac{k}{D}}$$

The smooth law applies when:

$$R_* < 0.3$$

The rough law applies when:

$$R_* > 60$$

(ii) Practical application of logarithmic laws

Moody took equations (8.1), (8.2) and (8.3) and the Poiseuille equation for laminar conditions and plotted logarithmically the friction coefficient, λ, against Reynolds number, R_e, for a family of values of relative roughness k/D. The result was the well-known Moody resistance diagram. The engineer, however, normally deals with the variables v (or the flow quantity Q), D and h_f and since the variables of the Moody diagram each include D, and v is included in λ and R_e, many problems are insoluble by methods other than trial and error. The usual way of overcoming this disadvantage has been to produce charts or tables for a specified material and fluid. For example, the ventilation engineer might use a chart associating h_f/L, Q, v and D for air at stated conditions in galvanized steel ducts. Correction factors are included for changes in temperature and for the use of other materials.

Work at the Hydraulics Research Station, now Hydraulics Research Ltd, has transformed the Colebrook–White equation into parameters using Q, v, D and h_f/L individually, and has produced a universal resistance diagram (Ackers 1969). Using this diagram it is possible, knowing two of the quantities given above and the properties of the pipe and the fluid, to determine the two unknown quantities. By the introduction of reference lines and overlays it is possible to eliminate most of the calculations for any given fluid. Formulae are available to determine the changes in k/D due to the effects of ageing.

(iii) Empirical formulae

Prior to the publication of the logarithmic formulae purely empirical formulae, usually of the exponential form:

$$v = \text{constant} \times D^x(h_f/L)^y$$

were used, and several were devised for the flow of specified fluids within pipes of certain materials. Simplicity is their chief merit and many are still widely used. Satisfactory results are obtained within the limits of their application, but the use of the logarithmic laws described above is now generally to be preferred.

Secondary losses

Valves, bends, junctions and other fittings cause local disruption to the flow conditions,

generating downstream eddies and separating the flow from the surrounding conduit walls, see figure 10.3. These eddies absorb available energy and are perceived as pressure losses. In both gravity- and mechanically-driven flow this has the effect of either reducing the deliverable flow or increasing the required energy input to maintain flow. These separation losses, also known as secondary or minor losses — misleading terms as in building services applications they can account for the majority of the system resistance, unlike the one bend in a 500 km oil pipeline — are best represented for practical purposes by a loss coefficient K_s defined for any particular fitting by

$$\Delta P_s = K_s \, \rho \frac{v^2}{2}$$

or if the loss is expressed in head of flowing fluid,

$$\Delta h_s = K_s \frac{v^2}{2g}$$

and determined by experiment. A wide range of fitting loss coefficient data is now available, CIBSE (1986, 2001), Millar (1978) and will not be reproduced here.

The magnitude of the fitting loss depends upon the shape and dimensions of the fluid path through the fitting. The significance depends upon the percentage contribution made to total system resistance by this and other fittings. As an example a simple 12 mm diameter stopcock or tap will have a K_s value of about 10, an elbow about 1.0 and a swept bend 0.4. These K_s values may also be represented by the equivalent length of pipework of that diameter that would generate the same frictional loss. Working from the Darcy frictional loss expression utilizing friction factor f yields

$$\Delta P_s = K_s \, \rho \frac{v^2}{2} = L_{eqv} \, \rho \frac{4f}{D} \frac{v^2}{2}$$

and hence if $L_{eqv} = nD$, where n is the number of pipe diameters necessary to generate the same frictional loss

$$n = \frac{K_s}{4f}$$

Both the Moody chart and Colebrook–White imply that, for turbulent flow, friction factor decreases with flow Reynolds number until a final value is reached at high Reynolds numbers dependent only on surface roughness. As friction factor falls the number of pipe diameters necessary to generate the equivalent pressure loss will rise.

The concept of an equivalent length to represent separation losses is extremely convenient as it allows the combined frictional and separation losses in any steady uniform flow condition to be determined from a single application of Darcy's equation, namely

$$\Delta P_{\text{pipe}} = (L + L_{\text{eqv}})\rho \frac{4f}{D} \frac{Q^2}{2A^2}$$

where the flowrate Q/cross-sectional area A replaces mean velocity, as in most building services applications Q is of more direct interest as a design variable.

Steady flow energy equation

While the presentation above is adequate for simple single pipeflow conditions under gravity as the driving force, more complex networks may be handled by application of the Principle of Conservation of Energy. Considering the flow from point A to point B illustrated in figure 8.2 and expressing the equivalence of energy at A and B yields the Steady Flow Energy Equation:

$$p_A + \rho g\, Z_A + \frac{1}{2}\rho\, v^2 + \Delta P_{\text{input}} - \Delta P_{\text{out}}$$

$$= p_B + \rho g\, Z_B + \frac{1}{2}\rho\, v^2 + \Delta P_{\text{friction}} + \Delta P_{\text{separation}}$$

where Δp_{input} and Δp_{out} refer to energy input by a fan or pump and energy extraction, for example by a turbine, respectively. For gravity-driven systems both terms are zero.

The $\Delta p_{\text{friction}}$ and $\Delta p_{\text{separation}}$ terms may be combined through the equivalent length approach to fitting losses and for a network made up of different diameter or surface roughness ducts in series between A and B may be expressed as

$$\Delta P_{\text{pipes}1 \to n} = \sum_{i=1}^{n} \left[(L + L_{\text{eqv}})_i \, \rho \, \frac{4f_i}{D_i} \frac{Q_i^2}{2\,A_i^2} \right]$$

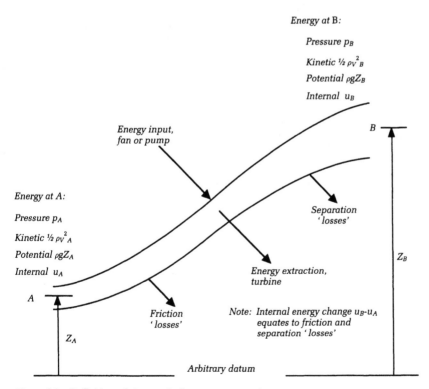

Energy at B:

Pressure p_B

Kinetic $\frac{1}{2} \rho v^2{}_B$

Potential $\rho g Z_B$

Internal u_B

Energy input, fan or pump

Energy at A:

Pressure p_A

Kinetic $\frac{1}{2} \rho v^2{}_A$

Potential $\rho g Z_A$

Internal u_A

Separation 'losses'

Energy extraction, turbine

Friction 'losses'

Note: Internal energy change $u_B - u_A$ equates to friction and separation 'losses'

Arbitrary datum

Figure 8.2   Definition of the steady flow energy equation terms

It should be noted that the choice of points A and B can materially simplify the application of the Steady Flow Energy Equation. Clearly conditions must be known at both locations, this data to include local flow velocity, to yield kinetic energy, local flow elevation, to yield potential energy, and the ambient pressure, to yield pressure energy. It is convenient to express energy in pressure terms. In many examples, such as ventilation from external air to a room, both A and B may be at atmospheric pressure and the local velocities may both be zero. Similarly, in the case of air the potential energy may be neglected. Hence the fan input has to overcome friction and separation losses only. In clean rooms or operating theatres held above atmosphere then the local pressure at B say would exceed atmospheric pressure and the supply fan would have to overcome this steady pressure differential in excess of the flow losses. Conversely the extract fan would gain from the 'free' energy input represented by the higher room pressure. In liquid pumping cases the potential pressure difference is often referred to as 'static lift', however, this may be negative — the case of pumping fuel 'downhill' from a tanker to a basement storage tank. (In this case note that the tank must be fully vented to avoid pressure build-up in the air above the incoming fuel adding to the pump loading due to a rise in pressure energy deficit across the system.)

The Steady Flow Energy Equation is a direct consequence of the Principle of Conservation of Energy and as such the use of the term 'losses' to describe the frictional and separation effects discussed is strictly incorrect. These factors affect the fluid internal energy which increases as a result of the flow conditions discussed. However, that change in internal energy is perceived as a pressure loss to be overcome by the fan or pump or the pipe gradient in gravity systems. Thus internal energy changes are commonly referred to as losses.

In general, turbines may be neglected, so the Steady Flow Energy Equation is most often used to deal with the interaction of fans and pumps with their connected networks. It will be appreciated that the input energy associated with a fan or pump is perceived as a pressure rise that depends upon fan speed and diameter. Thus a particular fan or pump in a particular network will deliver a particular flowrate, depending upon effectively the simultaneous solution of the Steady Flow Energy Equation and the machine pressure flow characteristic. Figure 8.3 illustrates such a solution for a supply and extract ventilation network

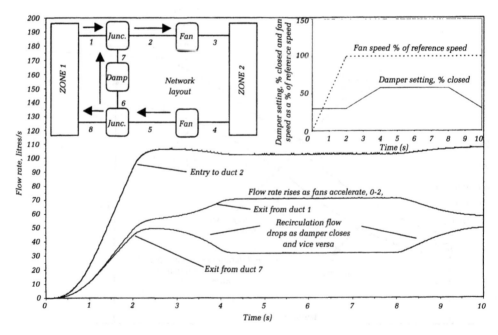

Figure 8.3 Air flow continuity at junction of inflow/recirculation ducts

incorporating recirculation. In this case it is necessary to represent the network loops by remembering that the pressure drop along parallel paths linking zones at a constant pressure are always equal, the flow distribution in each path ensures equivalence of pressure drop — the basic criterion for any flow balancing whether in low pressure hot water radiator heating systems or in ventilation ducts. The case illustrated in figure 8.3 is a further example of the finite difference modelling techniques introduced in chapter 9.

Flow in partially filled pipes

Within building drainage networks the most common regime is free surface flow in the horizontal branches linking appliances to vertical stacks and in the drain leading to the sewer. The Chézy equation applies in these cases provided that the flow is both steady and uniform, i.e. no changes with time, and the pipes remain of the same diameter and at the same slope over the length over which the Chézy equation is applied.

While these are very limiting constraints it is useful to consider the steady uniform flow in drainage networks as a precursor to dealing with the more usual unsteady flow.

The velocity terms are mean velocities over that cross-section. The actual velocity profile within the cross-section will show variations from the mean as the flow velocity increases from zero at the enclosing surfaces to a maximum some little way below the free surface — see the upper sketch in figure 8.4. This form of profile has implications in the study of solid transport velocities presented later; a solid close to the surface in relatively deep flow may have a velocity exceeding the flow mean value owing to its position on the velocity profile.

Resistance formulae for steady uniform flow in partially filled pipes

In steady uniform flow there is no acceleration in the flow direction and thus no change in flow depth; therefore the frictional resistance

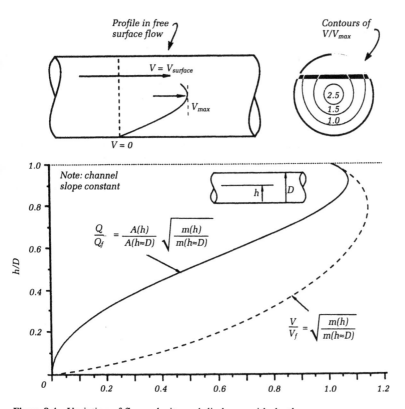

Figure 8.4 Variation of flow velocity and discharge with depth

to flow is balanced by the change in potential along the downward-sloping channel. Under steady flow conditions this is the same as the channel slope and leads to the Chézy equation (introduced earlier in its full-bore flow form):

$$V = C\sqrt{mS_0}$$
$$Q = AC\sqrt{mS_0}$$

where m is known as the hydraulic mean depth, i.e. flow cross-sectional area/wetted perimeter; $m = D/4$ for a full-bore flow. The Chézy loss coefficient C, is defined as $(2g/f)^{0.5}$ and is, therefore, dependent upon the system of units chosen, having dimensions $L^{0.5}T^{-1}$. The simplest formula for C, now widely used, is that due to Manning, dated 1860. Manning found that C varied as $m^{1/6}$ and was dependent upon a roughness coefficient, n, now known as Manning's roughness factor. The form of Manning's equation is thus:

$$V = \frac{1}{n}m^{0.67}S_0^{0.5}$$

It should again be noted that the dimensions of n are $L^{-1/3}T$ and thus the coefficient is dependent upon the unit system.

The Chézy coefficient C may be expected to depend upon Reynolds number, boundary roughness and possibly channel cross-sectional shape. While the frictional losses in full-bore pipe flow have been fully investigated, a similar investigation of Chézy C has not been undertaken. This is not only due to the extra variables involved in free surface flows but also because of the wide range of surface roughness met in practice and the difficulty in achieving steady uniform flow conditions outside controlled laboratory conditions. The behaviour of C may, however, be inferred directly from the full-bore flow friction factor as $C = \sqrt{2g/f}$. The Colebrook–White equation, with the pipe diameter term replaced by $4m$, becomes:

$$1/\sqrt{f} = -4\log_{10}(k/14.8m + 1.26/R_e\sqrt{f})$$

where $R_e = \dfrac{\rho V_m}{\mu}$

In applying this information the flow is assumed to be steady and fully developed so that the flow depth is 'normal', i.e. unchanging downstream in a constant slope drain. Further

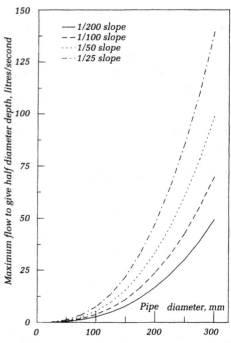

Figure 8.5 Permissible flow for a range of slopes with maximum depth equal to half diameter. Manning's $n = 0.009$

the flow is assumed not to exceed 50 per cent of the pipe diameter. Figure 8.5 illustrates the permissible flow rates based on these assumptions for a range of pipe diameters and slopes. Manning's roughness coefficient of 0.009 assumed is typical of clean plastic piping; see tables 8.1 and 8.2. Owing to the form of the Manning representation of the Chézy equation the permissible flows for any other roughness may be determined by multiplying the flow by (0.009/new Manning n). Examples of the application of the Colebrook–White equation to roughened drains are given in chapter 12.

The obvious difficulty with this technique is that it does not take into account any wave

Table 8.1 Values of Manning's n

Glass	0.009–0.011
PVCU	0.010–0.011
Cast iron	0.014–0.018
Concrete	0.011–0.014
Clay	0.014–0.018
Smooth metal flumes	0.011–0.015
Corrugated metal flumes	0.022–0.030
Steel, riveted	0.017–0.020

Table 8.2 Values of k appropriate to the Colebrook–White equation

	k (mm)
Cast iron (coated)	0.15
Cast iron (uncoated)	0.30
Concrete	0.15
Glazed paper	0.06
PVCU	0.06
Glass	0.03

attenuation that might occur nor any depth changes in the vicinity of pipe junctions. The 50 per cent diameter maximum depth rule actually works quite well to limit the wave and hydraulic jump effects at junctions; however, it inevitably means that between junctions the flow is artificially shallow with an enhanced possibility of solid deposition.

Flow depths and velocities under steady uniform conditions

Utilization of the Chézy equation with roughness represented by Manning's coefficient, together with details of pipe slope and cross-sectional shape and dimensions, allows the flow depth and mean velocity to be related under the special conditions of steady uniform flow. Thus if A is a known function of depth h, i.e. $A(h)$, and m is also a known function of h, i.e. $m(h)$, the depth of flow for a steady uniform flow rate Q is given when Manning's equation is satisfied:

$$S_0 - Q^2/C^2 A(h)^2 m(h) = 0$$

For building drainage networks the channel section is normally circular so that the solution of this equation requires an iterative solution, such as the bisection method. The resulting flow depth is known as the normal depth of flow and only occurs under steady uniform flow conditions, namely where the slope of the pipe provides a sufficient fluid mass force to overcome the surface shear forces. As mentioned this condition is rare in building drainage networks, but it is found to be the basis of many current design guides.

Manning's equation may be solved directly for rectangular channels where the flow area $A = hB$, B being the channel width, and the wetted

perimeter $P = B + 2h$. Thus $m = hB/(B + 2h)$ and:

$$S_0 C^2 B^3 h^3 - 2hQ^2 - BQ^2 = 0$$

is solvable for h, the normal depth.

For partially filled pipes there are optimum depths of flow for maximum velocity and discharge. As the flow level rises with increasing flow a point is reached beyond which the wetted perimeter increases very rapidly in comparison with the flow area, merely as a consequence of the pipe's circular cross-section. This causes the hydraulic mean depth, and hence the velocity, to decrease. As flow rate is the product of flow cross-sectional area and velocity it follows that eventually the flow rate will also decrease as the depth rises, as the drop-off in velocity outweighs the increasing area.

The depth of occurrence of these phenomena may be determined for circular section channels. It may be shown that the maximum flow velocity occurs when the flow depth, for constant slope and Chézy coefficient, is equal to 81 per cent of the full-bore diameter. For maximum discharge, the result depends upon the choice of resistance formula. Choosing the Chézy equation, the depth for maximum discharge may be shown to be 95 per cent of diameter, a depth at which the hydraulic mean depth, m, is $0.277D$. The lower sketch of figure 8.4 illustrates this effect. It therefore follows that the maximum discharge may be expressed as a percentage of the full-bore flow rate as $h \to D$ and $m \to 0.25D$:

$$\begin{aligned}Q_{full}/Q_{max} &= \sqrt{m_{full}/m_{max}} \\ &= \sqrt{0.25/0.287} = 0.933\end{aligned}$$

Steady, non-uniform, free surface flow

Steady uniform flow in building drainage networks, characterized by the attainment of a normal flow depth, is exceptional. Where changes in pipe slope, diameter or roughness occur the flow will experience changes in depth and velocity in the flow direction, while the overall flow rate remains constant. The flow specific energy, E, at any section may be defined in terms of the local depth and flow rate, from Bernoulli's equation, as:

$$E = h + Q^2/2gA^2$$

where A is a function of h, i.e. $A(h)$. There are two possible real solutions for h at a given Q and E and one depth, the critical depth, h_c, where the two roots coincide. At this depth the specific energy is a minimum for a given Q and the flow rate is a maximum for a given E. It may be shown that this critical depth is given by:

$$1 - Q^2 T_c / g A_c^3 = 0$$

where T_c and A_c are the flow surface width and cross-sectional area respectively when $h = h_c$, the critical depth (figure 8.6). The flow velocity when flow Q has a depth h_c is given directly as $V_c = Q_c / A_c$:

$$V_c = \frac{Q_c}{A_c} = \sqrt{g \frac{A_c}{T_c}}$$

This velocity becomes important in the classification of flows. It should be noted that h_c is independent of pipe slope and roughness and depends only on the flow rate and the channel cross-sectional shape. Critical flow conditions occur at the minimum specific energy for a given flow rate and hence divide the possible flow regimes into subcritical, or tranquil, flow and supercritical, or shooting, flow. These two regimes are characterized respectively by deep, slow-moving and shallow, rapid flow. The latter is the norm in building drainage networks, with the exception of those pipe lengths upstream of junctions where the restriction to flow presented by the confluence of streams imposes a region of forced subcritical flow terminating in an upstream hydraulic jump as a transition from supercritical conditions.

The terms subcritical and supercritical also refer to the ability of small surface waves to propagate against the flow. It may be shown (Swaffield and Galowin 1992) that the velocity of propagation of surface waves is:

$$(V_1 - c) = \sqrt{g \frac{A \Delta h}{T \Delta h}} = \sqrt{gA/T}$$

Taking velocities as positive downstream, the wave speed c relative to the pipe is $(gA/T)^{0.5} + V$ downstream and $(gA/T)^{0.5} - V$ upstream. Thus if $V > (gA/T)^{0.5}$, waves cannot propagate upstream. It will be noted that this value of V corresponds to the critical velocity. Similarly if $V < (gA/T)^{0.5}$, then the wave may propagate in either direction. In terms of the subcritical and supercritical flow regimes it follows that waves cannot propagate upstream in supercritical flow but may do so in subcritical flow. These flow regimes are therefore defined by a non-dimensional coefficient, the Froude number, F_r, where:

$$F_r = V / \sqrt{gA/T}$$

Thus if $F_r > 1$ the flow is supercritical; if $F_r < 1$ the flow is subcritical.

The term A/T is defined as an average depth, not to be confused with the hydraulic mean depth, m. For rectangular channels the analysis above would have yielded $F_r = V/(gH)^{0.5}$ where H was the flow depth in the channel.

The definition of flows into subcritical and supercritical regimes will become important in the later treatment of steady non-uniform and unsteady flows.

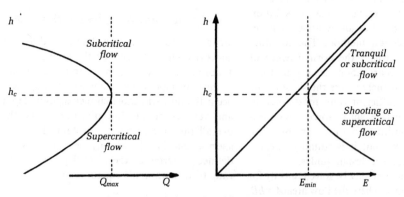

Figure 8.6 Alternate flow depths for constant specific energy E and flow Q

Gradually varied flow

In drainage hydraulics it is necessary for flows to undergo transition from subcritical to supercritical, or vice versa, as pipe slope, diameter or roughness change, or as pipe junctions and terminations are traversed. Similarly it is necessary to predict changes in depth within either flow regime as the flow depth responds to control conditions at the pipe boundaries, e.g. a restriction at exit caused by joining flows at a junction or the reduction in flow depth experienced in subcritical flow at discharge to a vertical stack. The changes in flow depth in such cases, extending several pipe diameters upstream or downstream of the control section, are referred to as backwater profiles, if the flow decelerates, or drawdown profiles if the flow is accelerated, in gradually varied flow.

Gradually varied flow may be considered as a special type of steady non-uniform flow where the flow parameters are assumed to change so slowly that it is reasonable to approximate the local frictional forces over any small length, ΔL, by the equivalent steady uniform flow losses based upon the local flow velocity and depth and calculated via the Chézy equation. It may be shown that the relationship between flow depth and length within such a backwater profile within either sub- or supercritical flow is given by the integral:

$$\Delta L_{h_1 \to h_2} = \int_{h_1}^{h_2} \frac{1 - Q^2 T/gA^3}{S_0 - Q^2/A^2mC^2} \, \mathrm{d}h \qquad (8.4)$$

The numerator and denominator of this integral are respectively the expressions for critical and normal depth, introduced above.

If the numerator is zero, then the flow is critical and there is no change in L for a change in h, i.e. the condition for an hydraulic jump. If the denominator tends to zero then the flow is steady uniform and no change in depth occurs, i.e. the flow is at normal depth for that pipe slope, diameter and roughness.

Again the fact that building drainage systems employ circular pipes indicates that the A, m, T terms are all functions of depth h and the integration must be approached numerically.

Discharge to a stack or free outfall

Critical flow conditions must occur if flow is to change from subcritical to supercritical. In the development of the Chézy and related equations, the pressure distribution down through the fluid at any section has been taken as hydrostatic, $p = \rho g h$ at A (see figure 8.7). This is taken to be the case as the streamlines are assumed parallel with no substantial accelerations perpendicular to the mean flow direction. With a free outfall the accelerations at the brink are large and vertical and thus the hydrostatic assumption breaks down, resulting in a much lower pressure distribution, as shown at B in figure 8.7. Thus the flow accelerates between A and B, and as the flow rate is constant this results in a rapid decrease in flow depth. Critical depth should theoretically occur at the brink; however, as the flow over the brink is highly curved the necessary assumptions break down and the actual critical depth section occurs upstream. Experimentally the brink depth has been found to be some 70 per cent of critical; the critical depth section occurs 3 to $4h_c$ upstream of the brink. The specific energy decreases from its normal depth value at A to a minimum at the critical depth section. The drawdown curve, AC, may be

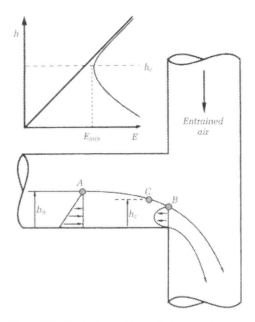

Figure 8.7 Typical transition – drain to stack

determined by numerical integration; the depth limits, to be subdivided, are the upstream normal depth at A and the critical depth at C in both cases. The integration commences at C and thus yields the distance upstream from C necessary for each equal Δh from h_c to h_n.

This illustrates an important point in gradually varied subcritical flow, namely that the control lies downstream.

Supercritical flow approaching a free outfall or a change in pipe slope remains 'unaware' of the change as information cannot be propagated upstream. Thus supercritical flow would exit to a vertical stack with only a local disturbance at the brink.

Gradually varied flow profile upstream of junctions or obstructions

The joining of· flows within a system will inevitably lead to some local depth increase as a result of the flow mixing process. The extent of this may be limited by good junction design; however, it is an ever present condition. The approach flows to such a junction will respond by establishing backwater profiles leading from the normal upstream flow depth to the imposed junction depth. Where both the imposed depth at the junction and the approach depths are either all less or all greater than critical, equation (8.4) may be utilized, the junction depth providing the downstream control and being determined experimentally for any junction geometry. However, in the majority of cases this will not be so and both profiles will include hydraulic jumps.

Rapidly varied flow and the hydraulic jump

The hydraulic jump (figure 4.9) is an important example of local, rapidly varying flow. The underlying assumption in the previous treatment of gradually varied flow was that the flow parameters changed slowly enough to allow use of the Chézy or Manning equations. In rapidly varying flow the rate of change of the flow parameters is too great to support this assumption. Theoretically the water surface profile from supercritical to subcritical flow is vertical as it passes through the critical depth. In practice this cannot occur and the transition takes the form of a steeply sloping water surface of finite length. This jump is associated with violently 'turbulent' flow conditions and a large concentrated energy 'loss', or dissipation by eddy formation; see figure 8.8 and figure 4.9.

As the flow is steady the jump will be stationary in the flow. Thus both the equation of continuity and momentum may be applied across the jump in order to determine the conjugate depths upstream and downstream of the discontinuity, h_1 and h_2:

$$\rho A_2(g\bar{h}_1 + Q^2/A_1^2) = \rho A_2(g\bar{h}_2 + Q^2/A_2^2)$$

where $\bar{h}_{1,2}$ are the respective centroid depths.

The loss of energy in the jump may be expressed by the change in specific energy from section 1 to 2 in figure 8.8.

The determination of drawdown or backwater profiles has already been introduced for cases where the appropriate control section depth is known. For the cases identified, i.e.

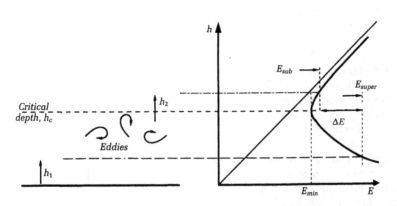

Figure 8.8 Hydraulic jump in a horizontal channel – conjugate depths and specific energy

where the upstream control depth is less than critical but the downstream depth is greater than critical, it is necessary to locate the jump position; the overall depth profile is determined in two segments, up to the jump and then on to the downstream control. Two cases will be considered, the first where the flow upstream of the jump is fully developed, i.e. its depth is the supercritical normal depth associated with the pipe slope and diameter and flow rate, and the second where the upstream depth is now known.

Figure 8.9 illustrates the first case, arising upstream of a junction. The downstream control is known as h_j, the depth at the junction, as a function of the combining flow rates and is above critical. The upstream depth, h_2, is the conjugate depth at the jump. As the overall depth change $h_j - h_2$ is known the backwater profile from the junction to the jump may be determined, integration commencing at the junction, and the location of the jump is thus found. In the second case, figure 8.10, the depth downstream of the jump may be assumed to be the normal depth appropriate to the steady flow and the pipe slope and dimensions; the jump equation may be utilized to yield the upstream conjugate depth, h_1. The upstream control depth, h_e, depends upon the annular flow in the stack. Integration of the gradually varied depth profile from h_e to h_1

thus yields the distance of the hydraulic jump from the base of the stack. The depth then rises abruptly to the downstream normal depth.

Application of the gradually varied flow profile integration technique, together with the location of any jumps in the pipe network, allows the steady flow depths and velocities to be calculated throughout a pipe network and serves as a basis for the unsteady flow simulation to be treated later.

Steady flow depths at pipe junctions

The presence of a level invert junction results in an enhanced depth at the pipe common coordinates, the backwater profile in each of the joining pipes then depending upon the steady flow carried by each; see figure 8.9. In order to predict the backwater profile, and the position of the hydraulic jumps normally positioned in each supply pipe, it is necessary to determine an empirical relationship linking h_j to the combined through flow, $\sum Q$. In building drainage networks the normal flow condition is supercritical so that jumps become inevitable in the vicinity of junctions. Experimental measurement over a wide range of Q_1 and Q_2 combinations for 90° and 135° level invert junctions showed that h_j may be described by a relationship of the form:

$$h_j = K \sum Q^a$$

where K has values of 0.29 to 0.35 for 135° to 90° junctions, with $a = 1.75$.

Critical depth as a variable is useful with top entry junctions, as shown in figure 8.11. The junction depth was investigated by introducing the critical depth appropriate to the main drain steady flow, h_{mc}, and the critical depth associated with the combined downstream

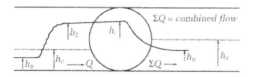

Figure 8.9 Jump location upstream of a level invert junction

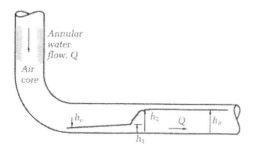

Figure 8.10 Hydraulic jump downstream of a stack

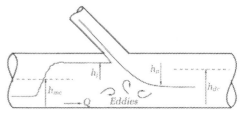

Figure 8.11 Schematic representation of a top entry junction flow combination

flow, h_{dc}, resulting in a defining expression of the form:

$$(h_j - h_{mc})/h_{mc} = K(h_{dc} - h_{mc}/h_{mc})^n$$

These critical depths are independent of pipe slope and roughness and depend only on flow rate and pipe diameter; typical results for a 90° top entry junction were $K = 2.15$ and $n = 0.56$.

Compressible flow: back siphonage test

Although as already explained the flow in building service installations does not normally involve density changes, one situation may be envisaged where the drop in pressure causing air flow may be sufficiently large for a density change to occur. This is under the conditions assumed for back-siphonage (chapter 3) — a substantial drop in pressure within the water supply installation causing a pressure below atmospheric at the point of use which may suck water back from an appliance into the supply piping. Consider the situation illustrated in figure 3.7(a) and suppose that there is a drop in pressure in the supply line which, with the outlet fitting open, draws air from the atmosphere into the piping, although no water. With a fall in pressure of a small fraction of a bar there is no change in density as the air flows through the outlet fitting, i.e. the air remains at substantially the same density as nearby in the atmosphere. Suppose, however, that the suction gradually increases; the velocity of flow will increase and there will also be a fall in air density as the air passes through the fitting into the pipeline. Eventually a stage is reached where the simple relationships outlined at the start of the chapter no longer hold. It is then necessary to introduce additional relationships between the density, pressure and temperature of the air as a basis for estimating flow rates. It is hardly necessary to discuss this matter in detail here — fluid mechanics texts deals with the matter for those interested — but it is useful to discuss one particular aspect.

Assume that the suction in the supply line continues to increase, with a resulting increase in the velocity of air flow through the fitting. Very high speeds may be reached but a limit is found to exist in practice, corresponding to the velocity of sound in air (340 m/s at 15 °C, for instance). This is the speed at which sound waves — small air pressure disturbances — are propagated in air. Once the suction has increased to give this air speed within the fitting, a further increase in suction cannot be propagated upstream, i.e. against the air flow at sonic velocity. The air flow is said to be 'choked', and a further increase in suction does not cause any further increase in speed. Theory shows that this stage is reached with air flowing from the atmosphere into a nozzle when the downstream pressure has fallen to 0.53 bar. This is the basis for the suggested test for back-siphonage in chapter 3, i.e. that the suction applied to a device should be 0.5 bar for a period of time, representing a worst condition.

For practical installation purposes it is necessary to consider what happens when, say, a leak occurs at S (figure 3.14) and the water column falls. Air is drawn in at A and, depending on the airway, a suction occurs at W. Practical tests and calculations, based on the fluid mechanics principles outlined at the beginning of this chapter, show how the suction is related to the airway size. Typically under these conditions anti-vacuum valves for domestic installations could be expected to have a flow resistance equivalent to a circular orifice about 9 or 10 mm in diameter and limit the suction in a 28 mm nominal size smooth riser to 300 mm of water. Similarly it is possible to calculate the equivalent orifice sizes for valves installed at the top of larger-size risers, again to limit the suction to 300 mm of water. This type of analysis forms the basis of protection described in chapter 3.

Waterhammer

Unsteady conditions are commonly met within the water and sanitary services of buildings. Chapter 9 will consider unsteady flow simulation in partially filled drainage, including rainwater gutters, air pressure transient propagation in vent systems as a result of entrained air flows and siphonic rainwater systems. However, the most easily recognized manifestation of unsteady flow within buildings is waterhammer — often characterized by the 'banging' of pipework following a rapid tap closure.

Waterhammer, or more correctly pressure transient propagation, is a well-understood phenomenon that occurs in all fluid transport networks, from hydroelectric power stations to aircraft fuel systems, and may be analysed and simulated by computational techniques. While a more detailed treatment may be found in Swaffield and Boldy (1993) or Wylie and Streeter (1983), the following presentation will illustrate the main factors involved.

Suppose a tap at the end of a pipe is closed instantaneously. The flow is brought to rest and there is an instantaneous rise in pressure on the tap upstream face. (Note that if the tap is within a pipe length that an equal negative pressure is generated on its downstream face.) The kinetic energy of the fluid, $1/2\rho V^2$, is converted into strain energy stored within the fluid and pipe system. If the pipe is assumed rigid then this strain energy per unit volume may be expressed in terms of the bulk modulus of the elasticity of the fluid, K, by the term $1/2\delta p^2/K$, where δp is the pressure rise. Thus the pressure rise experienced in the rigid pipe on tap closure becomes $\delta p = V\rho\sqrt{(K/\rho)}$. Since the speed of propagation of a pressure wave through the fluid in the rigid pipe may be expressed as $c = \sqrt{(K/\rho)}$, it follows that the pressure rise on tap closure is given by the simple formula $\delta p = \rho V c$, an expression first developed by Joukowsky in Moscow in 1904. Thus for a given fluid in a rigid pipe the pressure rise depends only on the fluid properties and the flow velocity destroyed.

Most pipes are not rigid. The cessation of flow causes a pressure rise that distorts the pipe cross-section. This may be easily taken into account by introducing an 'effective' fluid bulk modulus, K, that may be shown to be $K' = K/(1 + KD/Ee)$, where E is the pipe Young's modulus of elasticity, D its diameter and e its wall thickness. Thus for large diameter thin walled pipes the wave speed c is reduced, as it is for elastic pipes, resulting in a decreased pressure rise on instantaneous tap closure. It should be noted that the presence of even small quantities of air mixed into the flow will also have this effect.

Although only approximate and applicable only to near instantaneous closures, this analysis provides a useful basis for considering practical cases, for example a water flow of 3 m/s, density 1000 kg/m^3, when brought to rest instantaneously will generate a pressure rise of 4.3×10^6 N/m^2, comparable to a hammer blow. Flow stoppages in more than one pipe period, i.e. the time taken for the pressure wave to propagate to a major reflector boundary and return to its point of inception, are less than this value. (In entrained airflows a cessation of flow having an initial velocity of 1 m/s will generate a pressure rise of approximately 40 mm water gauge — at the other end of the spectrum but equally important.)

Figure 8.12 illustrates the pressure rise following a 0.1 second valve closure at the termination of a 20 m pipeline of 0.1 m diameter, carrying a water flow of 0.02 m^3/s and the subsequent oscillations of pressure as the waves propagated reflect within the pipe system. As the wave speed is 1000 m/s, the pipe period is 0.04 seconds and the valve closure is thus 'slow', i.e. greater than one pipe period so that the full Joukowsky pressure rise of 2546 kN/m^2 is not quite generated.

The valve closure generates a pressure wave that is reflected at the upstream end of the pipeline and is subsequently damped by friction. (In this case a value of $f = 0.01$ was used, sufficient to give damping and a small 'friction recovery' secondary rise in pressure at the valve once closed. Note that in high frictional head systems this secondary recovery, known as line packing, can exceed the transient pressure.) In a pipe of length L, and wave speed c, the time taken for a reflection to arrive back at the valve is $2L/c$, a travel time referred to as the pipe period. If the valve takes longer than this time to close then the peak pressure generated does not reach the Joukowsky value, $\rho V c$. Thus the Joukowsky analysis is only true for closures completed in less that one pipe period. The relationship is characterized by a rapid reduction in peak pressure, perhaps 50 per cent if the valve closes in five pipe periods but a much slower decrease beyond 15 pipe periods. However, care must be taken. As the pipe period depends on pipe elasticity, a plastic pipe has a lower characteristic wave speed than a steel pipe and hence a longer pipe period for any fixed pipe length. This results in any fixed valve closure time corresponding to a smaller number of pipe periods for the plastic pipe and hence a pressure rise on valve closure that may exceed that for the steel pipe at the same time of

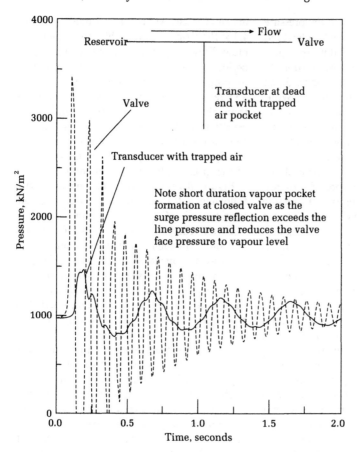

Figure 8.12 Pressure surge following a value closure, also illustrating the effect of trapped air on a transducer output

closure. It is therefore important to use pipe periods rather than time in assessing valve closure effects.

For extremely slow valve closures, perhaps taking longer than 15–20 pipe periods, the mass oscillation generated surge pressures may be calculated by considering the forces acting on the long plug of water coming to rest in the pipeline. The method is analogous to that used for the treatment of self-siphonage described later. Neglecting friction and assuming a horizontal pipe it is apparent that the pressure difference δp across the plug, length L, during deceleration, dV/dt, is given by the term $\rho L \, dV/dt$. The pressure rise thus depends upon the time of closure, dt, the length of the pipe, L, and the velocity destroyed, dV. The surge predicted by this technique is sometimes referred to as the inertia pressure. In the example above suppose that the valve closes

linearly in 0.5 seconds, then dV/dt becomes $(2.55/0.5)$ m/s^2 and the pressure rise for a water-filled 20 m long pipe is $102 \, kN/m^2$, much less than the Joukowsky value, and less than that predicted by a full unsteady analysis for a valve closure in 13 pipe periods. However, these techniques converge for very slow valve closures, confirming the later application of the inertia method to the study of mass oscillations, including trap self-siphonage.

The perception of waterhammer and its measurement also raises important issues. Chapter 9 will consider the effect of trapped air in utility systems such as fire sprinklers and dry risers. Trapped air may also affect the correct measurement of transient pressures. Figure 8.12 illustrates the measurement of the valve pressure rise by means of a pressure transducer mounted on a long branch connection unfortunately unvented. The attenuated transducer output is simulated by allowing the transient to compress the trapped gas. The transducer output dangerously underestimates the transient. It must be realized that this effect is wholly dependent upon the volume of trapped air assumed, however, it does reinforce the importance of understanding the phenomenon to be investigated.

While waterhammer does depend on initial flow velocity and the pipe/fluid properties, it is greatly dependent upon valve closure time. Fast-acting valves, particularly solenoid driven on water-using appliances, may be a source of troublesome noise.

Investigation and analysis of self-siphonage

The discharge–self-siphonage–seal loss sequence that occurs with a typical basin, trap and waste installation has been investigated by Wise (1954). The flow from a basin with pipe slopes of a few degrees is illustrated in chapter 4. Air is present when the basin overflow is open; the spiral formation results from the vortex in the basin. Blocking the overflow eliminates the air for most of the discharge but the vortex provides a little air at the end of the flow. Conditions at the end of discharge (figure 4.8)

are shown diagrammatically in figure 8.13. The water column breaks just beyond the trap. Some water is held in the trap but, owing to inertia, the plug of water in the waste pipe continues its motion. A partial vacuum thus develops and this removes some of the water held in the trap and that still entering from the basin. Air bubbles drawn through the seal during this movement assist the siphonage by their pumping action and the final result of this plug flow is considerable loss of seal.

Under the action of the pressure difference across the plug between the atmosphere at the stack end of the pipe and the reduced-pressure air pocket (assisted by frictional resistance) the plug slows down and comes to rest in the pipe. It then begins to move back towards the trap on account of the maintained pressure difference, while the water left in the trap drops back towards its equilibrium position. The net result is that, if the plug has come to rest sufficiently close to the trap, the return flow partly or completely refills it. The research provided test data for typical installations.

The negative pressure developed during the suction stage is also important; as shown in figure 8.14 it falls below atmospheric by an amount approximately equal to the trap depth and then fluctuates about this value. With a long plug movement, as in figure 8.14, the flow of air through the trap tends to restore the pressure to atmospheric and hence the suction

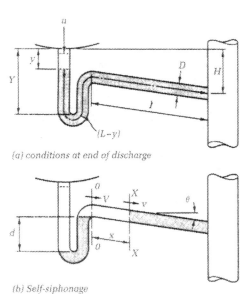

(a) conditions at end of discharge

(b) Self-siphonage

Figure 8.13 Definition sketch for self-siphonage at the end of a discharge

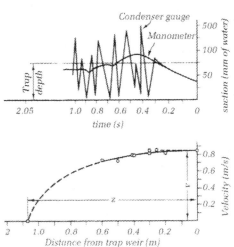

Figure 8.14 Example of velocity and suction during self-siphonage

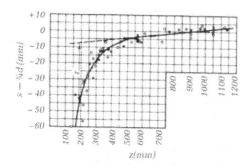

Figure 8.15 Seal loss due to self-siphonage against plug movement

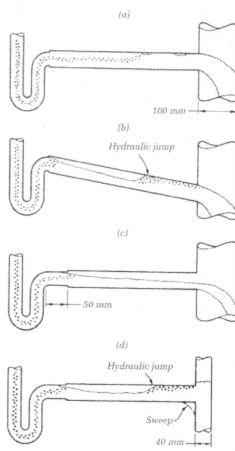

Figure 8.16 Examples of flow in enlarged waste pipes

towards the end of the travel is rather less than the trap depth. The pressure fluctuations correspond with the passage of air bubbles through the trap.

The general trend of the results was for any increase in total head or reduction in resistance to increase trap seal loss, because such changes enable the plug to move further from the trap, thus increasing the siphonage time and reducing the amount of refill into the trap from the waste pipe.

A theoretical analysis of the flow and plug movement was confirmed by experiment (Wise 1954) with the end results shown in figure 8.15. This gives seal loss in terms of trap depth d and plug travel z. Equations were developed to calculate z in terms of pipe parameters, figure 8.13, and discharge rate. The work provided a sound understanding to support design recommendations.

Conditions are different when enlarged waste pipes are used. Figure 8.16 illustrates the flow observed in 40 mm pipes receiving water from 32 mm bore brass traps. As indicated in sketch (a) the flow fills the pipe when the enlargement occurs immediately beyond the trap and the slope is about 7 per cent or less. This results in plug action at the end of the discharge and may cause seal loss. At slightly greater slopes and assuming the basin overflow to be open, an air pocket develops beyond the trap, as in sketch (b), ending in a hydraulic jump. In tests using a 1.25 m long pipe at 9 per cent slope, the jump became established about 250 mm from the stack, and the pressure in the air pocket was found to be about 50 mm (water gauge) below atmospheric. As the discharge ended, the jump

moved a little way upstream, not reaching the trap, and the negative pressure in the air pocket gave rise to a seal loss of 60 mm. At slopes greater than about 11 per cent, the jump formed as described above but moved slowly down the pipe and fell into the stack before the discharge ended. The partial vacuum was thus relieved and, as there was no suction at the end of the flow, the seal remained full. Under these conditions, blocking the overflow gave rise to full-bore water flow followed by plug action and seal loss.

With traps having tails 50 mm long as in sketches (c) and (d), conditions were different. The tail piece introduced water into the pipe at a depth of about three-quarters of the bore and the flow then depended on the pipe slope and on conditions downstream. With a 100 mm stack and with slopes steeper than 2 per cent, the profile in sketch (c) appeared when the waste pipe was straight. Seal losses were nil because siphonage could not develop at the end of the run. Under the same conditions but with a smaller stack, the profile in sketch (d) was obtained. The obstruction offered by the small-bore stack caused backwater with a hydraulic jump. Air pressure in the pocket rose about 40 mm (water gauge) above atmospheric as the backwater formed, but gradually fell to 60 mm below atmospheric. This led to considerable seal loss except at very flat slopes, when the jump ran back upstream to the trap and refilled it. Backwater also occurred when a horizontal right-angle bend having a radius of 75 mm or less was introduced into the waste pipe. Backwater could normally be avoided by sweeping the waste pipe into the stack as shown dashed in sketch (d), or by using a large-radius bend instead of an elbow. The flow then appeared as in sketch (c), and there was no seal loss.

With a large radius copper trap connected to a 40 mm pipe the flow shown in sketch (a) normally occurred and considerable seal loss resulted. It appeared that the larger curvature of the outlet bend of the copper trap tended to project the water upwards towards the top of the pipe, whilst the lower resistance of this trap tended to give a higher discharge rate; these two factors combined to make full-bore flow more likely.

Open-channel-flow theory (Wise 1954) can be used to predict the type of flow in pipes normally too big for plug movement to occur. The significant factors are those discussed earlier (see also figure 8.17); they are the 'normal' and 'critical' depths h_n and h_c and the critical slope θ_c. For calculation purposes the mean discharge Q may be taken as 0.7 l/s from tests of 32 mm brass traps. The discharge rate during early stages of the flow is 10 to 15 per cent greater than this. The pipe diameter D is 40 mm. Values of h_n, h_c and θ_c can be

calculated from these data using the gradually varied flow equations for circular conduits: $h_c = 37$ mm for the maximum discharge and 34 mm for the mean discharge; the corresponding critical slopes θ_c are 0.0191 and 0.0154. One illustration of theory is given in figure 8.17 which shows the profiles that occur when water is admitted at a depth less than $h_c = 37$ mm and the actual slope is greater than $\theta_c = 2$ per cent. Profile A corresponds to figure 8.16(c) for a straight pipe and a large stack; the surface is asymptotic to h_n, and at 9 per cent slope, for instance, for which $h_n = 17$ mm, the actual outlet depth was 18 mm. Profile B corresponds to figure 8.16(d); a tight bend or a small stack caused a jump through the critical depth to full-bore flow beyond. Conditions at less than the critical slope can be forecast in the same way.

Discharge from water closets

A knowledge of the water flow rates from sanitary appliances is basic to the design of drainage installations. The discharge from appliances such as washbasins was discussed in the foregoing section and we now consider the flow from water closets. Specially designed metering equipment has been used to determine such rates of flow and how they vary with time, i.e. to determine the discharge characteristics of the appliance. Figure 8.18 shows a typical example of the type of curve obtained. A peak flow is attained quickly and lasts for a short period; a slower rate of flow then continues for some seconds.

The values obtained depend on the rate of supply of water into the pan and on the detailed design of the pan itself, but not usually on the piping downstream since the flow is

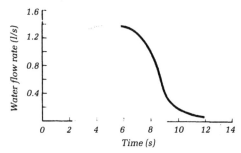

Figure 8.18 Example of discharge curve for low-level washdown WC with S-trap using 9 l of water

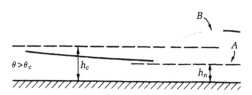

Figure 8.17 Definition sketch for open-channel flow

Table 8.3 Flow parameters for water closets using 9 l (Pink 1973(ii)) and 5–7 l (McDougall and Swaffield 1994)

Appliance	Trap	Peak flow rate (l/s)	Peak flow rate over 1 second duration (l/s)	Time of discharge for $Q_w \geq 0.1\,l/s$ (s)
Low-level 9 l washdown WC	P	1.4	1.4	9
High-level 9 l washdown WC	P	2.4	2.3	8
WC, 5.9 l		1.8	—	4
WC, 6.8 l		2.6	—	5

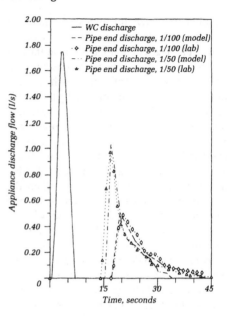

Figure 8.19 Attenuation of a 6 litre WC flush along a 12 m, 100 mm diameter, drain set at slopes of 1/50 and 1/100, compared to the predicted attenuation chapter 9, based on the measured WC discharge profile and drain parameters

nowhere full bore. Table 8.3 gives data for several different types of water closet, covering peak flow, peak flow over 1 second and time during which the flow exceeded 0.1 l/s. Such data guide the selection of representative values of flow rate and time for practical use in design procedures. Measurements have also been made of the discharge characteristics of individual appliances after the flow has passed some distance through pipework. Down stacks discharge rates remain sensibly constant.

Attenuation is most likely to affect appliance discharge flows in long horizontal pipework. Figure 8.19 illustrates the effect on the discharge from a 6 litre WC due to its passage along 12 m of 100 mm diameter drain set at either 1/50 or 1/100 slope. It will be seen that the peak flow is reduced from 1.75 l/s at the WC spigot to 1 l/s at 1/50 and 0.5 l/s at 1/100. The predicted attenuation, derived by the methods discussed in chapter 9, is also presented for comparison. It should be stressed that this form of attenuation depends on the appliance discharge profile as well as the drain

parameters. The WC discharge profile, featuring relatively rapid changes in flow, attenuates rapidly, as illustrated. Sink, bath, shower or basin discharges with long-period quasi steady flow will exhibit little or no attenuation to the peak flows, although some extension of the flow duration will occur as the tail of the discharge spreads; see chapter 9. Figure 8.19 also exhibits the other characteristic of wave attenuation, a clear steepening of the wave leading edge, while the tail of the discharge spreads.

Table 8.4 illustrates the peak flow rate and flow duration for the same WC, measured and predicted, at the end of 12 m of 75 mm, 100 mm and 150 mm diameter drain at the two slopes of

Table 8.4 Influence of pipe slope and diameter on the attenuation of a 6 litre WC discharge along 12 m of drain

	WC only, no drain	Pipe slope					
		1/100			1/50		
Dia. (mm)	—	75	100	150	75	100	150
$Q_{max}(l/s)$	1.75	0.55 (0.56)	0.52 (0.54)	0.50 (0.48)	1.08 (1.05)	1.03 (0.98)	0.93 (0.9)
$T_{flow}(s)$	10.0	45.0 (43.0)	45.0 (43.0)	45.0 (43.0)	20.0 (18.0)	22.0 (19.0)	28.0 (26.0)

1/50 and 1/100. These figures confirm that attenuation depends heavily on pipe slope and less so on pipe diameter. The predicted values are also shown, in parentheses, for comparison. These results were recorded at Heriot-Watt University as part of an evaluation of American low-flush toilets.

Investigation and analysis of stack flows and pressures and vent sizing

Satisfactory design of soil and waste pipe systems in large buildings requires a knowledge of the pressures that develop in vertical pipes running partly full. In practice, the vertical drainage stacks have to be open at the top to allow the escape of foul air and this means that, when water is discharged down the stack, air is sucked in at the top, travels downwards with the water and discharges into the sewers. The ratio of air to water flow may be as big as 100 to 1, with the actual air flow between 10 and 200 l/s. The air flow and the pressures depend on the drag between the water and air and on the frictional resistance of the components of the system. Research has sought a generalized scheme for the calculation of the maximum pressures above and below atmospheric that arise in practice. As with most fluid flow problems it has been necessary to combine theory with experimentally determined coefficients.

Features of the water and air flow

Whereas most hydraulic installations involve continuous water flow, soil and waste pipe systems in buildings are in use intermittently and the flows from individual appliances may last for only a few seconds. This poses the need to assess the load from a probability standpoint (chapter 1); but the fact that individual flows are only of short duration must also be considered in the hydraulic analysis. Many of the stack flow investigations conducted in the laboratory have used continuous flow to ease the problem of observation and measurement of relevant flow features and parameters. Such work has enabled a picture to be built up of the characteristics of flows in branches and vertical pipes and has provided some of the basic data needed for a generalized design method. Much work has also been done using normal

appliance discharges, both in the laboratory and in the field. This has provided data to supplement that obtained with continuous flow.

Several investigators have shown that, with the rates of water flow encountered in practice, nearly all of the discharge flows down the stack as a thin annular sheet on the inner wall of the pipe, the remainder of the stack being occupied by a core of air. This has been demonstrated with experiments in eight- and 32 storey buildings having 100 and 150 mm stacks in PVC and cast iron respectively (Pink 1973(i)). A composite probe was used to traverse a stack carrying various rates of water flow; a reduction of electrical resistance across the tip of the probe indicated the presence of water at the tip. Figure 8.20 shows typical results with water introduced into the stack through swept branch fittings. It was found that flow was generally annular; water flow within the air core was only a few per cent of that in the annulus. The estimated thickness is shown by the dashed lines in the figures.

Water entering a vertical stack from a branch is directed onto the wall of the stack and flows downwards, then, as an annular sheet. The water accelerates under the influence of gravity and, assuming a constant volume flow, the thickness of the sheet correspondingly decreases. After a fall of 2 or 3 m, a terminal velocity and thickness are

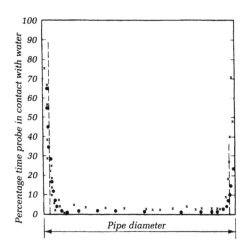

Figure 8.20 Traverses at right angles across 100 mm stack flowing at 2 l/s; discharge from 7.5 m above traverse

attained, beyond which the velocity and thickness remain substantially constant. For simplicity the sheet of water can be considered as a rigid body moving down a plane vertical wall, neglecting the radial velocity gradient that will, in fact, exist. To find the terminal velocity and/or thickness we make use of the relationship that accelerating force is equal to the gravitational force less the frictional resistance, noting that when the terminal velocity is reached the acceleration is zero and the gravitational force balances the frictional force.

For full pipe flow the friction factor, pipe diameter, Reynolds number and pipe roughness are related as shown earlier in the Colebrook–White equation for smooth turbulent and rough turbulent conditions and for the transition between the two. As given earlier, this equation is:

$$\frac{1}{\sqrt{\lambda}} = -2\log_{10}\left(\frac{k}{3.7D} + \frac{2.51}{R_e\sqrt{\lambda}}\right) \tag{8.3}$$

where λ is the friction factor, k is the pipe roughness, D is the pipe diameter and R_e is the Reynolds number of the fluid. Under full pipe flow conditions $\lambda = 8\tau_0/\rho v^2$ where τ_0 is the boundary shear stress, ρ is the fluid density and v is the fluid velocity. Substituting for λ in equation 8.3.

$$\sqrt{\frac{\rho V^2}{8\tau_0}} = 2\log_{10}\left(\frac{k}{3.7D} + \frac{2.51}{R_e}\sqrt{\frac{\rho V^2}{8\tau_0}}\right) \tag{8.8}$$

It may be shown that equation (8.3) can be applied to flow conditions with a free surface if the hydraulic radius R_H, defined as the ratio of cross-sectional area of flow to wetted perimeter, replaces the pipe diameter. For full pipe flow:

$$R_H = \frac{\pi D^2/4}{\pi D} = \frac{D}{4}$$

Replacing R_e by vD/ν and substituting for D in terms of R_H in equation (8.8):

$$\sqrt{\frac{\rho V^2}{8\tau_0}} = -2\log_{10}\left(\frac{k}{14.8R_H} + \frac{2.51\nu}{4VR_H}\sqrt{\frac{\rho V^2}{8\tau_0}}\right) \tag{8.9}$$

where ν is the kinematic viscosity of the fluid.

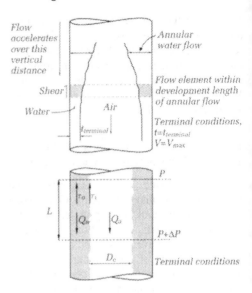

Figure 8.21   Definition sketches for annular flow

Considering annular flow in a vertical drainage stack (figure 8.21), a force balance equating frictional force to gravitational force on a length of the annulus ΔL, of terminal thickness t and moving at terminal velocity V, gives:

$$\pi D\tau_0\Delta L = \rho\pi Dt\Delta Lg$$

$$\tau_0 = \rho tg$$

In this balance the shear stress between the water and the air core τ_i is neglected since it can be shown to be small in comparison with τ_0. Also, by continuity Q_w equals πDtV and:

$$R_H = \pi Dt/\pi D = t$$

Substituting for V, for τ_0 and for R_H in equation (8.9) and rearranging:

$$\frac{Q_w}{4\pi Dt}\sqrt{\frac{1}{2gt}} = -\log_{10}\left(\frac{k}{14.8t} + \frac{0.31375\nu}{t}\sqrt{\frac{1}{2gt}}\right) \tag{8.10}$$

This equation can be solved iteratively for t.

Values of the annular thickness t predicted in this way are included for comparison with the experimental data in figure 8.20. The thicknesses calculated are indicated by dashed lines.

Equation (8.10) offers the opportunity to calculate both the terminal thickness and the velocity. It is useful to reduce this expression to link terminal velocity directly to stack diameter and flow rate. From figure 8.21 in the accelerating flow region the equation of motion indicates that:

$$\frac{dV_w}{dt'} = g - \frac{\tau_0}{\rho t}$$

where t' is time and:

$$\tau_0 = f\frac{1}{2}\rho V_w^2; \quad t = Q_w/\pi D V_w$$

Hence:

$$\frac{dV_w}{dt'} = g - \frac{1}{2}f\frac{\pi D}{Q_w}V_w^3$$

Thus terminal velocity, V_w, at $t = t_{terminal}$ becomes:

$$V_{terminal} = \sqrt[3]{\left(\frac{2g}{f}\right)\frac{Q_w}{\pi D}}$$

as $dV_w/dt' = 0$ once terminal conditions are reached.

The $(2g/f)$ term will be recognized from the open-channel-flow equation as related to the Chézy coefficient C:

$$C = \sqrt{2g/f} = m^{1/6}/n$$

where m is the hydraulic mean depth in a free surface channel flow where the channel roughness is expressed by Manning's coefficient, n.

Annular flow in a stack is limited, for practical purposes (chapter 5), to a terminal thickness of one-sixteenth of the stack diameter, i.e. one-quarter of the stack cross-section is water filled. Full-bore plugs of water tend to form with higher flows. Thus it is an appropriate approximation to consider the annular flow as if it were in a rectangular channel where the flow width to depth ratio was 16.

The appropriate hydraulic mean depth m has already been shown to be equivalent to the flow thickness, t. Thus, as $t = Q_w/\pi D V_w$ it follows that by substituting for C^2:

$$V_{terminal} = \sqrt[3]{\frac{1}{n^2}\left(\frac{Q_w}{\pi D V_t}\right)^{1/3}\frac{Q_w}{\pi D}}$$

Thus:

$$V_{terminal} = K(Q_w/D)^{0.4}$$

where

$$K = (0.2173/n^2)^{0.3}$$

For a typical, smooth, PVCU stack $n = 0.007$ and hence $K = 12.4$. This value compares closely with that quoted by Chakrabarti (1986) and other authors, including Wyly (1964). The value of K obtained by plotting V_t against $(Q_w/D)^{0.4}$ yields a coefficient value of 14.9 for smooth pipes, again a result in broad agreement with that calculated above. This agreement in effect confirms the use of the Chézy approximation for the annular frictional resistance.

The distance necessary to achieve terminal velocity may also be determined from the equation of motion for the accelerating annular flow as follows:

$$\frac{dV}{dt'} = \frac{dV}{dz}\frac{dz}{dt'} = V\frac{dV}{dz}$$

$$\frac{dV}{dz} = \frac{1}{V}\frac{dV}{dt'} = \frac{1}{V}\left(g - \frac{1}{2}f\frac{\pi D}{Q_w}V^3\right)$$

where V is the annular water film velocity and t' is time:

$$dz = \frac{V\,dV}{(g - \frac{1}{2}f\pi D\ V^3/Q_w)}$$

Substituting for the terminal velocity yields:

$$dz = \frac{V_t^2}{g}\frac{\theta\,d\theta}{(1 - \theta^3)}$$

where $\theta = V/V_t$. Integration yields the distance to terminal velocity and thickness at $z = z_{terminal}$:

$$\int_{z=0}^{z=z_t}dz = \frac{V_t^2}{g}\int_{\theta=0}^{\theta=1}\frac{\theta\,d\theta}{1-\theta^3}$$

However, the result will be infinite as the flow will approach the limiting condition asymptotically. Normal procedure is to calculate the distance to 99 per cent of terminal conditions, i.e. $\theta = 0.99$ as an upper integration limit.

The vertical fall required to attain terminal velocity, in SI units, is thus:

$$Z_t = 0.159V_t^2$$

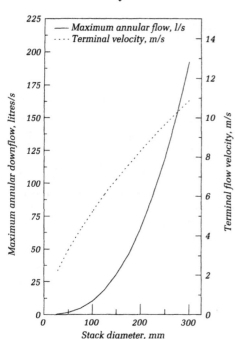

Figure 8.22 Annular flow terminal velocity and maximum allowable flow rate for a one-sixteenth terminal film thickness. Smooth stack

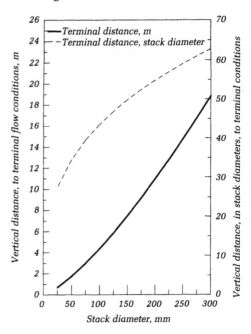

Figure 8.23 Vertical stack length required to attain terminal conditions at the maximum allowable flow, based on a one-sixteenth stack diameter film thickness

Figure 8.22 illustrates the maximum allowable flow in the stack based upon a one-sixteenth diameter, annular film thickness limit, together with the corresponding terminal velocity, for a stack material with a Manning n value of 0.007. Similarly figure 8.23 illustrates the vertical distance necessary to reach this terminal condition, both in metres and in stack diameters. It will be seen that the terminal lengths required rarely exceed one- or two-storey heights, a result that supports the use of terminal velocity over the greater length of a vertical stack in a multi-storey building. The equations determining terminal flow velocity and thickness, and the distance required to fall to attain these conditions, confirm the expected relationships, e.g. the dominance of stack diameter, in these results. The terminal velocity is lower in rougher pipes, the distance required also decreasing. The terminal thickness increases as the final velocity achieved decreases.

Water flow capacity of stacks

Using an empirical formula (Manning) for pipe friction, simpler than the Colebrook–White equation, Wyly and Eaton (1961) derived an expression for the terminal velocity in stacks.

They coupled this with the continuity equation to obtain an expression for flow capacity in terms of stack diameter and the fraction r of the cross-section of the stack occupied by the water. This is:

$$Q = 31.9r^{5/3} D^{8/3} \qquad (8.11)$$

where Q is in cubic metres per second and D is in metres, and the pipe is cast iron. It reduces to:

$Q = 1.6D^{8/3}$ with the stack flowing one-sixth full and

$Q = 3.15D^{8/3}$ with the stack flowing one-quarter full

The latter was suggested as an upper limit for capacity because full-bore plugs of water tend to form with higher flows. These expressions, coupled with simple rules about vent sizes, have formed the basis for the development of some of the sizing tables in codes and standards, including BS 5572 and the proposed CEN standard (De Cuyper 1993); see chapter 5. The figures for this purpose have been rounded. This simpler approach may be compared with, for instance, figure 8.22 giving values for 'quarter-full flow' in smooth stacks (see also figure 5.12).

Approach to vent sizing

An approach to the general calculation of venting was first attempted by Wise (1957a) and Lillywhite and Wise (1969). In the latter paper a scheme of calculation was taken to the point at which sizing tables for vents could be produced and they form a part of BS 5572 (1994), and see Appendix 1. We now describe the main aspects to be considered in this method.

The nearest approach to the stack being filled with water occurs at the branch inlets from the appliances. With a small branch discharging into a large stack, a cylinder of water about equal in diameter to the branch lies across the stack (figure 4.13) and this is seen on plan as a strip partly blocking the cross-section, as shown diagrammatically in figure 8.24(a). Figure 8.24(b) and (c) give an idea of the conditions that develop when the branch and stack diameters are equal, as with a WC. The air can flow downwards past the inlet through the free area on either side of the water. At changes of direction, such as at the bend at the base of the stack, water velocities are reduced and there is a corresponding increase in the proportion of the cross-section of the pipe occupied by water.

The picture, then, is of an annular flow of water down the stack which induces a flow of air downwards. At water inlets especially, but also at changes of direction in the pipework, more of the cross-section is occupied by water since water velocities are less than in the annular flow. At such points, and also in the dry part of the stack above the topmost branch discharging, there is a resistance to the flow of air.

(i) Pressure distribution in a stack

With one branch discharging the air pressure distribution in the stack is as shown diagrammatically as an example in figure 8.25. The top of the stack is at atmospheric pressure. Pressure decreases down the dry part of the stack AB, i.e. there is a suction, corresponding to the pipe friction loss in this length. At B, where the water is discharging, a marked drop in air pressure occurs. Pressure then rises from B to C within the annular flow and at C is above atmospheric pressure by an amount

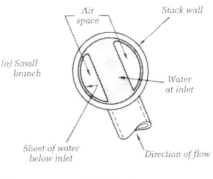

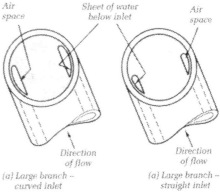

Figure 8.24 Diagrams to illustrate branches discharging into a stack

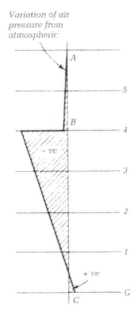

Figure 8.25 Diagram to illustrate form of pressure variation in a stack with one branch discharging

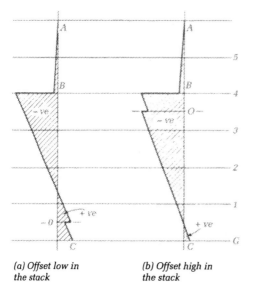

(a) Offset low in
the stack

(b) Offset high in
the stack

Figure 8.26 Diagrams to illustrate the effect of an offset in a stack

depending on the resistance of the bend and drain. The rate of change of pressure with distance is known as the pressure gradient, a term used with particular reference to section BC, i.e. the 'wet' part of the stack in the following analysis. With more than one branch discharging, a drop in pressure occurs across each stream of water entering the stack and an example was given in chapter 4. The behaviour of offsets is not fully understood but the kind of modification of the curve that is thought to occur when an offset is incorporated into a stack at O is shown in figure 8.26. An offset introduces an extra resistance to air flow. When high in the stack this tends to increase the suction and when low in the stack tends to increase the bottom pressure. The effect depends on the shape and dimensions of the offset and on the amount of water flowing through it.

(ii) Rates of air flow

The air flow is caused by the shear between the annulus of water and the air core. Early work on short stacks had seemed to show that the rate of air flow was independent of the height of the stack but later measurements on tall stacks (Pink 1973 (iii)) disproved this. Consider figure 8.21 which shows a short length of stack

with steady, annular water flow, assumed to have reached a constant terminal velocity. The diameter of the air core depends on the water flow rate, the stack diameter and the surface roughness of the pipe material. It is obtained by using the foregoing equations to calculate the terminal thickness of the water annulus, from which the diameter of the air core may be simply deduced. Figure 8.21 shows the shear stress τ_i that exists between the water and air, and the change of air pressure ΔP over the length ΔL. A force balance on the air core yields:

$$\Delta P \frac{\pi}{4} D_c^2 = \tau_i \pi D_c \Delta L$$

or:

$$\tau_i = \frac{\Delta P}{\Delta L} \frac{D_c}{4}$$

Supposing that the volume flow of air Q_a depends principally on D_c and τ_i, it follows that Q_a is principally a function of the water flow rate Q_w, the stack diameter D, the pipe roughness k and the pressure gradient $\Delta P/\Delta L$, i.e.:

$$Q_a = f(Q_w, D, k, \Delta P/\Delta L)$$

Experiments on tall cast iron stacks of 100 and 150 mm diameter provided data such as those shown in figure 8.27 based on Pink 1973(iii), where the air flow varies with pressure gradient in the wet part of the stack for a given water flow rate Q_w and stack diameter D and roughness k. Such data have enabled the development of the curves typified by figure 8.28. Here Q_a is plotted against Q_w and is seen to depend on $\Delta p/\Delta L$, values of the pressure gradient being indicated against the curves; the factors D and k are constant. As would be expected, for a given water flow rate the air flow is greater the smaller the pressure gradient acting against the flow.

(iii) Pipe and fitting losses

Calculation of the friction loss for air flow in the dry part of the stack is a matter of applying one of the formulae outlined earlier in this chapter. Pressure loss coefficients for air flowing within fittings partly filled with water have to be determined experimentally. The situation at a water inlet is shown, for example, in figure 8.24; air flows through the twin

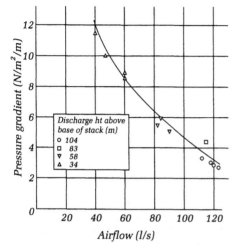

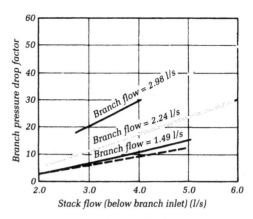

Figure 8.29 Variation of branch pressure drop factor with stack and branch flows for a 100 mm stack and branch

Note: Water flow rate 1.7 l/s
Stack diameter 150 mm
Height of discharge point above stack base between 34 and 104 m

Figure 8.27 Air flow rate against pressure gradient for a water flow of 1.7 l/s in a 150 mm cast iron stack

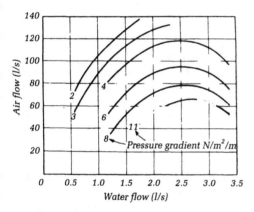

Figure 8.28 Variation of air flow with water flow in a 150 mm cast iron stack

segments on either side of the water. At a change of direction, air flows through the area left in the centre of the cross-section by water at the periphery. With a sudden change of direction a film of water may partly cover the central air core, reducing further the area available for air flow. The first measurements of pressure loss coefficients for air flowing in such circumstances were made some years ago (Wise 1957a). Figure 8.29 gives examples of what is now available and shows that the loss coefficients depend on the water flow rates as

well as on the dimensions and shapes of the fittings themselves.

(iv) General scheme of calculation

From the foregoing considerations a general scheme for the calculation of the maximum suctions and pressures that arise in discharge stacks may be formulated (Lillywhite and Wise 1969). Principal interest is in the maximum suction that can arise which, numerically, is generally somewhat greater than the maximum pressure. To simplify the procedure the following assumptions are convenient:

1. Any discharges into the stack are assumed to take place on adjacent floors; this has been shown in practice to give the greater seal losses.
2. Discharges from the branches are assumed to be steady and continuous at the maximum rates likely from the appliances.
3. Water flow is assumed to be truly annular.
4. The horizontal drain from the base of the stack can discharge freely with negligible resistance to air flow. Hence pressure at the base of the stack is taken to be atmospheric.

The maximum suction at any part of the stack may be written:

suction = pressure drop at entry to the stack + pressure drop due to bends and piping in the dry part of the stack + sum of pressure drops across the discharging inlets + velocity head

In simplified terms:

$$h_s = 0.974 \frac{Q_a^2}{D^4}\left(0.5 + \sum k_{bends} + \frac{fl}{D} + \sum k_{inlets} + 1\right) N/m^2 \quad (8.12)$$

The factors in parentheses on the right-hand side represent the various pressure loss coefficients of the kinds described, whilst Q_a is the air flow and D the stack diameter.

In practice, the equation is used in an iterative procedure which eventually provides h_s and computer programs have been written to do the arithmetic. For design purposes, the computed maximum suction for a given installation is compared with the peak allowable of 375 N/m^2. Tables of design information given elsewhere in the present publication, also in BS 5572 (1994) have been derived making use of this method and supported by a range of practical tests in the laboratory and in buildings in use.

Further analysis of annular downflow, entrained air and suction

This chapter has introduced several fundamental concepts of fluid mechanics, including the Steady Flow Energy Equation and the no-slip condition between a fluid and a conduit surface and between two fluids in contact with each other. When combined with the summation of the suction pressures down the length of a vertical stack (Lillywhite and Wise 1969), these offer a way into the relationship between stack annular water flow and the entrained airflow to be expected. The back pressure sometimes experienced at the base of the stack when added to the suction pressure total yields the load that has to be overcome by the entrainment action between the falling water and air core in the stack:

$$\Delta P_{total} = \Delta P_{entry} + \Delta P_{dry\ pipe\ friction} + \Delta P_{branch\ junction} + \Delta P_{back\ pressure}$$

In order for the Steady Flow Energy Equation to be complied with between the atmospheric air above the roof level entry to the stack and the atmospheric pressure in the sewer, it follows that there must be an energy input to balance this load. This has been described as a 'pseudo-

fan' effect, however, a more accurate description would be based around the continuous application of a shear stress at the annular water flow / entrained air core interface arising from the application of the 'no-slip' condition. Thus for the wet stack height where air and water are in contact at the annular interface the energy input required may be represented by an application of Darcy's equation:

$$\Delta P_{input} = \frac{4\rho f_{air/water}\ L_s(V_{air} - V_{water})^2}{2\ D_{air\ core}}$$

where L_s is the stack height and $f_{air/water}$ is an equivalent friction factor which will empirically define the shear stress between the annular water film and the entrained air core.

Equating the summation of losses incurred in maintaining the air core flow down the stack to the energy input implied by the shear stress allows an expression for friction factor $f_{air/water}$ to be developed as

$$f_{air/water} = \frac{2\ D_{air\ core}\ \Delta P_{input}}{4\rho\ L_s(V_{air} - V_{water})|(V_{air} - V_{water})|}$$

Note that the absolute value of $(V_{air} - V_{water})$ ensures that the sign of $f_{air/water}$ reflects the velocity differential between the air and water surfaces. Inspection of this expression indicates a methodology for determining $f_{air/water}$. For a particular building stack the pressure loss may be determined by measurement at a range of applied water downflows. The air core mean velocity may be determined by monitoring the entrained air flow at entry to the stack. The water downflow velocity may be taken as the annular film mean velocity and the height of the stack may be approximated assuming fully developed annular flows. Jack (1997, 2001) presents the outcome of an extensive experimental programme encompassing both site testing in a 20 storey public authority housing block in Dundee, Scotland and laboratory and campus building based measurements at Heriot-Watt University. Jack's work related the entrained air phenomenon to a range of non-dimensional groups so that the results may be widely applied, and proposed a set of empirical expressions to determine both $f_{air/water}$ and the reductions in stack height necessary to allow for the establishment of fully developed annular flow. These expressions have been

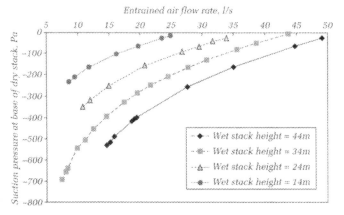

Figure 8.30 AIRNET model predictions for a variation in the wet stack height for 150 mm diameter 'smooth' pipe, when the discharge flow rate is 1.147 l/sec

incorporated into the AIRNET vent system simulation, chapter 9, that allows the entrained airflow for any range of applied water flow conditions to be determined, together with the propagation of low amplitude air pressure transients as a result of airflow establishment.

The ability to express the shear stress effects between the air core and the annular water film allows a range of previous observations to be explained. For example if the stack is segregated into wet and dry sections it will be seen that the air pressure will always fall in the dry stack due to friction between the airflow and the dry stack walls. In the wet stack the air pressure may rise or fall depending upon the relative magnitudes of the air and water velocities. In a single stack system with one discharging branch the air pressure will recover in the wet stack. For a stack with a multiplicity of discharging branches the combined flow in the lower regions may entrain an airflow greater than that appropriate in the upper floors where the annular water film represents a lower water flow rate. Hence air pressure may continue to fall in that section of the wet stack. Overall the input and extracted energy must balance.

The discussion above clearly allows the pressure profiles presented in figure 8.25 to be explained along with the observation from the Centre Point research that stack height influenced entrained airflow rates. Application of Jack's results within the AIRNET model allowed the prediction of airflows and hence total pressure drop in a stack dependent upon

both wet stack length and applied water flow. Figure 8.30 illustrates these predictions for the local authority multi-storey block already mentioned. It is clear that increased water downflow increases entrained airflow. However, the length of vertical stack subjected to the traction between the water film and the air core also determine the entrained airflow and the associated pressure requirement. These results are in all respects the equivalent of a set of fan characteristics where the available wet stack height replicates the effect of fan speed.

Similarly the inclusion of separation losses present at any discharging branch or offset allows the profiles in figure 8.26 to be understood. The water–air shear stress effectively drags air through the obstruction formed by the discharging branch or the water films that form at the offset entry — films analogous to those that form at the base of the stack. Thus in all cases an offset will reduce the local air pressure as illustrated.

The inclusion of offsets in a vertical stack are, in general, bad design. Historically, prior to a full understanding of the attainment of terminal annular flow conditions, it was believed that offsets could limit the acceleration of the water downflows. However, in practice, offsets may be the cause of severe transient problems as they may be the location of intermittent surcharge in response to appliance discharge patterns further up the building. Figure 8.31 illustrates the effect of an offset surcharge in a four-storey building. It will be seen that the cessation of airflow

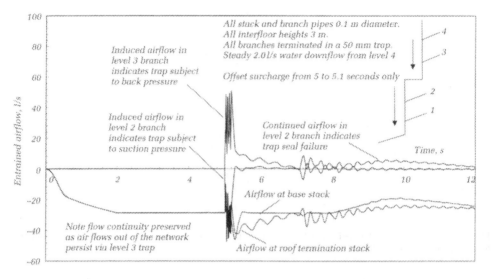

Figure 8.31 Entrained airflows within a network subjected to a short duration offset surcharge

results in a positive transient propagated up the stack, effectively displacing trap seals due to positive pressure. Below the offset a negative transient of equal magnitude depletes trap seals by induced siphonage. These effects are illustrated by the predicted airflows established in the appliance branches at the floor levels immediately above and below the offset. The continued airflow out of the network through the trap on the second level indicates that the trap seal has failed due to induced siphonage.

The establishment of full bore flow in a group branch that includes a vertical section, figure 8.32, may also lead to trap seal failures. Figure 8.32 indicates the likely pressure variations upstream and downstream of a vertical branch section that momentarily runs full bore due to combined appliance discharge. The piston effect applies induced siphonage failures upstream and positive back pressure failures downstream. The inclusion of air admittance valves may protect against the induced siphonage, however, these provide no protection against the positive pressure transients as they close to prevent odour ingress to the surrounding habitable space.

Effect of detergents on entrained airflow in vertical stacks

Current drainage and vent system design guides and codes are based on empirical data gathered across a wide range of international laboratories. The base data relate to cold clean water as the discharge fluid responsible for the entrainment of an airflow within the drainage vertical stack and hence for the air pressure regime experienced by the drainage network. However most 'grey' and 'black' water sources in buildings are dosed with detergent and are often warm. These additives and temperature changes affect the air entrainment as they, acting as surfactants, contribute to the shear stress active between the falling water annulus and the entrained air core. These effects were investigated by Campbell and McLeod (2000, 2001), the results indicating that the effects were dependent upon the chemical composition of the detergent. The detergent types used included anionic detergents — found in soap liquids and bars, cationic detergents — found in softeners and conditioners and non-ionic detergents found in fabric cleaners.

Significantly higher induced airflows were generated in a two-storey test rig for anionic and non-ionic surfactants when compared with clean water — up to 40% in some cases. These detergent types are associated with washing, cleaning and bathing activities where large water discharges are involved. These discharges may well be warm. Cationic detergents resulted in slightly reduced entrained airflows and are associated with the smaller discharge volumes, for example at the end of a washing

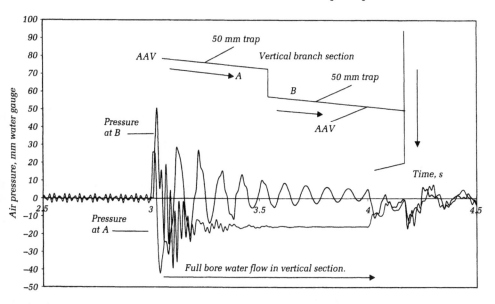

Figure 8.32 Pressure predicted upstream and downstream of a surcharged vertical section in a branch serving several appliances

machine cycle. These results suggest that for water discharges comparable to several bath, shower or sink discharges or WC discharges in a two-storey building configured according to the current codes, greater entrained airflow will be generated with consequently greater suction pressure within the network. This research will continue and will be included in future simulations of drainage and vent system air pressure and entrainment.

Influence of fluid mechanics theory on building drainage and water supply

This chapter has confirmed that the study of air and water flows within building drainage and water supply systems must be firmly based upon a wider understanding of fluid mechanics. These flow conditions are an application of phenomena capable of analysis via the mathematical tools and criteria developed through extensive research in fluid mechanics. The central importance of this understanding is clear in both the presentation of vent system air pressure conditions, which led to the introduction of the single stack system in the UK, and their impact upon trap seal retention and the work on unsteady flows. The application of frictional representations, together with the Steady Flow Energy Equation allows a rational explanation of the steady state pressure profiles experienced in vent systems, while the introduction of well-established methods for treating waterhammer lays the foundation for a later treatment of unsteady flows in general. While this chapter has, in the main, concentrated upon the description of system flows and pressure phenomena in terms of steady state relationships, it is clear that in building drainage, vent and water supply systems that condition rarely applies. All flows are dependent upon appliance usage patterns and are hence inherently time dependent. In the case of free surface flows wave attenuation will also add to the unsteady nature of all the flow phenomena encountered. Chapter 9 will continue the theme of the application of techniques from the wider field of fluid mechanics to both solve problems within building systems and to generate system simulations that will be of direct benefit to system designers, code bodies and appliance manufacturers.

9 Unsteady Flow Modelling in Water Supply and Drainage Systems

The fluid flow phenomena encountered in building water supply and drainage systems are inherently unsteady in nature. Whether considering the passage of an appliance discharge through a drainage network or the entrainment of air by the annular water film in a vertical stack, the observed flow parameters are clearly time dependent. In the past the available techniques were unable to handle these conditions at a level accessible to the average designer. However, the advent of fast available computing and a computer literate society makes the application of more comprehensive simulation techniques both practical and desirable. This chapter will consider the application of a particular family of simulations to the prediction of unsteady flow conditions within both building water supply and drainage/vent systems. The systems to be considered will include free surface flows in the horizontal branches of drainage networks and open gutter rainwater gutters, the entrained air flows and associated low amplitude air pressure transients in vertical stack and vent systems and the pressure transients generated in water supply services and siphonic rainwater systems. All these conditions will be shown to belong to the same family of unsteady flow phenomena soluble by the application of finite difference schemes based on the method of characteristics and the use of currently available computing.

The term 'unsteady' has an accepted definition that at any chosen location in the pipe network the dominant flow parameters for each flow mechanism will vary with time. In the case of drainage and vent systems the flowrate and depth, or the annular downflow thickness, entrained airflow and suction pressure, will vary with time. Similarly, rainwater gutter depths will depend on storm intensity and duration, while the pressures within siphonic drainage networks will be similarly time dependent. Pressures in water supply systems will depend on demand and valve closure rates. Water flow within a building drainage network may be described as unsteady due to both the variation of outflow at each appliance during its discharge, the random usage of the appliances connected to the network and the modifications to the appliance discharge due to wave attenuation and combination with joining flows at junctions during its passage through the network.

Current design methods have incorporated the usage pattern of system appliances by the introduction of probability of use functions, beginning with the Fixture Unit Method (Hunter 1940). However, the design guidance offered is conservative and assumes steady flowrates related to a summation of the network fixture units. As the expected steady flowrate in a branch rises, the design guidance available proposes increases in pipe diameter and/or gradient. It may be shown, chapter 8, figure 8.5, that the flowrate at which such changes are recommended correspond to steady flow depths reaching 50 per cent of the pipe diameter.

As a result of multiple appliance discharge the downflow in the vertical stacks connecting each floor network to the building collection drain will also be unsteady. This implies that the entrained airflow in the stack, and any associated vent network, will also be time dependent, and, more importantly, the pressure transients generated by these changes in annular water downflows will depend upon both the downflow and the rate at which it changes. For example, an annular downflow rising from zero to 4 l/s in 1 second will give a much greater transient suction pressure than the same change in water flow accomplished in, say, 5 seconds. National codes are generally based on steady flow testing that limits the downflow in any stack diameter to that sufficient to generate a given suction, in the UK – 375 N/m^2, is used, or to the annular downflow sufficient to give a particular fully developed annular thickness; 1/16 the diameter or 25 per cent of the cross-sectional area is commonly quoted.

In parallel to the US / UK design guidelines outlined in chapter 1, all other developed national codes have their own particular methodologies and limiting values. As these are entirely based upon the translation of steady state empirical data into 'safe' design guidelines, it is extremely difficult to compare their relative merits or suitability to meet changing norms and usage patterns within building drainage.

The introduction of water conservation, the rise of 'sustainability' as a predominant research agenda and the installation of low water using appliances set up design requirements not necessarily covered by the steady state empirical data utilized to generate current design guides. Rapid changes in the water consumption of appliances, together with water conservation considerations and future metering and costing strategies imply that improved modelling methods are required to ensure that systems are not over designed for their likely loadings. In building drainage system design, overdesign cannot be regarded as a safety factor as increases in pipe diameter at any given flow rate automatically lead to an enhanced likelihood of solid deposition and maintenance costs. The introduction of the Water Regulations 1999, and in particular their re-introduction of dual flush WC operation at the reduced maximum flush volumes of 6 and 4 litres (see chapter 11) reinforces the necessity to consider water conservation in building water supply and drainage design. The sizing of drains in general and in relation to WC operation will become important to efficient system operation.

Modelling unsteady flow conditions

The water flow and entrained airflows within building water supply and drainage systems belong to a family of unsteady flows that may be described by the full equations of continuity and momentum, whose solution was first investigated by d'Alembert in the 1750s and generally now referred to as the St Venant wave equations. The following text will deal in a general way with unsteady flow modelling by developing the general form of the equations of momentum and continuity that apply to all the unsteady cases defined above and in particular will define the importance of wave propagation velocity in each case.

The momentum equation may be stated (Swaffield and Boldy 1993, Douglas, Gasiorek and Swaffield 2000), as

$$\frac{1}{\rho}\frac{\partial p}{\partial x} + \left[\frac{\partial V}{\partial t} + V\frac{\partial V}{\partial x}\right] - g\sin\alpha + \frac{\tau_0 P}{\rho A} + \frac{qV}{A} = 0$$

(9.1)

reducing to

Term I^m + Term II^m − Term III^m + Term IV^m + Term V^m = 0

Each of these terms has its own significance, e.g. Term III^m represents gravitational forces while Term IV^m represents the frictional resistance acting to oppose the local flow, Term V^m represents the acceleration of any lateral inflow.

The continuity equation may also be derived in a general form. The conduit is assumed to be linearly elastic, only subjected to small deformations. The fluid is assumed to undergo small changes in density compared with the magnitude of its density.

$$\frac{\partial V}{\partial x} + \frac{1}{A}\left[\frac{\partial A}{\partial t} + V\frac{\partial A}{\partial x}\right] + \frac{1}{\rho}\left[\frac{\partial \rho}{\partial t} + V\frac{\partial \rho}{\partial x}\right] - \frac{q}{A} = 0$$

(9.2)

where q is the lateral inflow per unit length of conduit,

Table 9.1 Identification of relevant terms for each unsteady flow regime considered

Flow regime	Equation of motion	Continuity equation
Waterhammer (closed conduit flow, siphonic rainwater systems, all terms relevant except lateral inflows)	I,II,III,IV	I,II,III
Free surface building drainage networks, i.e. no density changes	I,II,III,IV	I,II only
Free surface rainwater gutter systems, i.e. as for drainage networks but including lateral inflows	I,II,III,IV,V	I,II,IV only
Air pressure transients in drainage vent systems, i.e. no changes in flow cross section, lateral inflows or longitudinal duct extensions	I,II,IV only	I,III only

reducing to

$$\text{Term I}^c + \text{Term II}^c + \text{Term III}^c - \text{Term IV}^c = 0$$

Term II^c represents the effect on transient propagation of a change in flow cross-sectional area and includes, via the Poisson's ratio term, the effect of longitudinal extension of the conduit wall. Similarly Term III^c represents density changes due to the passage of the transient. Having understood the function of each term in the equations of motion and continuity it is possible to identify which are relevant for each of the unsteady flow regimes to be considered (see table 9.1).

The method of characteristics

The unsteady flow equations of momentum and continuity for each of the flow cases identified above are examples of quasi-linear hyperbolic partial differential equations and as such may be solved by numerical techniques, provided that they are transformed into a total differential format. Following the initial work in the 1960s (Lister 1960), the method of

characteristics has emerged as the industry standard simulation technique. This method re-casts the equations as total differential equations in two dependent and two independent variables.

The combined total differential equation may thus be written as

$$\frac{\mathrm{d}u_1}{\mathrm{d}t} \pm C_2 \frac{\mathrm{d}u_2}{\mathrm{d}t} + C_3 = 0 \qquad (9.3)$$

provided that the Courant Criterion, expressed as

$$\frac{\mathrm{d}x}{\mathrm{d}t} = u_1 \pm c \qquad (9.4)$$

is maintained. The appropriate values of u_1, u_2, C_1, and C_2 are given in table 9.2.

Particular points emerge from this analysis:

1. In pressure transient or waterhammer analysis the pressures generated are sufficient for the elasticity of the pipe wall to affect the wave speed calculation. Hence wave speeds will be much lower in plastic as opposed to steel pipes, however, the actual pressure rise following a valve closure may

Table 9.2 Identification of the coefficients in the finite difference equations applicable to the building drainage applications considered

	u_1	u_2	C_2	C_3
Waterhammer (inc. siphonic rainwater systems)	V	p	$1/\rho c$	$-g \sin \alpha + fV|V|/2m$
Drain	V	h	g/c	$g(S - S_0)$
Gutter	V	h	g/c	$g(S - S_0) + q(V \pm c)/A$
Vent	u	c	$2c/(\gamma - 1)$	$fu|u|/2m$

Note that the friction factor f in the vent system application can under some conditions of annular water downflow and entrained airflow act as the driver for the entrainment process

Table 9.3 Wave speed calculations for a range of flow conditions, together with the pressure levels generated where appropriate for a 1 m/s instantaneous reduction in flow velocity

Rigid pipe	$c \to 1500\,\text{m/s}$	$c = \sqrt{\left(\dfrac{K_f}{\rho_f}\right)}$
Elastic pipe	$200 < c < 1500\,\text{m/s}$	$c = \sqrt{\dfrac{K_f}{\rho_f\left(1 + \dfrac{DK_f}{Ee}C_1\right)}}$
Gas/fluid pipeflow	$150 < c < 600\,\text{m/s}$	$c = \sqrt{\dfrac{K_{\text{eff}}}{\rho_{\text{eff}}}}$
Low amplitude gas	$c \to 350\,\text{m/s}$	$c = \sqrt{\gamma\dfrac{p}{\rho}}$
Free surface wave	$0.25 < c < 5.0\,\text{m/s}$	$c = \sqrt{g\dfrac{A}{T}}$
where	$K_{\text{eff}} = \left(\dfrac{1}{\dfrac{1-y}{K_f} + \dfrac{y}{K_g} + \dfrac{D}{Ee}C_1}\right),$	$\rho_{\text{eff}} = (1-y)\rho_f + y\rho_g$

Pressure surge values for a 1.0 m/s reduction in flow velocity vary from 1500 kN/m² for water in a rigid pipe to 40 mm water gauge for a 1.0 m/s reduction in entrained airflow velocity in a building drainage vent stack. For partially filled circular cross-section pipeflow the half–diameter depth wave speed reduces to 2.776D. For air density, ρ, may be taken as 1.3 at 0 °C, atmospheric pressure, p, as 101 kN/m² and γ as 1.39 at 1 atm and 0 °C, so that wave speed, $c = 329\,\text{m/s}$.

be comparable as the reduced wave speed effectively means that changes in flow conditions occur in a shorter time as measured in pipe periods.

2. With air pressure transients the amplitude is normally insufficient to cause pipe wall distortion so that the wave speed may be determined from density and pressure considerations only. However, it should be noted that in this case the characteristic equations are expressed in terms of fluid velocity and wave speed as pressure and density are not independent of each other. This implies a further calculation to determine pressure once the characteristic equations have been solved as

$$p_P = \left[(p_o/\rho_o^\gamma)(\gamma/c^2)^\gamma\right]^{1/(1-\gamma)} \tag{9.5}$$

3. In free surface flows, either in horizontal branches, sewer connections or gutters, the wave speed is the surface wave propagation velocity, dependent upon flow depth and velocity. Flows may be characterized into subcritical, where the flow velocity is less than this wave speed, and supercritical, where the velocity exceeds the surface wave speed. Hence, in supercritical flows disturbances cannot propagate upstream leading to the formation of hydraulic jumps to separate

these flow conditions. (Note that the defining Froude number is effectively a form of Mach number as it relates local flow velocity to the wave propagation velocity.)

The finite difference expressions only exist provided that the dx/dt relationships are adhered to; this is known as the Courant criterion and links simulation time step to

Table 9.4 Values of Young's modulus of elasticity, Poisson's ratio and shear modulus for a range of common pipe wall materials

Material	Young's modulus $\times10^{-9}\,\text{N/m}^2$	Poisson's ratio	Shear modulus $\times10^{-9}\,\text{N/m}^2$
Aluminium	70.0	0.3	27.6
Cast iron	80.0–110.0	0.25	40.0–80.0
Concrete	20.0–30.0	0.1–0.3	–
Copper	107.0–130.0	0.34	–
Glass	68.0	0.24	–
GRP	50.0	0.35	3.92
Polythene	3.1	–	1.10
PTFE plastic	0.35	–	–
PVC plastic	2.4–2.8	–	–
Reinforced concrete	30.0–60.0	0.15	–
Rubber	0.7–7.0	0.46–0.49	–
Steel	200.0–214.0	0.3	82.7
Titanium	103.4	0.34	44.8

Table 9.5 Values of bulk modulus, and density for a range of common fluids

Fluid		Bulk modulus $\times 10^{-8}$ N/m^2	Density kg/m^3
Carbon tetra- chloride	at 15°C	11.2	1590.0
Ethyl alcohol	at 15°C	12.0	779.0
Kerosene		13.6	814.0
Oil s.a.e. 10		16.7	880.0–940.0
s.a.e. 30		18.6	880.0–940.0
Sea water	at 10°C	22.0	1026.0
Water	at 0°C	20.5	1000.0
	at 20°C	20.5	998.0
	at 80°C	20.5	972.0

distance increments or inter-node pipe lengths. It will be seen that the Courant criterion depends upon flow velocity and wave speed; it is useful to note that for free surface flows these are comparable, commonly < 5 m/s, while for air pressure transients the wave speed is of the order 340 m/s with entrained airflow velocities an order smaller. Much higher wave speeds, up to 1500 m/s, may be encountered in waterhammer applications. Table 9.3 illustrates the wave speed calculations necessary for each of the cases discussed, while tables 9.4 and 9.5 present Young's Modulus and fluid Bulk Modulus values required in the waterhammer case. These dx/dt equations define lines drawn in the x–t plane, known as characteristics.

These characteristics may be thought of as the mode of communication of information from one location in the flow at the current time to another location one time step in the future. Figure 9.1 illustrates this general concept for either free surface or full bore flow system applications.

At the boundaries to any system only one characteristic can exist and it is necessary to represent the boundary by some developed equation linking flow depth, or air pressure to flow rate and/or time. It is the ability to develop such boundary conditions that endows the method of characteristics with its particular suitability as a simulation technique for unsteady flow in building drainage and vent systems. Boundary conditions include WC discharge to a drain; pipe junctions, including displaced hydraulic jumps; entry to vertical stacks; the transformation of the stack annular flow back into free surface flow at entry to the lowest level collection drain; solid transport within a pipe network and the interaction between stack annular downflow and the vent system.

The development of these boundary conditions requires both an empirical, laboratory-based input, as well as a study of the mathematics required to transfer the observed boundary effects into a set of relationships compatible with the method of characteristics model. A fuller description of the model mathematics and the

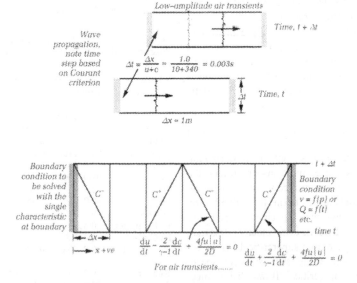

Figure 9.1 Summary of the method of characteristics applied to unsteady flow within building utility systems. Note importance of the boundary equations representing duct section entry and exit

validation undertaken may be found in Swaffield and Galowin (1992) and Swaffield and Boldy (1993). At nodes within a pipelength the simultaneous solution of the available characteristic equations is sufficient.

Application of computing models to building water supply and drainage

The method of characteristics formulation of the equations of continuity and momentum defining unsteady flow in building systems may be used to address a wide range of phenomena including pressure surge in appliance water supplies, mechanisms and interrelationships inherent in horizontal free surface branch flow, including the effects of defective drainage and solid transport, air pressure transient propagation in vent systems and trap seal retention, gutter flow and capacity and siphonic rainwater system priming. Each of these cases will be developed with examples of simulation output and the relevant boundary conditions necessary to fulfil the requirements of a method of characteristics application.

Drainage systems as an example of an unsteady flow regime

The definition of unsteady flow clearly applies to a building drainage system fed by a random selection of appliance discharges. An appliance discharge to the head of a drain may be described by an inflow vs. time curve. This generates a wave that propagates down the drain towards its termination at a vertical stack or at a junction with other branches, which may naturally be carrying their own discharges. Figure 9.2 illustrates such a wave, where the deeper zones tend to move faster than the shallower, effectively redistributing the water mass in the wave forward, steepening the leading edge and spreading the tail of the wave. Friction reduces the difference between the flow velocities along the wave, so that not all waves 'break'. The overall effect will be to attenuate the wave, lowering the peak flow depth as the wave moves through the network, with a consequent reduction in both flow depth and peak flowrate, and an increase in flow duration at any location. None of these effects can be predicted by the current design methods and this provided the impetus for the development of the methodology and numerical modelling described.

The degree of attenuation experienced by any appliance discharge is dependent upon pipe diameter, roughness, slope, the shape of the initial flow vs time profile and the baseflow over which it must pass. Figure 9.3 also illustrates why sewer flow may be described

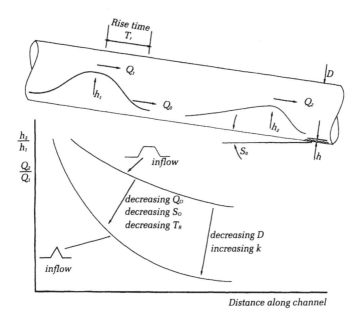

Figure 9.2 Dependence of wave attenuation on pipe and flow parameters

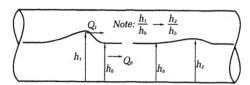

Figure 9.3 Wave attenuation becomes insignificant further downstream as the base flow Q_0 increases due to combining flows

as quasi-steady. The driving force for attenuation is the difference in depth and velocity within the wave itself; as these differences decrease due to attenuation, so does the attenuation decrease. Far downstream the wave becomes 'almost' a steady flow, as the summation of many such attenuated waves within the sewer leads to conditions that may be approximated by the equations governing steady free surface flows.

Any simulation must also be capable of dealing with multiple discharges to a horizontal drain network that will include junctions, either level invert or top entry, as well as increases in diameter in the flow direction. The schematic figure 9.1 emphasized the need to develop boundary conditions at such junctions. An essential capability is therefore to be able to predict the effects of back-wash at junctions — only possible with the method of characteristics model. A sustained research effort has developed the DRAINET model that is capable of simulating multi-appliance, multi-storey drainage networks, delivering flowrate and depth at chosen locations at time intervals

determined by the user and the application of the Courant criterion. This simulation also includes solid transport prediction and the ability to determine the influence of defective drainage installation on flow conditions.

System simulation

Two examples of the versatility of the method of characteristics follow. The introduction of an above floor collection branch serving a series of WCs was found to cause user annoyance due to the back-wash of WC discharge into adjacent toilets. A combined experimental and simulation study was undertaken on behalf of the code authority involved. Figure 9.4 illustrates the level invert network of five WCs considered. Figure 9.5 illustrates the predicted backwash into adjacent units. For a range of test cases close agreement was found between the measured back-wash volume, simply collected at the drain to unflushed WC connection, and that predicted by the DRAINET simulation. As a result of this study the code was changed to ensure a minimum of one pipe diameter as a vertical drop between the WC discharge and the level invert junction. This code initiative was positive as above floor collection would ease any future refurbishment or change of use of the space. A further implication of this study was that as WC flush volumes decrease internationally, e.g. the UK Water Regulations allow 6/4 dual flush, 6/3 dual flush is standard in

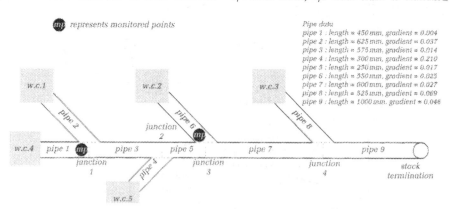

Note - all pipes UPVC (internal roughness = 0.06) and 104mm internal diameter

Figure 9.4   Drainage network modelled

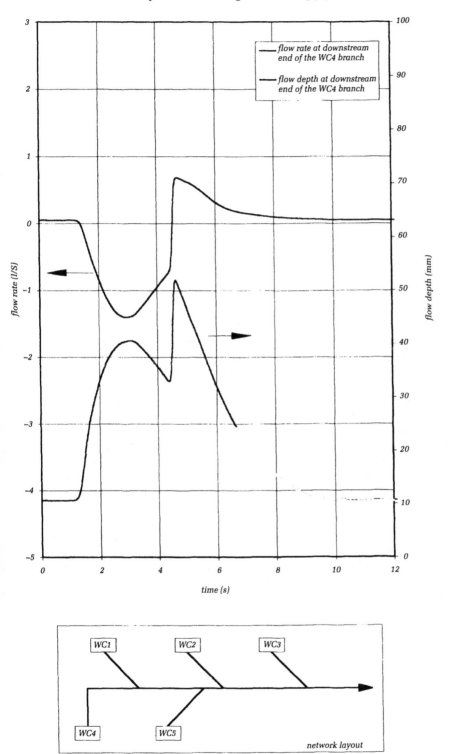

Figure 9.5 Predicted variation in flow rate and depth with the operation of WC 1 (flush volume = 6 litres) at 0 seconds

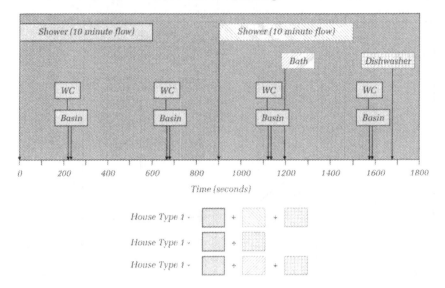

Figure 9.6   Assumed dwelling usage pattern, 'morning rush hour'

Australia and Scandinavian WCs are available at 4/2 litres dual flush, the use of reduced size horizontal branches must also be considered. A reduction to 80 mm in the case discussed here would have made both the above floor collection drain and the single diameter vertical drop more acceptable and practical.

On a larger scale the question of the appropriate house to collector drain diameter and the number of dwellings that may be safely serviced by a particular drain diameter has also been the subject of application of the DRAI-NET model. Figure 9.6 illustrates the assumed appliance usage in single family dwellings on an estate. In order to determine the drain diameter suitable for a particular number of houses it was assumed that the peak loading would occur during a morning 'rush hour' and a statistical analysis suggested that a simulation of 9 houses over a 30 minute period would provide the required data. Figures 9.7 and 9.8 illustrate the house-drain layout chosen and the flowrate and depth at entry and exit from the network. It is clear that reducing the drain diameter or increasing the number of dwellings served is possible. Probably the most interesting fact to emerge from this study was the extremely low flow depths associated with even the peak loading assumed. This, as will be seen later, has a direct impact upon solid transport and deposition, with consequences for maintenance, as flush volumes are decreased further and water conservation in the design of other appliances becomes a reality.

Inclusion of solid transport within a method of characteristics unsteady flow model — the solid velocity/flow velocity decrement approach

The introduction of a moving solid into the method of characteristics model requires a

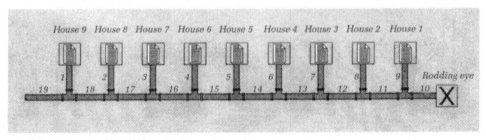

Figure 9.7   Layout of house and collection drains

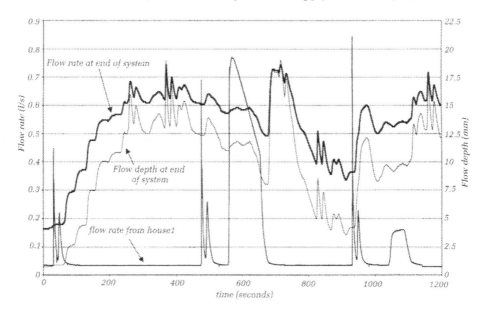

Figure 9.8 Cumulative flow rate and depth at end of system for 100 mm diameter pipes, together with flow rate from House 1

suitable boundary condition to be developed that links solid velocity to the flow conditions at the solid location at any particular time. Figure 9.9 illustrates the flow conditions in the vicinity of the solid. Observations of actual drainage system operation, Bokor (1982), McDougall (1995), McDougall and Swaffield (2000) indicate that for a large proportion of discharged solids, i.e. paper towels, toilet paper, faecal material, the solid velocity may be approximated by a relationship based on local flow velocity. This model would retain all the advantages of the method of characteristics prediction of wave attenuation effects in the determination of local flow velocity at the solid location, but would in all probability slightly underestimate solid transport distance as the solid interference with the local flow depth would be disregarded. (In the latter stages of transport, as the solid decelerates, a flow depth differential builds up across the solid that would enhance the forces encouraging continued motion, however, it may be argued that these are effectively included in the empirical solid velocity – local flow velocity relationships developed by McDougall).

McDougall (1995) developed a solid velocity model that was compatible with the method of characteristics simulation of unsteady free surface flows in horizontal branches. The method of characteristics solution accurately predicts local flow velocity and depth along the drain at any time in response to any combination of appliance discharges. If the relationship between these local flow conditions and the velocity of a solid whose mass and dimensions were known could be determined then a prediction of solid transport performance would be possible. Experimental work provided data linking the flow depth and velocity in drains of 75 to 150 mm diameter and a range of slopes appropriate for both steady subcritical and supercritical flows to the measured velocity of test solids defined in terms of their diameter, 20 to 50 mm and specific gravities, 0.4 to 1.6. The solids chosen were derived from those used in earlier National Bureau of Standards research (Swaffield and Galowin 1992), being hollow closed-ended plastic cylinders whose sg. was controlled by filling with appropriate fluids, referred to as NBS solids. The dependence of solid velocity on local flow depth, local flow mean velocity and local flow specific energy were assessed as a base for an empirical relationship. Depth as a sole determinant was discarded as an increasing depth around the solid would not necessarily imply increasing solid velocity, for example, approaching an obstruction or in the subcritical flow region upstream of a junction (chapter 8).

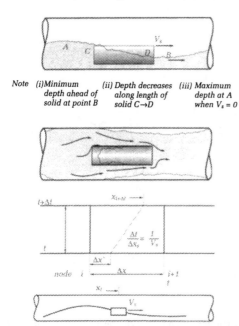

Note (i)Minimum   (ii) Depth decreases   (iii) Maximum
      depth ahead of   along length of   depth at A
      solid at point B   solid C→D   when $V_s = 0$

Figure 9.9 Flow conditions surrounding an NBS type solid and an illustration of the calculation technique used to track the solid path through the method of characteristics specified time interval x–t grid

It was decided that solid velocity as a function of local mean flow velocity was the best determinant during solid motion, with flow depth being a determinant of the initiation of motion for solids deposited in the drain.

Figure 9.10 illustrates this relationship for a particular NBS solid/drain diameter/slope combination as applied flowrate increases. A number of regions may be identified. This may be thought of as a velocity decrement approach as the solid velocity is linked to local mean flow velocity. Initially the solid remains deposited on the drain invert, region 1, until the flowrate is sufficient to initiate motion. In region 2 the solid velocity increases with increasing flow velocity but lags until the flowrate increases to a value identified as the start of region 3 — note change in the slope of the solid velocity vs flowrate line fit — corresponding to a flow depth roughly equivalent to the solid diameter. Within region 2 the solid velocity increases with the surrounding water velocity, however, the solid generally remains in sliding contact with the pipe invert. The relative importance of each region varies with the solid properties. As expected the

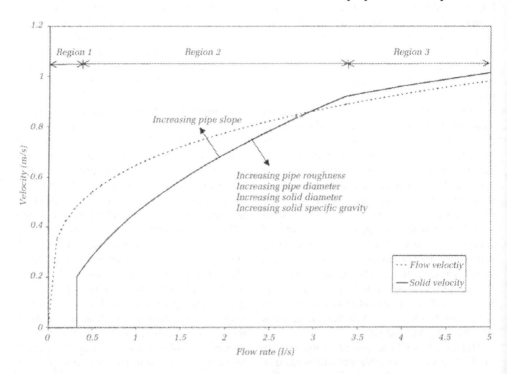

Figure 9.10 Illustration of the format of the McDougall solid transport model for the case of a 38 mm diameter NBS solid of 1.05 specific gravity in a 100 mm diameter cast iron drain at a 0.01 slope

diameter of the pipe has little effect, it is the local flow depth surrounding the solid that determines the buoyancy forces. Figure 9.10 also indicates the changes in the relationship if drain slope, diameter and roughness or the solid parameters of diameter and specific gravity were altered. It will be appreciated that the form of the solid velocity–surrounding flow relationship indicates that in region 2 the surrounding water passes the solid so that its relative position in the flush moves back. Once in region 3 it is possible, due to the velocity distribution at any section in the flow (figure 8.4), for the solid to travel faster than the mean surrounding flow due to its buoyancy and position on that velocity distribution. This analysis confirms the necessity for solids to be discharged early in the flush from WCs. Current international testing standards, including the Australian and DIN Standards and the UK Water Regulations define 40 per cent of the flush volume as a trailing discharge.

The local velocity of the solid may then be calculated from the local flow depth and velocity calculated by the full method of characteristics solution (figure 9.9). The model can be made to deal with multiple solids that close up or merge by introducing a change in the form of the relationship to conform to some larger merged solid, however, the simulation is currently being developed to allow the treatment of strings of solids, as would be expected in a WC discharge or during the merging of discharges from adjacent WCs. This model presents the designer with a rapid methodology for identifying both possible solid deposition, and design solutions by

adjusting pipe slope, diameter or appliance choice. Returning to the dwelling collection drain example, figure 9.11 illustrates the solid deposition data predicted for the dwelling to sewer drain diameter assessment illustrated in figure 9.7.

Multiple solid transport mechanisms and the effect of transient deposition

It is wholly misleading to imagine that solids discharged from a WC exit a typical network as a result of a single WC operation. The mechanism of solid transport is rather as shown schematically by figure 9.12 and by figure 9.13, drawn from Boker's observations of an actual WC drain operation in the interfloor voids of Greenwich Hospital in the early 1980s, and discussed further in chapter 12. Note that figure 9.13 uses an x-axis defined as the root of solid transport distance over drain slope, a group found to facilitate the comparison of solid transport performance between pipe materials and slope (Swaffield and Galowin 1992). Solids decelerate following discharge and may become deposited, as demonstrated for the heavier faecal matter, third solid in figure 9.13. Similarly waste discharged late in the flush may be prone to deposition, demonstrated by the tissue data, fourth solid, included in figure 9.13. A series of solids discharged in the same flush will interact, with trailing solids often closing with those ahead and causing short term re-acceleration, as illustrated by the interaction between the first and second solids in figure 9.13. Later WC operations, or the discharge of other appliances have the effect of moving solids on. The final travel distance depends upon the design of the drainage network, the properties of the solid and most importantly upon the attenuation of any subsequent discharge to the drain. Later discharges carrying solids have a particular effect as trains of solids interact as shown.

The effect of multiple WC discharges on the final deposition of a particular solid is illustrated by the simulation illustrated in figure 9.14. In the United States the introduction of a mandatory 6 litre flush in 1992 opened the market to several innovative flushing mechanisms. The traditional drop valve cistern was joined by a pressurized tank cistern which

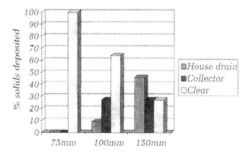

Figure 9.11 Distribution of depositions predicted in the drainage network following solid discharge from WCs in each of the houses modelled, figure 9.7

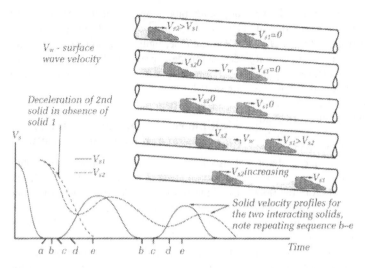

Figure 9.12 Interaction between a series of solids transported in an attenuating appliance discharged wave or subject to successive appliance discharges

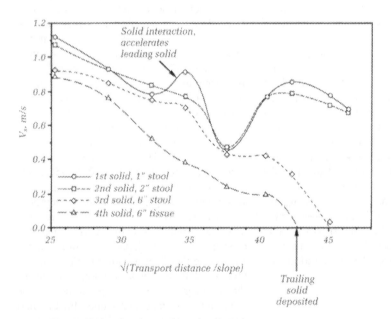

Figure 9.13 Measurements of faecal and tissue transport in an installed drainage network. Solid-to-solid interaction affects solid velocities while trailing solids are deposited by the attenuating discharge wave

consists of a tank that is filled from the mains so that the pressure available on discharge is the mains pressure. While this is seen as innovatory such devices were historically common in France but without the pressure tank being 'hidden' within a traditional ceramic cistern. Laboratory test results were considered to be in error as they suggested that the final travel distance for a single solid achieved by the gravity cistern could, after many flushes, exceed that for the pressure cistern despite the fact that on the first or second flush the deposited solid travelled further in the system fitted with the pressure cistern WC. As shown by figure 9.14, the simulation confirmed the laboratory results. The explanation lies in the

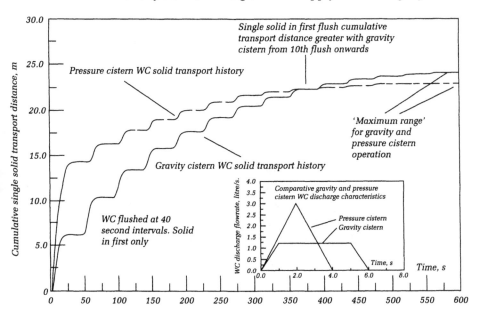

Figure 9.14 Comparative transport performance for a single solid within the first flush from either a gravity or pressure cistern flushed 6 litre WC discharging to a 100 mm diameter drain at 0.01 slope. Simulated 38 mm diameter 1.05 sg NBS solid

differential attenuation of the two WC discharges (figure 9.14). The pressure cistern having a steeper rise of flowrate against time attenuates more rapidly in the drain, hence its effective transport range is decreased. This confirmation and explanation of these apparently dubious results confirms the applicability of the method of characteristics simulation.

Defective drainage systems

Building drainage networks are rarely perfect. Installation conditions are often difficult and may lead to defects in terms of pipe slopes that diverge from the design requirement. Similarly obstructions may be encountered, possibly emanating from poor jointing techniques or below ground root penetration; see also table 12.5. Persistent stranding of WC discharged materials may also lead to effective obstructions. The DRAINET simulation was enhanced, following an extensive series of site surveys, to include boundary conditions representing both these defect categories (Swaffield, McDougall and Campbell 1999).

Two main defect types were addressed, namely slope defects away from the design requirement and localized obstructions that reduce the pipe diameter. A slope defect survey

was conducted over a kilometre of the 24 kilometres of glass pipe that make up the interfloor drainage network at the Queen's Medical Centre Nottingham. The assumed drain slope was determined by reference to the elevations of consecutive branch connections and then compared to the local pipe section slopes making up the overall pipe length between junctions. Further slope defect data were collected from the Lothian Regional Council George IV Bridge building below ground car park and Roodlands Hospital where the drainage system was installed in a below floor 'crawl' space. The same measurement methodology was utilized at all the sites surveyed.

The results of the slope defect survey are summarized in figure 9.15, which illustrates a distribution plot for all pipe sections considered regardless of diameter or material and may be seen to form a normal distribution for the error between actual section slope and the overall design slope that could be used to define the actual slopes in a multi-pipe section drainage run between two fixed elevations. The above ground data collected suggests that drain section slope defects during installation are corrected for overall by the operatives and may be represented by the error distribution

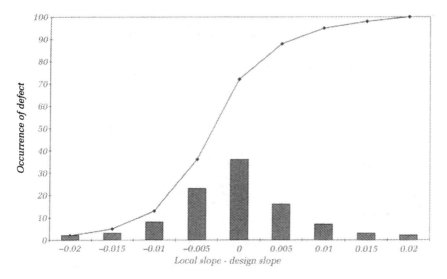

Figure 9.15   Slope defect occurrence distribution

illustrated. Therefore it would be possible to predict the effect on solid transport and wave attenuation in any such installation by introducing randomized local slopes between junctions into the simulation data set, provided that the overall fall between junctions was maintained — effectively the same criteria demonstrated by the actual drain installation. This would increase the number of pipes considered within any simulation and would introduce the appropriate boundary conditions for slope change discussed in chapter 8, however, if required this would not be an impediment to the simulation.

The obstruction survey reviewed CCTV surveys conducted by London Underground Ltd (LUL) as part of the station refurbishment programme. The data presented represents (tables 9.6 and 9.7), some 6 kilometres of mostly buried drainage pipework, with an age range of several decades. The data is typical of drains that have been in use for some time, however, as some of the footage referred to specific investigations of known defects, the defects to be found in existing systems would be represented by the less extreme of the defects presented.

Table 9.7 presents a summary of the defects found by pipe material and diameter. Defects affecting more than 50 per cent of the pipe diameter or cross-sectional flow area were excluded and this affects the percentages presented.

Table 9.6   Abbreviations used in Table 9.7 to describe drain defect identification

CI	Cast iron
CON	Concrete
PF	Pitch fibre
PVC	Plastic (polyvinyl chloride)
VC	Vitreous clay
B(j)	Pipe broken (at j m)
C(j)	Crack (at j m.)
CX	Defective connection
D %	Pipe deformed, % loss of diameter
DE(j)	Debris, % height loss, (over j m)
DI %	Dropped invert, % of diameter
E %	Encrustation, % loss of cross-sectional area
F (j)	Fracture at j m
I	Infiltration
JD	Joint displaced
JX	Defective junction
OB %	Obstruction, % diameter loss
RF(j)	Fine roots (at j m)
RM %	Mass roots, % loss of cross-sectional area
RT	Tap root
X	Pipe collapsed

Some defects occur more often on some pipes than others, for example, encrustation occurs almost exclusively on cast iron pipework, appearing on the CCTV records as a combination of rust and dried waste. Also some defects only affect underground pipes and may be discounted in simulating above ground suspended systems. Jointing defects clearly apply to both types of installation, however, there are different types of jointing to be considered dependent on application.

Table 9.7 Summary of drain obstruction survey

Pipe type	CI100	CI150	CI225	CI300	CON300	PF100	PF150	PVC100	VC100	VC150	VC225	VC300	Total
Length (m)	960	293	36	3	52	5	19	36	1423	2253	232	575	5887
B		1							23	34	2	3	63
B (m)	0.6											1.5	2.1
C	4	1							133	158	13	36	345
C (m)										11.5	36.6	35.6	83.7
CX	1	4							7	5			17
D 6%									3	4			7
d 11% (m)										19.6	23.7	1.5	44.8
DE 21%	29	17						1	49	58	7	1	162
DE 15% (m)	3.8	2.3						14.3	13.4	148.1	34.3	27.4	243.6
DI 35%	2	1								2			5
E 18%	47	25							93	230	5	144	544
E 14% (m)	493.9	85.1	2.8	3.0					42.2	145.4	3.0	46.3	821.7
F									72	71	5	2	152
F (m)										67.3	36.2	15.4	118.9
I										2	2	7	11
JD	42	5	1				7	2	757	1015	61	235	2125
JX											2		2
OB 25%	1	1							5	2			9
OJ	3	2						1	34	17		22	79
RF									6	68	9	1	84
RF (m)										20.7		1.9	22.6
RM 26%	1	1						1	1	53	11		68
RT									1	6			7
X	1	1							3	1			6

Figure 9.16   Local depth increase caused by a poorly aligned junction or coupling

Incorporating the defect into the method of characteristics simulation is simply a case of introducing a suitable boundary condition to define the obstruction or slope defect. In the case of an obstruction, experimental work has identified a series of expressions linking local depth at the obstruction to local flow conditions defined by the flow critical depth, chapter 8, and drain slope (figure 9.16). In the case of a slope defect it is necessary to predict the backwater profiles applicable through the defective drain section and to identify the location of any hydraulic jumps (figure 9.17). With these boundary conditions in place the simulation will predict the effect of the defect. Figure 9.18 illustrates the comparison of predicted and measured solid transport in a 75 mm diameter drain with either form of defect.

As a WC flush attenuates along any drain, the location of the defect relative to drain entry becomes a factor determining its effect on solid transport. The position of the solid within the WC discharge and the form of the WC discharge profile also affect transport. These interrelationships explain the need for a simulation such as DRAINET to address solid transport. Figure 9.19 illustrates the effect of a particular defect on solid deposition in a 75 mm diameter drain at a slope of 0.01 following discharge of a solid from an idealized 6 litre flush volume WC, the solid leaving at the 50 per cent flush stage. As the position of the defect relative to drain entry is changed, its effect on solid deposition is seen to change. While the defect is close to the drain entry it reduces considerably both the solid transport distance and the solid velocity, with

subsequent deposition occurring beyond the defect at an overall distance considerably shorter than that achieved in a defect-free drain. However, as the defect location becomes further away from the drain entry, a point is reached when the solid deposits at or upstream of the defect due to interaction with the associated hydraulic jump and backwater profile. In this region the local flow depth increases and reduces solid velocity leading to deposition either in the 'downhill' drain section carrying the hydraulic jump or in the 'uphill' section of the defect, figure 9.17.

Gutter flow considerations

The flow in a roof gutter during and following a storm is unsteady free surface flow, however, design guides concentrate on the determination of 'safe' flow capacities based on steady state data. The development of partially filled unsteady pipeflow simulations aimed at building drainage design also make possible simulations capable of defining the roof gutter flow response to both time and location dependent lateral inflows by means of a method of characteristics solution of the governing St Venant equations. Lateral inflows were included in the general derivation. The solution requires the definition of boundary equations at the upstream and downstream terminations of the gutter. In addition the simulation requires the calculation time step to be dependent upon local flow velocity and wave propagation velocity, the Courant criterion.

At the downstream end of the gutter the boundary is supplied by the flow vs. depth relationship assumed for the outlet. The most common assumption is of a free outfall at critical depth such that the downstream boundary may be expressed as

$$1 - \frac{Q^2 T_c}{g A_c^3} = 0 \qquad (9.6)$$

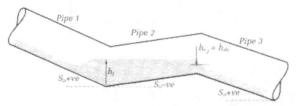

Figure 9.17   Backwater profile and depth calculations in the event of a pipe section having a backfall

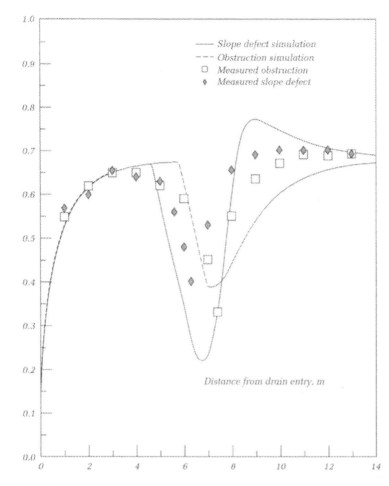

Figure 9.18 Comparison of measured and predicted solid transport velocities in a 75 mm diameter drain at 0.01 slope with either a 20 mm obstruction or a 10 mm slope defect 7 m from drain entry, carrying 1.0 l/s flow

where A and T are functions of h. Alternatively if an experimentally based relationship of the form

$$h = K(Q) \, Q^{n(Q)} \qquad (9.7)$$

is available then that may be utilized. It is also possible for the form of this relationship to be variable with depth, the simulation storing values of the defining coefficients, K and n, and utilizing whichever pair is appropriate to the approaching flowrate.

At the upstream gutter termination the boundary condition is supplied by a zero flow condition

$$V_{x=0} = 0.0 \qquad (9.8)$$

Frictional representation in the momentum equation was provided via an appropriate shear stress. In practice this has to be transformed into one of the two available frictional representations for free surface channel flow — namely the Chézy equation utilizing the Manning coefficient or the Colebrook–White equation, chapter 8. Generally the latter is now accepted as the more accurate for open channels under 1 m in diameter. Whichever methodology is used the equivalent steady flow resistance terms, Manning n or Colebrook–White generated friction factor will underestimate the total resistive forces due to the disruption to flow caused by the lateral inflows (Mein and Jones 1992, May 1982), see chapter 13.

The simulation, GUTTER, was run in comparison with historic data attributed to Beij

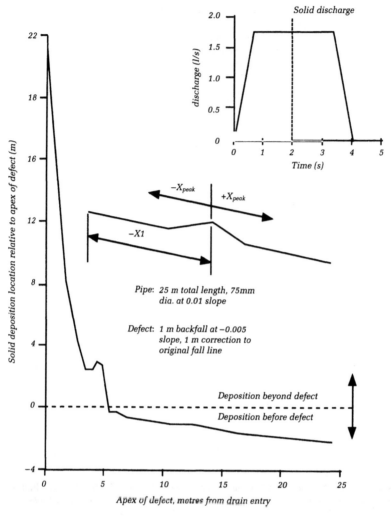

Figure 9.19   Influence of slope defect position on solid deposition downstream of an idealized 6 litre flush volume WC in a 25 m long, 75 mm diameter drain at 0.1 design slope

(1934) (figure 9.20) for a rectangular section gutter at two slopes and utilizing both the Manning n and the Colebrook–White friction factor. Following Mein and Jones (1992) it may be seen that adjusting the Manning n between 0.009 and 0.012 allowed close agreement with the data recorded by Beij (1934). However, for a simulation to be accurate it should be based on a knowledge of the gutter conditions rather than post-measurement curve fitting. Utilizing the Colebrook–White expression with a surface roughness of 0.3 mm appropriate to a cast iron gutter material yields acceptable predictions for flow depth, i.e. within 5 per cent at any location at either

slope (figure 9.20), confirming the superiority of the Colebrook–White approach already identified for other partially filled pipeflow or free surface flow conditions (Swaffield and Bridge 1983, ASCE 1963 and Ackers 1978). In each case close agreement is achieved over the full length of the gutter, confirming the applicability and stability of the model. Figure 9.21 illustrates the application of GUTTER to predict the flow depths in a gutter for the storm intensity shown. The time-dependent nature of the gutter flow conditions is well illustrated. The application of simulation techniques also allows the representation of longitudinal as well as temporal variations

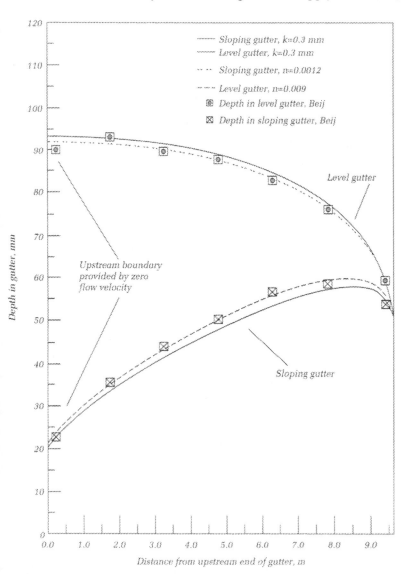

Figure 9.20   Comparison of the predicted depths with those measured by Beij for a 153 mm wide, 9.63 m long, rectangular section gutter carrying 330 l/min

in lateral inflow along the gutter, as may be the case with complex roof structures.

Method of characteristics applied to the modelling of low amplitude air pressure transients in drainage vent systems

The fundamental mechanism determining the air pressure regime within building drainage and vent systems is the air entrainment generated by the falling annular water film in the system vertical stacks. Under steady conditions the water flow generates a steady entrained airflow resulting in frictional pressure drop in the dry portions of the stack and separation losses as the airflow is drawn through the water curtains that form at discharging junctions and at the base of the stack, chapter 8. Any variations in annular water downflow will result in changed demand for entrained airflow and this will be communicated to the whole system in the form of low

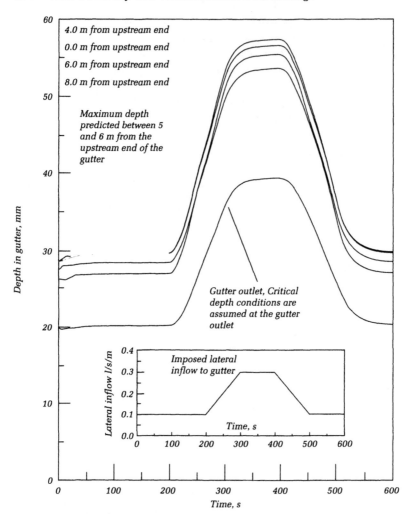

Depth in gutter, mm

4.0 m from upstream end

0.0 m from upstream end

6.0 m from upstream end

8.0 m from upstream end

Maximum depth predicted between 5 and 6 m from the upstream end of the gutter

Gutter outlet, Critical depth conditions are assumed at the gutter outlet

Lateral inflow l/s/m

Imposed lateral inflow to gutter

Time, s

Time, s

Figure 9.21 Time-dependent flow depth in a trapezoidal section channel, 0.1 m base, 45 degree side slope, 10 m long at gradient 0.001. Manning n 0.009

amplitude air pressure transients. Any obstruction to the airflow by, for example, water flow surcharge of the stack will generate larger transients that propagate throughout the network, communicating this change in system conditions. The main concern in all air pressure transient analysis is the preservation of the appliance trap seals connected into the drainage network. Once propagated any transient will be transmitted and reflected within the system until it attenuates or until a change in system condition negates its effect.

The propagation of air pressure transients may be modelled by the application of the method of characteristics. The methodology is identical to that already discussed, depending

on the simulation of transient propagation throughout the system and the representation of system boundaries, again the key to successful modelling. Boundaries can include pipe ends open to atmosphere, with and without local losses, junctions of two or more ducts, with or without an annular water film present, terminations featuring a water trap seal, pressure relief valves, and the essential linkage between stack annular water downflow and the entrained airflow.

(i) Network boundary conditions

Figure 9.1 illustrated the general method of characteristics solution, namely internal nodes being handled by simultaneous solution of the two available characteristics while at a boundary there is only one characteristic equation which must be solved with a suitable boundary equation. Boundary equations normally link local flow velocity to pressure or time, for example, an open termination is represented by either atmospheric pressure or a concentrated loss coefficient:

$$p_1 = p_{atm} - K\,0.5\rho u|u| \qquad (9.9)$$

where the absolute air velocity ensures that the pressure differential is always negative in the positive flow direction.

Junctions of two or more ducts are represented by flow continuity and noting that, in the absence of any local pressure loss, the air pressure at the junction will be the same for all joining pipes. If the junction features a horizontal branch actively discharging to the stack, the air pressure loss due to the water film is incorporated based on empirical expressions linking the joining water flows, the junction geometry and the entrained airflow. The solution of these relationships with the characteristics available at each pipe termination allows the prediction of all the unknown pressure and velocity terms.

In the case of a dead-ended branch, zero velocity at the dead end provides the boundary condition.

(ii) Moving boundary conditions — trap seal response

A trap seal capable of displacement in response to pressure fluctuation at the air to water interface is a common boundary. The trap response is described by

$$p(j, n(j) + 1) - p(ref) + \rho_w[H_1 - H_0 \\ - L(4f U_w | U_w |/2D - A d U_w /dt)] = 0 \qquad (9.10)$$

where A is the trap cross-sectional area, diameter D, L is the liquid column length, H_1 and H_0 are the water surface elevations above an arbitrary datum, p_{system} and p_{ref}, are the air pressures exerted upon the trap liquid, of density ρ_w, f is an appropriate friction factor and U_w is the liquid column velocity. Note that it is necessary to reduce the length of

the water column, L, if the trap water level overflows the entry to the drain. p_{system} depends on the transient conditions at the pipe-trap interface, introduced via the branch exit characteristic.

(iii) Air admittance valves

The air admittance valve (AAV) is a recent development intended to be fitted at the top of the stack to avoid the need for roof penetration or as a means of protecting groups of appliances from induced siphonage. In operation it allows air to enter the stack as long as the stack pressure is slightly less than atmospheric pressure, and closes if stack pressure rises above a small negative value to prevent air, and odours, escaping from the stack. Theoretically an air admittance valve solves many of the traditional trap seal depletion problems, however, it does represent a mechanical device that may over extended periods of use fail to seal. Also, as the AAV shuts to prevent odour ingress into habitable space, it cannot vent positive pressures generated within the network as a result of surcharge. This has been a major obstacle to its wider use and code acceptance.

As a boundary the valve may be described as a variable entry/exit loss, the value of the loss coefficient being dependent upon the degree of opening of the valve, which in turn is determined by the pressure differential across the valve.

Figure 9.22 illustrates the mechanism commonly found. The valve remains closed as long as the pressure in the vent stack is insufficient to overcome the mass force of the diaphragm. Once the diaphragm lifts, it is free to rise until the valve is fully open. Equally the diaphragm may 'hover' at some intermediate setting, dependent upon the pressure–time history in the stack. Thus the boundary condition representing the value may be expressed by a series of three simple equations

$$\text{Closed valve}: \quad u_{exit} = 0.0 \qquad (9.11)$$

$$\text{Partially open}: \quad p_{atm} - p_{exit} = K 0.5\, u_{exit}^2 \quad (9.12)$$

$$\text{Fully open}: \quad p_{atm} - p_{exit} = K_{min} 0.5\, u_{exit}^2 \quad (9.13)$$

that may be solved with the appropriate characteristic equation (K_{min} is the appropriate

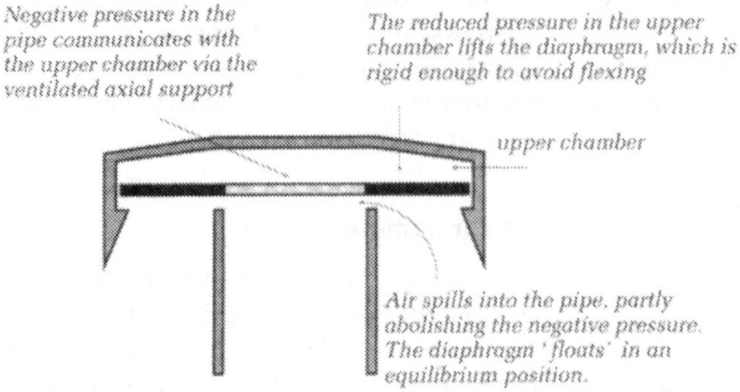

Negative pressure in the pipe communicates with the upper chamber via the ventilated axial support

The reduced pressure in the upper chamber lifts the diaphragm, which is rigid enough to avoid flexing

upper chamber

Air spills into the pipe, partly abolishing the negative pressure. The diaphragm 'floats' in an equilibrium position.

Figure 9.22   Air admittance valve — open position — diagrammatic

loss coefficient for the fully open valve). Figure 9.23 illustrates the variation of K values with differential pressure for a range of valve types (Swaffield and Campbell 1992), defining a 'good' valve characteristic. Loss coefficients should be low to avoid excessive suction pressures within the stack, however, valve diaphragms should not be too light as this will lead to flutter, i.e. rapid alternating opening and closing actions in response to system transients. The valve opening pressure should be close to atmospheric.

A similar analysis may be applied to other innovative devices, such as the waterless trap illustrated in figures 9.24 and 9.25. Here water pressure would hold the sheath open to allow appliance discharge. A positive air pressure in the connected branch would collapse the sheath and no odour escape would be allowed. However, if the branch pressure fell below atmospheric and if the appliance drain plug was not in place, the sheath would open allowing air into the drainage network. Thus the form of waterless trap illustrated is also a highly efficient air admittance valve whose loss coefficient curve

(figure 9.23) would be in the lower right sector of the diagram.

(iv) Positive pressure transient propagation

Positive air pressure transients may be generated within a drainage and vent network whenever there is a full or partial closure of the air path through the network. Such closures can occur as a result of surcharging the collection drain at the base of the vertical stack or as a result of overloading a branch to stack junction following appliance discharge. The Joukowsky relationship defining pressure change as a result of an instantaneous flow stoppage applies:

$$\Delta p = \rho_{air} \, c \, u_{air} \tag{9.14}$$

A 1 m/s air flow velocity with density 1.2 kg/m³ and a wave speed of 325 m/s will generate on stoppage a transient at the closure location of 390 N/m² or 40 mm water gauge. This transient will propagate throughout the network and will be reflected or transmitted at each boundary. An open termination will have a −1 reflection coefficient, thus generating a relieving transient. A closed end, for example, an air admittance valve which remains closed in response to the arrival of a positive transient, will have a +1 reflection coefficient that generates an enhanced positive pressure in the stack. It should, however, be noted that these Joukowsky values will only be attained if the flow closure is achieved in less than one pipe period, defined as the time taken for a transient to travel from the closure location to the first major reflecting boundary and back, in this

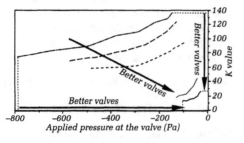

Figure 9.23   Change in loss factor K with applied pressure for a selection of typical air admittance valves

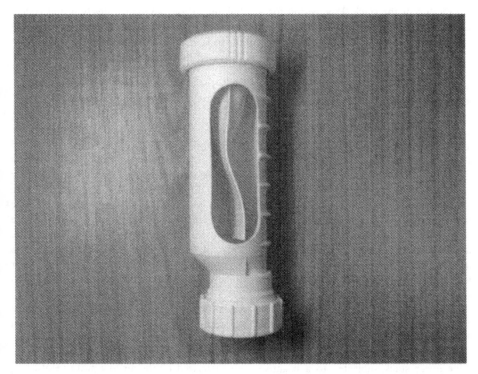

Figure 9.24 Hepworth 'Hygienic Self Sealing Waste Valve' waterless trap which also acts as an air admittance valve (Courtesy Hepworth Plumbing Products)

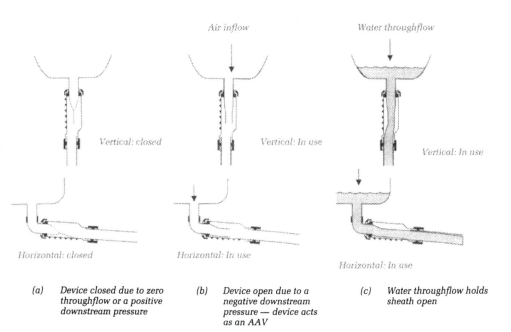

(a) Device closed due to zero throughflow or a positive downstream pressure

(b) Device open due to a negative downstream pressure — device acts as an AAV

(c) Water throughflow holds sheath open

Figure 9.25 Operation of the Hepworth waterless trap, illustrating the internal mounting of the sheath and its response to both water throughflow and the air pressure regime within the drainage network

case to the top of the stack. Pipe period for a stack height L may be thus defined as

$$T_p = 2L/c \qquad (9.15)$$

The effects of an air path closure at a branch to stack junction or at a stack offset are more complex. The upper section of the stack is subjected to a positive pressure transient while the lower portion of the stack is subjected to an equal but negative transient. The simulation can deal with this condition by redefining the air velocities on either side of the closure as zero for the duration of the blockage. This implies separation of the network into sections above and below the closure, however, if the closure is assumed to occur at an already modelled junction this presents no increase in model complexity. Once the closure is considered to have ended, indicated by the duration of peak water flows in the branch or the collection drain at the base of the stack, the local boundary conditions revert to those in place prior to the closure.

In the lower stack subjected to a negative transient it is possible that trap seal depletion occurs immediately. This effectively transforms the trap boundary to an open pipe and as such this becomes a relieving vent. As the method of characteristics deals with each boundary separately the model is automatically capable of dealing with this eventuality. This point will be demonstrated in later simulations. Similarly once the positive pressure transient has displaced the trap water up into the appliance this boundary also becomes a possible vent, however a vent with a high loss coefficient and as such not explicitly modelled.

Network modelling

The central issue in developing a vent system simulation was the linkage between annular downflow and entrained airflow as this determines the reaction of the network pressure regime to changes in system usage. The characteristic equations contain a friction term that normally has the expected function of opposing flow and eventually contributing to transient attenuation. However, in this case the frictional representation may be used to generate the airflow provided a relationship

linking airflow, annular water flow and system parameters has been developed. In chapter 8 the basis for this was laid when the total loss in the system was represented as

$$\Delta P_{total} = \Delta P_{entry} + \Delta P_{dry\ pipe\ friction}$$
$$+ \Delta P_{branch\ junction} + \Delta P_{back\ pressure}$$
$$(9.16)$$

The 'motive force' to entrain this airflow and compensate for these 'pressure losses' is derived from the shear force between the annular water layer and the air in the wet portion of the stack. Hence a 'negative' friction factor may be postulated that generates an equal pressure rise to that determined from equation (9.16). Ongoing research (Jack 2000) has identified the format and relationships governing this shear force representation and allows the prediction of the transient response of the stack network to variations in applied water downflows. The appropriate friction factor for each zone within the stack thus depends on the relative water and entrained airflows, as illustrated by figure 9.26, characterized by the airflow mean velocity V_a and the annular water film terminal velocity V_{tl}. The concept of increasing or decreasing air core pressure dependent upon the relationship between V_a and V_{tl} is discussed in further detail in chapter 8.

The identification of the 'negative' friction linkage allows the full development of the vent system simulation, AIRNET. Figure 9.27 illustrates a four-storey drainage network subject to simultaneous WC discharge on each floor with the subsequent combined flow at the stack base. The falling annular flow will generate pressure transients and may be sufficient to deplete an appliance trap seal — which then acts as a relief airflow path. Alternatively the water flow may be sufficient to generate a surcharge at the stack base. Figure 9.28 illustrates the first case. As the water flow increases so the entrained airflow rises and the system pressure falls. At the peak water flow the trap seal on level 2 is depleted by induced siphonage and the traces illustrate the trap acting as a relief valve. In the case of surcharge at the stack base (figure 9.29) positive airflow is recorded through the depleted trap. (Note that negative airflow in these results is flow down the stack from atmosphere to sewer.)

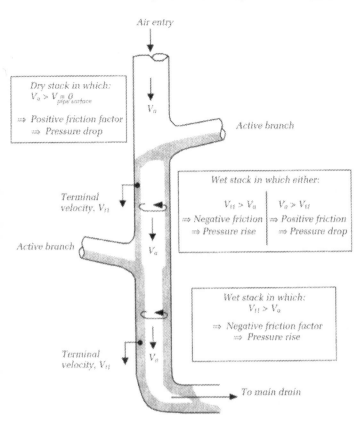

Figure 9.26 Definition of friction factor application through single stack system subject to multiple simultaneous discharge

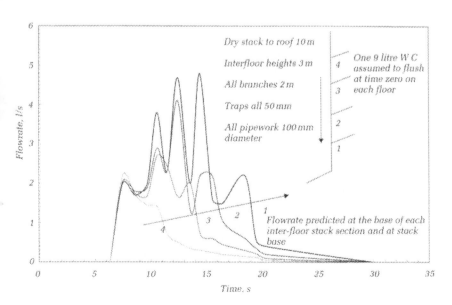

Figure 9.27 Annular downflows at each floor level for a four-storey single stack system

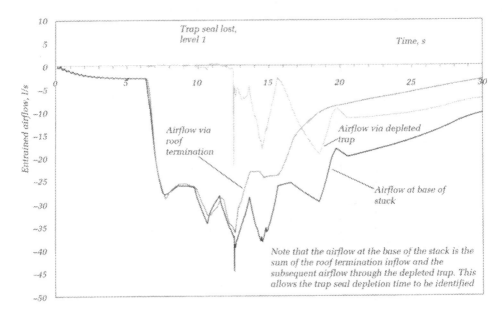

Figure 9.28 Entrained airflow in the vertical stack and through a level 2 branch following trap seal loss

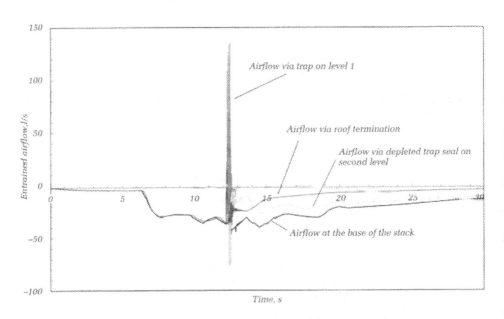

Figure 9.29 Entrained airflows in the network, illustrating the effect of a surcharge at the base of the stack

Pressure surges in water supply networks

Pressure surge occurs as a normal event within all fluid-carrying systems when a steady state condition is interrupted. Surge propagation cases range from specialist examples, such as

in-flight re-fuelling, to transients that damage or negate the operation of fire fighting systems by damaging dry riser or sprinkler networks on the rapid opening of isolation valves connected to the water supply system. The potential for excessive surge pressures in cases where there is uncontrolled compression of air or free gas by

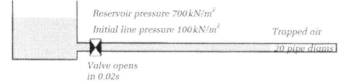

Reservoir pressure 700 kN/m²

Initial line pressure 100 kN/m²   *Trapped air*

20 pipe diams

Valve opens
in 0.02s

Figure 9.30 Trapped air at end of a closed conduit to represent transient boundary condition in dry riser and sprinkler head systems

a water column arise from the form of the gas laws which allow major reductions in volume prior to a final rapid increase in gas pressure. The surge pressure generated is enhanced due to the compression of the initial air within the dry pipework, even if the system is fitted with outwards relief valves. Gas compression allows the advancing water column to accelerate before finally being brought to rest as the trapped gas pressure rises and the final pressures generated may be destructive of the local fluid network. The literature includes dry riser and sprinkler investigations (Lawson *et al.* 1963, Hope and Papworth 1980) while a more specialist system failure, namely the violent fracturing of water closets due to the interaction between slugs of entrained air in the water supply and the flush valve mechanisms commonly found in US domestic and commercial installations, has been reported and simulated using the techniques discussed here (Ballanco 1999).

Figure 9.30 illustrates the source of the transient for the dry line priming case, which encompasses the vertical dry riser and the horizontal sprinkler or hydrant-fed hose application. Opening the supply valve to connect the high-pressure source to the empty line, or to an empty line terminated by an open valve, effectively allows the supply pressure to accelerate a water column through the system against little or no resistance. The air expelled through the open valve, or through any termination relief valve, will be subject to a low loss factor at the termination device due to the relatively low density of air compared to the approaching water column. Eventually the water column is decelerated rapidly due either to compression of the trapped gas or due to its arrival at the terminating valve, whose effective loss coefficient will increase by almost a thousand times when the transiting fluid is water and not air. The resulting positive

pressure surge may be sufficient to cause catastrophic failure of the supply system — most probably at any hose couplings. Figure 9.31 illustrates the form of the pressure transient propagation to be expected in such cases.

The description above also applies to the WC fracturing investigation that addressed the transient propagation as a result of slugs of unvented air passing through a WC on flushing with a mains connected flush valve. Figure 9.32 illustrates the form of the problem with the method of characteristics grid and boundary conditions superimposed. Figures 9.33 and 9.34 illustrate the transient propagation expected in three cases — water only flushing and water plus air slug with and without a relief valve set to open at a pre-determined multiple of the line pressure. The possibility of catastrophic failure of the WC bowl is clearly indicated and the likely transient condition following fracture is illustrated for the no relief valve case. The laboratory investigation reported by Ballanco (1999) included high-speed photography of a fracturing WC under controlled conditions. Figure 9.35 illustrates a catastrophic failure of the WC as the water column accelerated by the passage of unvented air through the unit is rapidly decelerated on entering the WC rim. Initially the air driven out of the system through the WC rim 'blows' bowl water out of the WC. The arrival of the water column leads to catastrophic failure of the ceramic — pieces may be identified in the later images.

Siphonic rainwater drainage systems

Over the past 30 years increasing amounts of industrial and commercial roof area have been drained using siphonic drainage. Examples of siphonic drainage may be found in many prestige or status buildings internationally, for example the Sydney Olympic Stadium,

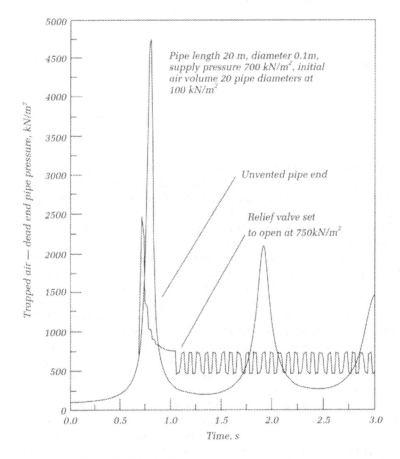

Figure 9.31   Effect of relief valve in reducing peak pressure by venting trapped air

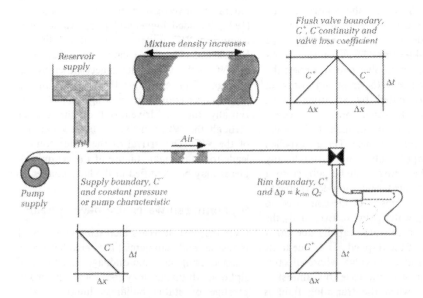

Figure 9.32   Method of characteristics simulation of the WC fracture event

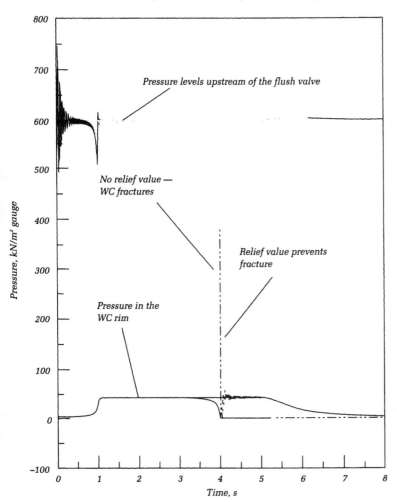

Figure 9.33 Simulated pressure traces upstream of the flush valve and in the WC rim for water only and entrained air flush examples. Effect of relief valve shown more clearly in figure 9.34

Hong Kong Airport and in the UK at Stanstead Airport and Murrayfield Stadium in Edinburgh.

The increase in usage has been largely due to the many advantages perceived over conventional systems. There are, however, still uncertainties regarding the operating mechanisms of these systems, particularly during the initial air expulsion and priming phase. A lack of fundamental understanding of the transient nature of the system operation makes failure identification difficult. In addition, as system design requires a higher level of expertise than conventional gutter/downpipe systems, their performance can be prone to failure due to design inaccuracies and erroneous assumptions.

Conventional roof drainage systems generally consist of open gutter outlets connected to vertical rainwater pipes designed to operate effectively at atmospheric pressure with a continuous entrained air core. The flow capacity of a conventional system is determined by the size of the outlets and the gutter depth — normally a maximum of around 100 mm. Additionally, any horizontal pipework must have gradients that assure self-cleansing action. An equivalent siphonic system will have a significantly higher flow capacity as the pipework flows full bore and the driving head becomes equal to the vertical distance between the roof gutter outlet and the point of discharge. A siphonic system draining a roof

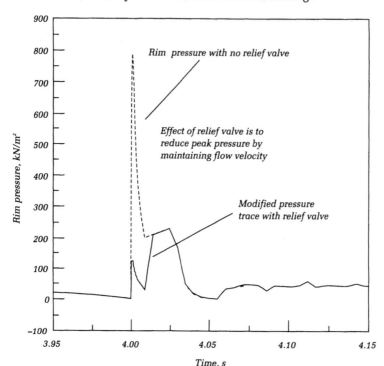

Figure 9.34   Effect of a relief valve on the pressure surge in the WC rim as flow reverts to water only

normally consists of a specially designed gutter outlet (figure 9.36) connected to a discharge point at or below ground level. The connecting pipework will be designed to run full bore in response to a specific rainfall condition, defined in terms of intensity and possibly duration. Any storm below this threshold will result in unsteady flow conditions within the system. Storms which exceed the design condition will result in flooding unless excess run-off is re-directed.

The higher capacities achieved by siphonic systems have the following benefits when these systems are compared to conventional systems:

- fewer outlets required;
- smaller diameter pipework;
- several outlets may be connected to each vertical discharge pipe;
- full bore flow driven by a high head differential yields route flexibility and removes gradient and self-cleansing flow restrictions;
- pipework may be routed at high level to pick up sequential gutter outlets, reducing the groundwork requirement.

The disadvantages mainly concern the establishment of primed conditions — restricted to one design storm intensity, and the need to ensure that siphonic conditions are not lost in multiple outlet systems if the rate of flow from each outlet is not dynamically balanced to ensure no air ingress as different roof sections are drained. If the design storm is not severe enough flooding may occur; if the design storm is too severe then the network may never prime and as siphonic action will then not occur the traditional concerns of self-cleansing and blockage due to leaf and other detritus deposition returns. Laboratory studies have suggested that below 40 per cent of the primed system capacity, single outlet systems act as conventional roof drainage. Above 40 per cent unsteady flow conditions prevail. Generally there is a need for more accurate design and an understanding of system operation when defining maintenance standards.

Partial priming and failure of the siphonic condition may also be a difficulty as this can lead to system vibration and possible noise problems. Under-specification of pipe wall thickness may lead to catastrophic failure due

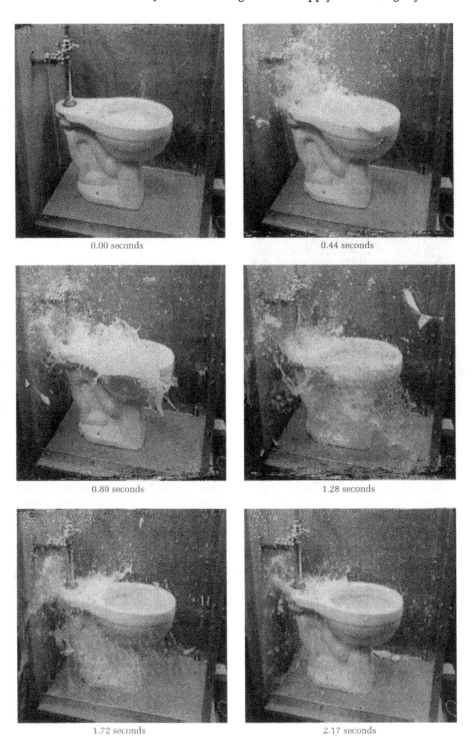

0.00 seconds

0.44 seconds

0.89 seconds

1.28 seconds

1.72 seconds

2.17 seconds

Figure 9.35 High-speed video reveals the sequence of failures for a ceramic WC subject to surge pressures. The initial expelled air displaces water from the bowl followed by ceramic fracturing on the arrival of the water column (Courtesy J. B. Engineering)

Figure 9.36   Typical, but different, siphonic roof outlets (both graphics reproduced with permission)

to implosion of the pipework (Bowler and Arthur 1999). It is often not appreciated that siphonic systems operating correctly will feature pressures well below atmospheric.

The challenges in modelling siphonic rainwater drainage systems therefore revolve around the issue of establishing and retaining siphonic conditions within fully primed systems. Figures 9.37 and 9.38 illustrate both a siphonic rainwater system two-roof-outlet test rig and the mechanisms required for priming. The sequence of events during a simulated storm may be summarized as follows (Wright *et al.* 2001):

1. Free surface flows in both the gutter and the system pipework with annular flow in the vertical stack.
2. As flow rate increases a hydraulic jump forms upstream of the pipe junction in both pipes.
3. Increasing inflow leads to full bore flow at the junction that propagates downstream towards the vertical stack.

4. When the full bore flow reaches the stack de-pressurization of the system occurs, increasing the inflow to the system from the gutter outlets and the establishment of full bore flow at the head of each branch — trapping air pockets in both pipes.
5. Both air pockets are swept downstream. In the laboratory case when the pocket in branch 1, figure 9.37, reached the vertical stack it caused a momentary re-pressurization which lasted until the air exits the system at the stack base. In the case of the branch 2 pocket, this became broken up in passing through the junction and had no effect on system pressure levels on reaching the stack.
6. Once the air had exited, the system pressure remained steady and the system operated at its design condition.

Figure 9.39 illustrates typical gutter depths and pipework pressure levels during the priming process discussed. These results are

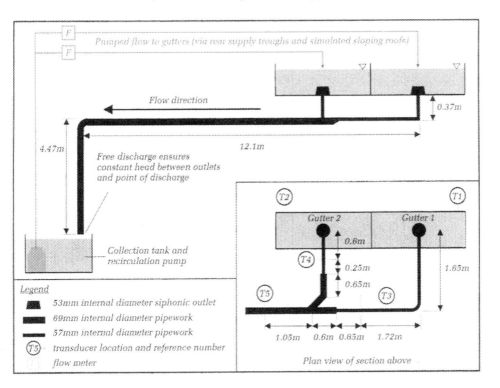

Figure 9.37   Test rig for siphonic rainwater systems

similar to those derived in an earlier single outlet investigation that included the development of a method of characteristics simulation of the events discussed for the simpler single outlet case. Figure 9.40 presents the system pressures recorded at an inflow corresponding to 42 per cent of the design capacity that confirms the oscillatory nature of the flow at these storm conditions, while figure 9.41 illustrates the agreement between measured and predicted pressures during the priming event, including the re-pressurization phase corresponding to the trapped air pocket reaching the vertical stack (Arthur and Swaffield 2001).

It is essential that laboratory measurements and simulation development is matched by site testing to validate and confirm both the models and the understanding of the flow mechanisms involved. As an example of what is required figure 9.42 illustrate system flow rates and pressures at the National Archives of Scotland building fitted with a siphonic rainwater system that uses pipework as small as 50 mm diameter to drain roof areas up to 3000 m^2. As mentioned siphonic systems require careful maintenance and figure 9.43 illustrates typical outlet blockages that may be encountered.

Impact of computer-based simulations

This chapter has concentrated on the application of one modelling technique to describe the unsteady flow conditions within building utility systems. The flow conditions met in full and partially filled pipe flow and in the transport of solids and entrainment of air have all been shown to be members of the same family of flow phenomena that may be addressed via the method of characteristics. While the mathematical techniques deployed have had a long history, it is the advent of cheap, fast and available computing that has made these accessible to the building services designer. The techniques presented have already also been used to address other flow conditions and systems within buildings. Examples lie in the modelling of air pressure transient propagation within the drainage vent systems appropriate to large underground structures

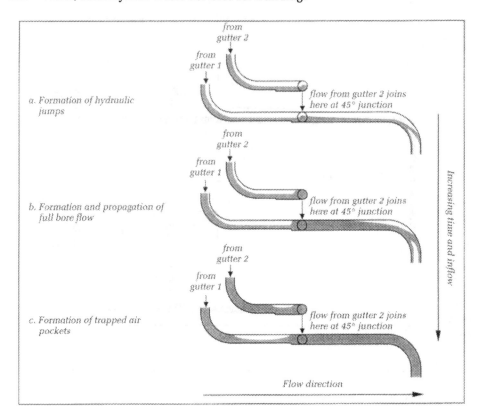

Figure 9.38 Priming process of the siphonic rainwater test rig

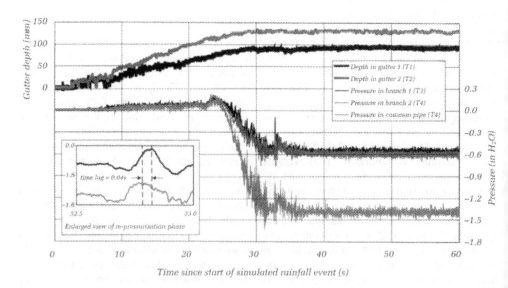

Figure 9.39 Measured gutter depths and system pressures for the design criteria rainfall event (gutter 1 inflow= 5.85 l/s, gutter 2 inflow= 7.78 l/s)

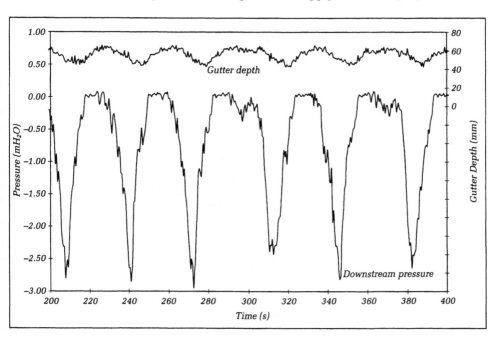

Figure 9.40 Ambient pressures in the system for a steady gutter inflow rate of 42 per cent of the measured capacity, illustrating the cyclic nature of the partially primed system performance

and the dynamic balancing of airflow delivered to various building internal zones by air conditioning or mechanical ventilation systems. In the first case, work for London Underground proved that the AIRNET type simulation was wholly appropriate for the study of the complex ventilation networks necessary to provide both venting and odour removal from drainage networks and collection sumps at low levels within the stations. In the latter

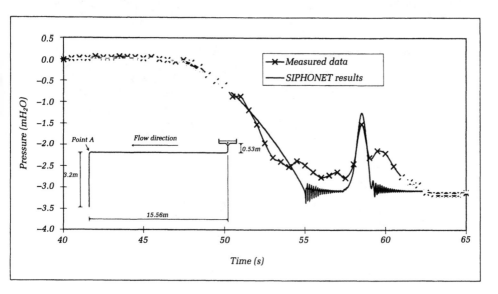

Figure 9.41 Computed and measured pressures at Point A during the priming of the siphonic roof drainage system illustrated

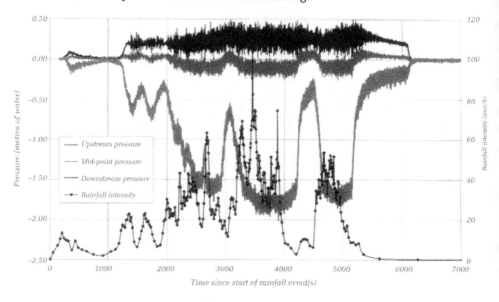

Figure 9.42 Measured rainfall intensities and system pressures for the main siphonic system at Thomas Thompson House (Edinburgh) for the rainfall event starting at 16:22:21 on 2/8/00

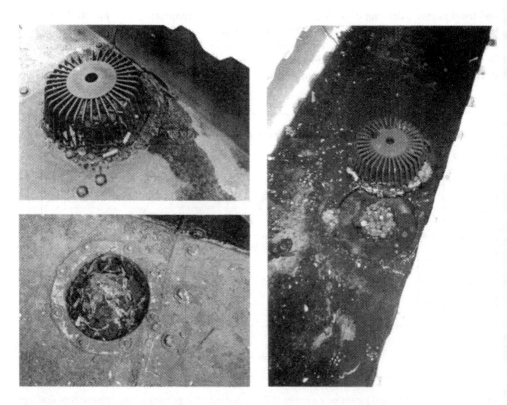

Figure 9.43 Blocked outlets — materials may vary from screw caps to leaves (Heriot-Watt University)

example current work is developing network models that will act as simulated test beds for control strategies to ensure balanced air quality and supply to building internal zones.

While these are exciting developments, the models presented continue to have as their prime objective the provision of design advice, both to building services and plumbing consultants and to the code bodies that determine permissible system installations. The models, based on a fundamental description of the flow conditions met in building water supply, drainage and vent systems, are international, crossing the boundaries of national codes. The replacement of national codes based on traditional rules of thumb and experiential knowledge will provide a basis for international codes of practice based on performance that will open markets and raise the overall level of provision, while at the same time allowing innovative appliance and system design. The examples developed in this chapter are merely the precursor to the design support that will become commonplace.

10   Noise

Noise is accepted as one of the factors to be considered in the design and installation of building services. Ventilation and air-conditioning systems as producers of noise have been dealt with extensively in the literature to provide a theoretical and practical basis for noise control. Water and sanitary services have been less well covered although the results of various researches have been reported in scientific and technical journals. The aim of the present chapter is to assemble theoretical and practical findings from this work as a basis for design and installation.

The extent to which noise from services is troublesome to the occupants of a building depends on a range of factors including the nature of the noise, the sensitivity of the people themselves, their occupation at the time the noise occurs, the way in which noise sources are separated and insulated from the hearer, and the background noise present. In a situation where, for example, traffic noise is loud and persistent the hum of a small pump transmitted into the room may go unnoticed. The same noise penetrating a quiet library where readers are concentrating or into a bedroom at night may, on the other hand, be highly disturbing. The variability of people in their response to noise — a common finding in surveys — is also important. Social surveys of noise nuisance have not dealt primarily with building services but certain surveys have provided some information of interest. One study (Chapman 1948) of people in some 2000 dwellings in the UK identified cistern noise as disturbing to sleep for some 6 per cent of occupants, whether the noise was from within their own dwelling or from next door. Table 10.1 reports relevant results from a survey of 800 semi-detached local authority houses (Parkin *et al.* 1960). Another survey of some 1500 flats in the UK (Gray *et al.* 1958) indicated a higher degree of disturbance by

Table 10.1   Disturbance by plumbing noise

Type of noise	% that hear	% disturbed
WC flush	28.0	2.3
Cistern filling	18.0	2.1
Water running	7.6	1.3
Bath filling	12.9	1.1
Bath emptying	10.3	0.9

WC flushing and other plumbing noises, with percentages exceeding 10. There have been broadly similar findings from surveys in continental Europe and the USA; a comprehensive American study (National Bureau of Standards 1976) identifies plumbing as a substantial cause of noise nuisance. Fuchs (1993) comments on the effect that sudden plumbing noise may have on sleep. A considerable number of people clearly are aware of the noise from water and sanitary services, whether from inside or outside their own home, and some are disturbed by the noise in one way or another.

Measures of noise

Subjective information of the kind outlined above can be of real use to the designer only if it is related in some systematic way to objective measures of the noise. With the increase in air and road traffic since the 1950s extensive studies of these relationships for noise from these sources have been made. Certain conclusions from such work are useful in providing a simple basis for quantitative assessment here. Background to the terms used in the following paragraphs is given in the final section of the present chapter.

A full expression of the noise caused by a sanitary appliance or fitting is given by its noise spectrum exemplified by figure 10.1 in a simplified form. For present purposes of

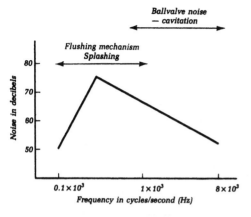

Figure 10.1 Representation of a noise spectrum for a WC suite, indicating ranges for sources

broadly classifying different equipment in terms of noise levels, a single measure is preferable. One simple measure is the dBA, the reading of a sound-level meter containing a weighting network that discriminates against low-frequency sounds, as does the human ear. This unit is particularly useful because in various fields it has been found to be quite a good indicator of subjective response, a steadily increasing annoyance being observed with increase in the dBA level. Better correlations have been obtained using more complex measures which take account of the fluctuating nature of the noise as well as its mean level. For present purposes that concept can give no more than a useful reminder of factors that may be significant. For quantitative assessment in what follows the dBA measure obtained using a sound-level meter is used. There are few indications of numerical values related to subjective response. In one study, however, complaints of excessive plumbing noise occurred when the sound level within dwellings lay in the range 50–60 dBA.

Noise sources

The results of measurements of noise levels from various appliances and equipment used in buildings are assembled in figure 10.2. Each line represents a study, drawing on the sources indicated. The range of noise produced is wide and the upper limits can exceed 90 dBA. These figures generally relate to points close to the source and the level of exposure is, of course, less away from the source — perhaps

10–20 dBA less in an adjacent room. The first four examples in the figure cover commonly used appliances which may serve as useful points of comparison with the remainder of the data which are more directly of interest. Food waste grinders can cause the highest levels of noise in the range of equipment listed here. Water closets are seen to give widely varying results with a range from some 45 to 85 dBA. This is associated, however, with a wide range of appliances from the noisier high-level suites still used in the UK and mains-fed flushing valves on some American suites, to the quieter low-level or close-coupled cistern and pan fitted with a good-quality flushing mechanism and quiet supply valve. Fuchs (1993) quotes values up to 45 dBA for some products, noise measured in an adjacent room. Pumps and fans also vary widely in their noise, with size, speed and type important. Pipework generates less noise as exemplified by the range given in figure 10.2. In comparison, the general background noise level in a quiet room in an urban area might be 30 to 40 dBA.

In none of the appliances, equipment and fittings in figure 10.2 is there a single origin of the noise which, if treated, will render the equipment quiet. This is obvious for such appliances as a WC suite where noise can arise in various ways during flushing, discharge and refilling. It is the case, also, for apparently simple devices such as valves: cavitation, as described later, can arise at several locations in the waterway, each of which must be treated if the valve is to be effectively silenced. Again, with pumps, predominant tones arise from the pressure fluctuations associated with the passage of the impeller blades past the fixed blades and from harmonics of this frequency; bearings can cause noise, and cavitation may sometimes occur. Effective noise control in any equipment thus requires a systematic study to identify all the possible causes and their likely effects and then to prescribe the necessary design measures, also any special requirements to be observed during installation, commissioning and use of the equipment.

The origins of the noise in the appliances, equipment and fittings of particular interest here may be summarized as follows:

1. *Flow noise.* With pipes and fittings running full (water supply), turbulent conditions,

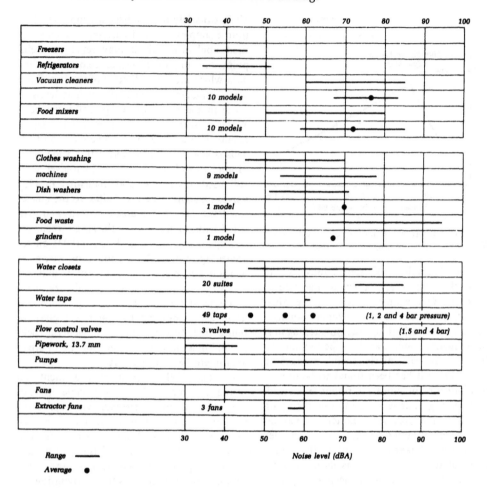

Figure 10.2   Noise levels for selected domestic appliances

Sources: Bugliarello 1976, Jackson and Leventhall 1975, Anzou 1967, Strumpf 1967, Ball and Webster 1976
Notes: Measurements generally at 1–1.5 m from source. Noise from taps (49) and valves (3) measured in adjacent room

separation at corners and bends, cavitation, waterhammer, splashing from terminal fittings. Air bubbles in supply piping may cause flow noise.

With pipes and fittings running partly full (drainage), turbulent conditions, splashing, air entrainment during flow, siphonage of water seals.

2. *Mechanical noise*. Flushing mechanisms; pump impellers, especially noise from bearings. Vibrations, rattling induced by fluid flow and by pump rotation.

3. *Temperature effects*. Noises due to expansion or contraction of pipework. Bubbling, steam formation and collapse.

We now consider some of the more significant of these factors, noting that chapters 4 and 8 in dealing with the fluid mechanics aspects of soil and waste drainage have touched on the noise implications and this information will not be repeated here.

Noise from valves and fittings — effects of cavitation

In most practical fluid flow situations the flow is turbulent (see chapter 8) and the movement of water in supply systems is no exception. One can visualize the motion in pipework and components as involving the passage of an

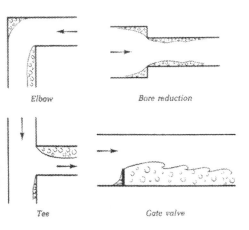

Figure 10.3 Diagrams to illustrate local flow separation and turbulence

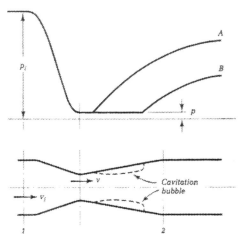

Figure 10.4 Illustration of pressure and velocity changes at a change in waterway

endless series of eddies or vortices. In certain situations, as exemplified in figure 10.3, the main flow separates from the surface and a particular form of local turbulence occurs which is the cause of the so-called secondary pressure losses in fittings described in chapter 8. These general conditions, but more particularly the local turbulence and vortex shedding, are realized in vibrations and hence noise that may be detected by the human ear. Additionally there is the special factor of cavitation that may arise in bends and fittings when conditions are favourable.

Cavitation means the generation of cavities or bubbles of vapour in a liquid as a result of a fall in pressure to the vapour pressure value, about 0.02 bar with cold water. The process is aided by the presence in the liquid of tiny nuclei of undissolved gas, usually air. With liquid in motion, a suitable drop in pressure may occur at a change in the boundary geometry in pipework; and it may also occur at the back of an impeller blade in a pump. The cavitation process itself is not thought to be noisy, but the subsequent collapse of the bubbles as they are carried away by the flow may give rise to an audible crackling, rattling or hissing.

The conditions shown in figure 10.3 are all suitable for the occurrence of cavitation, a change in cross-section or direction causing a reduction in pressure. As so far described the effect of such a change is one of local separation and turbulence but cavitation can develop if the transition is sufficiently marked.

This can be illustrated with reference to figure 10.4 which depicts the normal pressure and velocity changes at a reduction in cross-section of waterway, an application of the Bernoulli equation (chapter 8). Flow up to the contraction is at velocity v_1, and the corresponding pressure is p_1. Just beyond the throat, where the flow separates from the pipe wall, the velocity increases to v and the pressure falls correspondingly to p. If the rate of flow and the pipe sizes are such that this pressure p nears the vapour pressure of the water, vapour bubbles will be liberated in and near the region of eddies just beyond the point of smallest bore. The higher-pressure flow necessary to collapse the bubbles occurs towards p_2. For a given flow and inlet pressure and assuming that the pressure p at the throat equals the vapour pressure, the extent and intensity of the cavitation depends on the pressure downstream — compare curves A and B.

A further by-product of cavitation is erosion. Bombardment of the adjacent parts of the valve or pipe fitting takes place on a minute scale as the cavitation bubbles collapse. Serious damage has been caused by the action in large conduits, and the common corrosion of tap and ballvalve seats is in many cases a product of cavitation.

As examples of cavitation in water supply components it is useful to refer to two published case studies. Prior to a more extensive study of noise from pipework (Ball

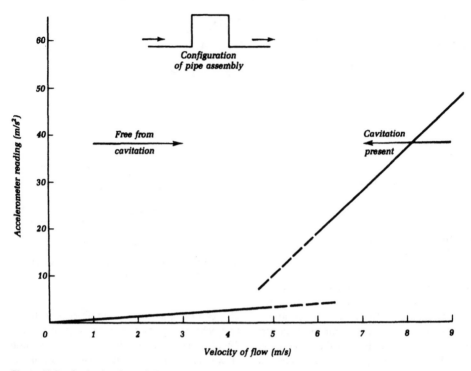

Figure 10.5   Cavitation in a 13.7 mm copper pipe assembly as a function of water velocity

and Webster 1976) the potential for cavitation in some standard small-bore pipe fittings was first investigated. Using the reverberant room to be described later the investigators supplied fittings with cold water at a range of velocities in a special rig which also enabled operating pressures to be varied widely. This work showed no evidence of cavitation in typical individual fittings at speeds up to 7 m/s (the maximum available), although the flow caused noise as shortly to be outlined. It was not until three elbows were assembed close together in series that sufficient change in the flow and consequent pressure loss occurred for cavitation to develop at lower speeds. Cavitation was then induced at velocities of about 5 m/s, as shown in figure 10.5, and the effect increased markedly at higher speeds. In this case the vibrations were recorded by means of an accelerometer attached to the fittings, but noise could be heard by the observer when close to them. For comparison it may be noted that American research using pipe diameters from 25 to 100 mm showed no cavitation in individual bends and elbows at velocities below 10 m/s, with water temperatures ranging from

10 to 55 °C (Rogers 1959, Stonemetz 1965). It may be concluded that cavitation does not present a major problem with pipe fittings in normal water supply installations and it can be readily prevented by limiting the velocity of flow. The effects of turbulence and separation remain and are dealt with in a subsequent section.

Cavitation can be much more of a problem with taps and valves and its influence may best be described by reference to an early study (Sobolev 1955) of a ballvalve. The valve investigated, the so-called Portsmouth type of 13.7 mm nominal size shown in figure 10.6, has

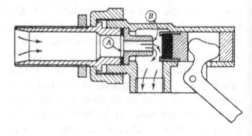

Figure 10.6   Diagram to illustrate the waterway through a conventional piston-type ballvalve

been very commonly used in the UK over many years. Whilst reasonably reliable it is noisy in operation and the brass seatings have been found to erode, sometimes in a matter of months. The use of phosphor bronze for the seating was not found to be a solution to the erosion. A flow visualization study using a valve of this type machined from transparent Perspex showed that cavitation occurred somewhere in the valve throughout the range of discharge. The effect had its origin at different points in the waterway at different stages of the piston travel. It occurred over most of the range at the inlet to the seating where there was a sharp reduction in cross-section (A); a plume of bubbles extended downstream from A along the bore of the seating. Cavitation occurred at the seating outlet both with small and large clearance between the seating face and the washer (B), related to the sharp change in the direction of flow as the water emerged from the nozzle. With large clearances between seating and washer, bubbles were observed also to develop at the washer face. In these two locations substantial erosion had been found to occur in practice.

The pressure measured in the throat during cavitation was reported at about 0.1 bar, which is well above the vapour pressure of the water. Thus cavitation appeared to begin somewhat before the vapour pressure was reached, although this measurement may not necessarily have been very precise. A theoretical prediction of the pressure in the throat cannot easily be made exact because the flow conditions are not known with precision. With the discharge rates likely to occur in practice velocities of flow in the bore of the seating are considerable — perhaps 25 to 50 m/s. As may be shown very simply from the Bernoulli equation, the conversion of a pressure of several bars in the supply pipe, where the flow velocity is 1 or 2 m/s, to a high velocity in the bore of the seating, leads to low pressures within the bore. This is the prerequisite of cavitation.

By rounding the inlet to the throat it was found possible to eliminate cavitation in this region, but modifications to the seating and washer were only successful over part of the range of discharge. Cavitation could be eliminated substantially only by redesigning the valve, providing a nozzle with rounded inlet

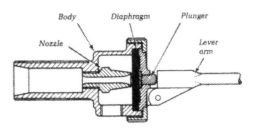

Figure 10.7 Diagram to illustrate the waterway through a diaphragm (Garston) ballvalve

and outlet and using a diaphragm instead of a piston with washer, so improving the overall flow characteristics of the waterway (figure 10.7). This work provided the basis for some designs of quiet ballvalve now on the market in the UK. Similar types of design are available abroad.

Overall it may be seen that there are three possible courses open in seeking to eliminate cavitation: reducing the discharge rate; increasing the hydrostatic pressure (downstream) under which the device operates (cf. figure 10.4); and improving the waterway. The last means making any change in flow direction or pipe bore as gradual as possible. Design of the waterway is, of course, of fundamental importance in planning new taps, valves or fittings for quiet operation. Equally it is important to establish the range of pressure and discharge conditions under which the device may be used.

Noise from pipework — effects of turbulence and separation

Several investigators have studied the general flow noise in piping and connections as distinct from the noise due to cavitation associated particularly with taps and valves. The latter tend to dominate but the study of the pipework itself is necessary if the basic principles of noise and its control are to be understood and applied. Before describing two studies it will be an advantage to state the conditions under which data were obtained. For the reasons given later the pipe assemblies to be studied were mounted in specially built reverberant rooms (of a size roughly the same as a small living room) rather than in a room containing the usual absorbent surfaces, carpets and other furnishings. The noise levels in a reverberant

Table 10.2   Noise levels, dBA, in a reverberant test room from pipework of 15–16 mm bore

Velocity of flow (m/s)	Pipe material			
	Copper	Steel	Lead	Plastics
0.1	24	26	25	29
0.55	27	30	29	30
3.4	46	38	39	41
5.2	46	38	38	41

room are higher than in a normal room but, as explained later, may be adjusted to correspond to the practical case. In what follows the reverberant test data are given first and then some adjusted values are given for typical practical situations.

Turning first to the results obtained in reverberant conditions, table 10.2 summarizes data of Marseille (1965) using a rectangular configuration of piping with elbows at the corners, all 15–16 mm bore. The circuit was 1.5 × 1.3 m in size, with rubber flow-and-return connections, and was suspended in the room from rubber hangers. The noise increases with water velocity as might be expected, although cavitation was hardly likely to be

present under the relatively low speeds used. The increase is associated with an increasing level of turbulence; indeed a significant increase in noise emission was noted at what was probably a transition from laminar to turbulent conditions (see chapter 8). In checking the effect of the type of pipe clip, it was found that the noise levels in the room were somewhat higher when steel instead of rubber clips were used (giving firm rather than resilient fixing). The noise levels in table 10.2 appear to be more or less independent of pipe material. Overall a noise level of about 40 dBA was obtained at a velocity of 3 m/s which might imply a level of, say, 33 dBA with the pipes mounted in a normal room.

The results obtained from studies by Ball and Webster (1976) are summarized in figures 10.8 to 10.11. As with the French work described above, the speeds were too low for cavitation to occur — none was detected — and the noise is associated with the turbulent flow conditions. The Reynolds number in these studies ranged from about 10 000 to 80 000 (see chapter 8). The noise levels increased with velocity of flow as shown. Particular points of interest follow.

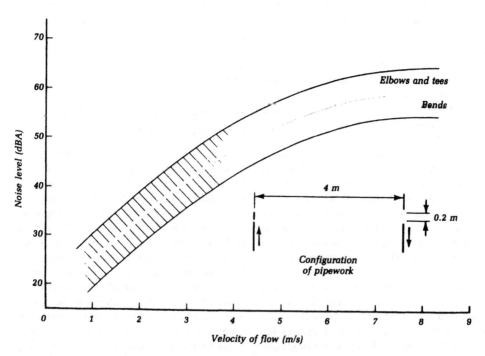

Figure 10.8   Noise levels for 13.7 mm copper pipework in reverberant test room. Pipework mounted in brackets on test room wall

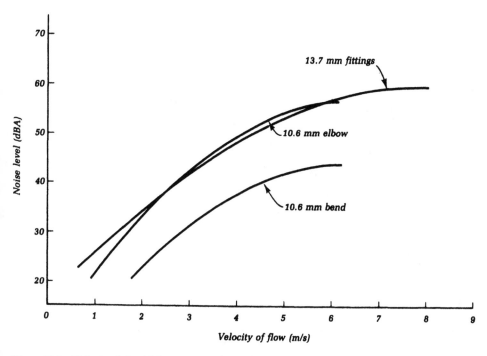

Figure 10.9 Noise levels for 10.6 mm copper pipework in reverberant test room, with curve from figure 10.8 for comparison

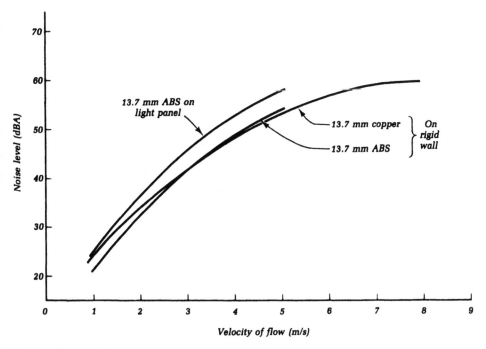

Figure 10.10 Noise levels for 13.7 mm ABS pipework in reverberant test room, with curve from figure 10.8 for comparison

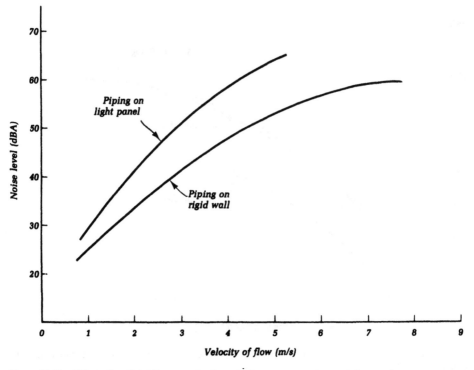

Figure 10.11   Effect of wall rigidity on noise from 13.7 mm copper pipework in reverberant test room

Figure 10.8 is based on the results of a large number of measurements on 13.7 mm bore, commercial-type copper fittings — elbows, tees and bends — and also pipe bends made in a bending machine. The configuration tested in each case was as shown in figure 10.8, a total length of pipe of about 4.4 m containing two changes of direction. The spread of results is indicated by the outer curves, a range of some 10 to 12 dBA overall associated with small differences in the shape and size of waterway. The effect of the radius of curvature of the fitting was not distinguishable at velocities below about 3.5 m/s, but it was at higher speeds. Whereas the results for elbows and tees tended to lie in the upper half of the scatter, those for bends tended to be in the lower half. These results are broadly comparable to the French data (table 10.2).

Figure 10.9 shows comparable results for 10.6 mm bore elbows and pipe bends. Here the elbow is noisier than the bend at all speeds of flow, an effect due to a combination of reduced curvature and an observed restriction in bore of the elbow.

Figure 10.10 shows results for an ABS copolymer pipe with elbows of bore similar to that of the larger copper piping, i.e. 13.7 mm. With the assembly fixed to the test room wall the results are similar to those for the copper pipework.

Figures 10.10 and 10.11 show the effect of mounting the assembly on a lightweight panel as compared with mounting it directly onto the test room wall. The panel in these studies, 18 mm plywood about 2.4 × 1.2 m in size, was fixed firmly to the wall of the room by six bolts but was separated from it by 40 mm spacers and was thus free to vibrate. It will be seen that this change in mounting increases the noise level in the test room. Whereas the heavy wall of the test room is not excited by the vibrations in piping fixed directly to it, a lightweight panel is more capable of being excited and thus increases the effective acoustic coupling between the piping and the air. The stiffer copper piping (figure 10.11) appears to be more effective in forcing the panel into vibration than the ABS piping (figure 10.10) — an increase of about 9 dBA compared with 5 dBA at 3 m/s was observed.

Table 10.3 Noise levels, dBA, in a room from pipework with water flow at 3 m/s

	Piping mounted on rigid wall	Piping mounted on light panel
13.7 mm bore copper piping with elbows or bends	33	43
10.6 mm bore copper piping: with elbows	35	41
with bends	24	30
13.7 mm bore ABS piping with elbows	34	39

Note: Reverberation time of room 1 second

As already explained, the noise levels to be expected in a normal room are less than those in a reverberant room and table 10.3 summarizes the findings of all this work converted to a more practical situation — a room with a reverberation time of 1 second — and assuming a water velocity of 3 m/s. The highest noise level with the pipe assembly fixed to a rigid wall is 35 dBA and does not exceed 43 dBA even with the assembly mounted on a light-weight panel capable of augmenting the sound radiated from the pipe. These levels may be compared with the commonly assumed general background levels of 30–40 dBA in the quieter rooms of dwellings in urban areas.

Whilst the foregoing data aid the general understanding of the factors that are important in noise control and contribute to the recommendations made later, they have particular significance in relation to the velocity of flow to be assumed for pipe-sizing calculations. Design speeds of 1 m/s have been used in the UK whilst values of up to 2 m/s have been suggested in some European countries. The results reported here suggest that noise levels generated in the pipework itself are not likely to be troublesome at speeds higher than normally assumed: a value of 3 m/s might reasonably be taken as an upper limit for design purposes in respect of noise control. Suitable fittings should be used.

Waterhammer

When water flowing in a pipe is stopped suddenly by rapid closure of a valve or tap, a high impact pressure is produced leading to a surge or wave which rebounds from the fitting and passes back down the pipe, continuing its motion until damped out by friction. The magnitude of the pressure depends on the velocity of the water before closure, and on the pipe bore, length and material. It can be shown that the pressure in small rigid pipes may rise to several times the normal running value when a valve is shut with sufficient speed. The usual result of the action is an appreciable knocking sound or hammer — waterhammer — whilst attendant troubles may be vibration of pipes and damage to the pipe system. When waterhammer occurs in a system fitted with a ballvalve, an oscillation of the ball float may be set up which itself tends to produce noise. Alternatively, the ballvalve discharging into a cistern may set up waves on the water surface and so develop a self-induced oscillation and waterhammer. A self-induced waterhammer may also occur with taps having a loose washer plate intended to act as a non-return valve. An alternating pressure — positive and negative — may occur at the seating at some flow rates and cause the loose plate to vibrate. The risk of waterhammer arising can be substantially reduced by careful choice of taps and valves, and through design of the system as a whole, avoiding excessive velocities of flow, whilst careful fixing of pipework and appliances can help to avoid vibration and rattling. Waterhammer is further analysed in chapter 8.

Noise transmission

Noise generated in a room is attenuated as it spreads through the rest of the building, with the walls, floors and ceilings providing a measure of insulation. A pipe passing through such a barrier is a potential cause of reduced insulation by facilitating transmission. It is important to distinguish between noise transmission along the pipe itself and transmission through any air space round the pipe where it passes through the barrier. Noise generated in the air external to the piping system can penetrate the air space much more easily than it can pass along the pipe. It is the noise originating in the appliances and fittings themselves and in pumps that can pass readily along the piping. Indeed the attenuation afforded by metal pipework is minimal and noise caused, for example, by cavitation in a valve or by waterhammer easily travels throughout the whole pipe system when metallic. Hence the noise levels in buildings are likely to be higher, for instance, than those set out in table 10.3 which relate to the pipework only. With taps and valves as the dominant noise sources, and with little attenuation in the pipe system, noise levels in dwellings, for example, are commonly in the range 35–55 dBA and may be higher.

The degree of attenuation in a pipe system depends both on the pipe size and material and on the fixing of the piping to the structure. There is little difference between copper and galvanized steel tube in this respect, both affording minimal attenuation, but plastics water pipes show a greater loss (Ball 1974, Wassilieff and Dravitzki 1992). As a guide for small pipes the following figures are reasonable, according to Ball:

PVC-U and ABS: 1 to 1.5 dB/m
Polythene: 2 to 2.5 dB/m

the precise value depending on the grade and size of pipe and the type of noise source. The overall length of the piping itself may be a factor, with some evidence that the attenuation per unit length is somewhat less for short than for long pipes.

The type of fixing also plays some part in the attenuation loss. In principle the best conditions for attenuation are when the pipe is fixed firmly at short intervals to the supporting wall, when energy is lost in setting the wall into vibration. As already noted this may increase the sound in a room adjacent to a noise source, especially when the pipes are mounted on a lightweight partition, but the increased attenuation loss may then reduce the noise transmitted further afield. Attenuation tends to be reduced when the pipes themselves are 'insulated' from their supports by wrapping in felt or rubber.

A short length of flexible tubing inserted between a noise source such as a pump or valve and the pipe installation offers the potential for noise reduction. The effectiveness of such a device is limited because sound travels through a pipe system both along the pipe wall and through the water. Introducing a pipe material less favourable to sound transmission does not reduce transmission through the water and

thus sound is able to excite the pipe again after the discontinuity. Nevertheless the device has been shown by several investigators to effect a worthwhile noise reduction. Table 10.4 summarizes the results traced in the literature in which there is a reasonable measure of agreement. The flow noise generators referred to are devices for producing 'white noise' — noise at a given level covering a wide frequency range — and were used to represent a standard source roughly equivalent to a noisy tap. The 'silencers' tried were intended to interrupt the noise transmission through the water, e.g. by a gauze or by a change in pipe bore, to supplement the effect of a change in pipe material. Attenuation ranging from 10 to 30 dB has been achieved by the methods indicated.

Reference was made at the beginning of the chapter to the pump noise associated with the passage of impeller blades over fixed blades. This action may excite pipelines into their bending frequency modes corresponding to the blade passage frequency in the pump. German research gave appropriate intervals of support for piping to keep fundamental bending frequencies below some value such as 50 Hz,

Table 10.4   Attenuation caused by pipe inserts

Conditions	Method	Attenuation (dB)
Noise generator with 13.7 mm pipework (Ball 1974)	Rubber steam hose, length over 100 mm	20–25
	Rubber tubing	10–20
	Metal bellows	5–15
Noise generator with 15–16 mm pipework	Short rubber coupling + simple silencer	20–30
Water tap with 15–16 mm pipework	Rubber tubing 150 mm long	8–10
Circulating pump with 15–16 mm pipework (Marseille 1965)	Rubber tubing 150 mm long	9
	1.5 m long	22
	Silencer — 200 mm long enlargement to 130 mm bore + gauze and tapered outlet	11
Circulating pump with 25 mm piping (Rogers 1959)	Silencer — 500 mm long enlargement to 200 mm bore	12

with the following intervals between supports related to the inside/outside diameters of metal pipes:

53/60 mm: 2.4 m
105/114 mm: 3.2 m
207/219 mm: 4.0 m
203/219 mm: 4.3 m

Prevention of noise

The solution to noise control problems generally involves three components: the noise source, the receiver, and the pathway between the two. The opportunity for preventing or limiting noise thus lies with various parties — the designer and manufacturer of appliances and fittings; the architect responsible for the planning and design of buildings and equipment selection; the builder; and the designer and installer of the service system. Code writers play their part in specifying requirements, design methods and testing procedures. The owner and occupants have an influence through the way in which they define the brief and determine the use of the building and its services. In what follows there are both specific points of attack and aspects of wider relevance that derive from the foregoing paragraphs.

In the UK, and in most other countries, there are no general mandatory standards relating to noise control for services. A few countries state requirements; Fuchs (1993) summarized these and stated that they limit the noise of household appliances and equipment to 30 dBA maximum in a building. In general, however, there are no legal obligations, and design is based on recommendations of what are reasonable noise levels for the particular type of building in question, based on research and practical experience of buildings in use and on the cost penalties normally associated with reduced noise levels.

As a guide we may note that complaints of excessive plumbing noise have arisen when the levels in dwellings in the UK, i.e. in living rooms and bedrooms, have been in the range 50–60 dBA, with a few complaints arising in the range 40–50 dBA. More general measurements in dwellings, not linked to complaints, have given sound levels in the range of 35–60 dBA when water installations were in

use. Again, the general background noise in living rooms and bedrooms in a quiet neighbourhood is commonly in the range 30–40 dBA. Finally we may note again the reference above to a limit of 30 dBA applied to the service installations in buildings. Typically (figure 10.2) the noise levels at source from water and sanitary appliances and fittings lie in the range 45–85 dBA. Attention to both sources and pathways is, therefore, likely to be necessary if the levels in occupied rooms are to fall in the range 30–40 dBA. This information is summarized in figure 10.12.

(i) Planning considerations

The basic rules for planning as an aid to noise control have long been understood and the following recommendations relating to noise from services derive from them:

1. Separate noise sources as far as possible from areas intended to be quiet.
2. Arrange rooms not particularly needing quiet as screens or baffles between sources and areas needing quiet.
3. Group noise sources together, away from quiet areas.
4. Locate equipment likely to be noisy in basements, if possible, where the general structure will probably be heaviest and thus more sound insulating and where vibrations may to some extent be absorbed into the ground.

The application of these rules may be seen in the location of bathrooms and WCs away from living rooms and bedrooms where space permits; the location of bathrooms directly above kitchens rather than above living rooms in small dwellings, a practice having other design and cost advantages; the location of service shafts near bathrooms and kitchens rather than near living rooms and bedrooms; and the location of plant rooms in basements and in service areas away from working or living accommodation.

(ii) Constructional considerations

Weight is important in determining the sound insulation properties of a wall or floor. It is much more difficult for a heavy wall of brick or concrete to be set into vibration than it is for

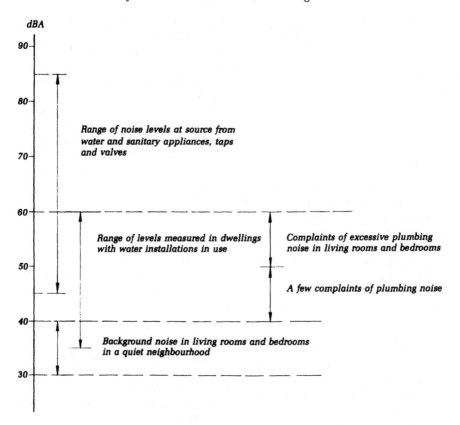

Figure 10.12   Complaints related to noise levels

a lightweight panel. For single walls the insulation is almost entirely determined by its weight per unit area and the so-called 'mass law' applies. The insulation afforded by a 230 mm brick wall (400 kg/m^2), for example, is about 50 dB, that by a 115 mm brick wall 45 dB, and that due to a partition weighing, say, 50 kg/m^2 about 35 dB on average. As a general rule, therefore, heavy walls radiate less sound than light ones, and this applies to the excitation caused by attached pipes. 'Indirect' or 'flanking' pathways for sound are also important in many circumstances, i.e. means by which sound can bypass a dividing wall or floor. The openings around pipes constitute a potential bypass and must be treated effectively as a first line of defence against noise transmission. A non-setting waterproof caulking compound might be used, for example, to fill the space.

In general, and assuming that all the living and work accommodation in a building is to have as little noise as possible, pipework should be insulated from supports and walls by means of rubber, neoprene, felt or mineral wool sleeves. Fixing should be to heavy rather than light walls or floors. Lengths of flexible piping, or one of the commercially available noise-reducing couplings, should be used close to the noise sources, and the pipework encased in substantial ducts. Appliances should be fixed where possible on flexible mountings and isolated from supporting walls by resilient gaskets. If noise in a room close to the source does not matter so much, any pipework passing through that room might be firmly fixed to a wall so as to lose energy to it, avoiding walls from which sound radiation should be avoided for other reasons as in a party wall between dwellings.

In small houses or flats the bathroom and WC is usually next to a bedroom or living room. The dividing partition should here be of at least 75 mm concrete plastered both sides and the sanitary appliances should not be fixed to it. The bathroom door should be as heavy as practicable and well fitting. Where the soil stack passes down through a kitchen it should be enclosed in a reasonably airtight covering such as blockboard or wood wool slabs plastered. If the soil stack is carried though a living room (to be avoided if at all possible) it should be enclosed. A recommendation for lagging and boxing in of soil stacks in such circumstances is made in Approved Document E in support of the Building Regulations 1991. The Regulations contain several other requirements relating to protection against noise from services in specific situations which should be observed.

(iii) Noise sources

Clearly it is important to select appliances and fittings that have been designed and manufactured with quiet operation as one of the main requirements. Even though not specifically designed with this objective, certain equipment has characteristics than tend to quiet operation. In general this implies the use of low-level rather than high-level WC suites; the quietest WC suites are as a rule those with close-coupled cisterns and double-siphonic traps, but they are more expensive than the ordinary washdown pan with low-level cistern. Pump noise depends strongly on the speed of the impeller and thus there is merit in choosing as low an operating speed as possible, commensurate with meeting the required duty.

Noise prevention in the design of taps, valves and pipe fittings requires attention to the principles set out earlier for the avoidance of flow separation and cavitation. Gradual changes in flow direction and bore are advantageous, to be employed as far as possible in the design of taps and valves and in the choice of pipe fittings. It is equally necessary to establish the range of pressure and discharge conditions under which the device is to operate. Taps provided with aeration or anti-splash devices cause less noise than the conventional types. The splashing of water from ballvalves is largely prevented by use of a simple silencing pipe dipping below the water surface in a cistern but its use runs up against water regulation requirements in respect of back-siphonage. Some designs of device are available that overcome this objection.

(iv) Design of the system

Excessive velocities in piping and excessive pressure drop at terminal fittings are, as shown in the foregoing paragraphs, factors likely to give rise to noise. An upper limit of 3 m/s might reasonably be taken as a guide for noise control in piping, (BS 6700), with this limit not necessarily applying to short lengths of small bore piping serving combination tap assemblies. The judicious use of pressure-reducing valves in supply pipework operating under substantial pressures is likely to be worthwhile. Some conventional UK practice has been that fittings should not be subjected to a head of more than 30 m (about 3 bar). American experience suggests that the static pressure of main supply lines in buildings of three storeys or less should be regulated so as not to exceed about 3.5 bar, and that the pressure in branch lines servicing individual apartment units should not exceed about 2.5 bar. More generally pipe layout should be as simple as possible, with a minimum of changes in flow direction; suitable fittings should be used, e.g. see figures 10.8 and 10.9.

Experimental methods

The levels of noise caused by pipework are not high and in establishing methods for their measurement it is necessary to seek for reasonable accuracy whilst providing conditions that are as realistic as possible. This requirement may conveniently be met by using a reverberant room for the measurement of the airborne noise levels. A reverberant room has smooth, hard, sound-reflecting walls constructed of brick or concrete and with the angles a little way from $90°$. It ensures an even distribution of sound inside and levels for measurement that are higher than when a room contains a good deal of absorbent material — carpets, furnishings and so on. The levels determined under these conditions may be readily converted to more realistic conditions by using the relationship between sound pressure level and reverberation time (rt). The reverberation time in a typical, small, reverberant room is several seconds whereas in a normal living room or bedroom it is 1 second or less. The change in sound pressure level when the reverberation time is changed from T_1 to T_2 is given by the term $10 \log_{10}(T_1/T_2)$. Suppose that the average *rt* in a test room is 5 seconds over the range of frequency concerned. The noise level in a room with an *rt* of 1 second would be about $10 \log_{10}(5/1)$, i.e. 7 dBA, less than in the test room. Similarly for a room with an *rt* of 0.5 seconds a value of $10 \log_{10}(5/0.5)$, i.e. 10 dBA, should be deducted from the test room results.

Thus in a study of the noise from pipework, the pipe assembly might be erected inside a reverberant test room of about the same size as a small room in a dwelling. It would be fed with water from and discharge to the outside, taking care as far as possible to avoid noise generation in the pipework outside that might be transmitted to the inside. In this way the pipe assembly for study is effectively isolated within the room where the noise measurements are to be made. This basic method was used in work already described and the data given in table 10.3 were converted from the reverberant to a more realistic condition by using the relationship stated above.

The dominant noise in water installations is that caused by taps, valves, pumps and sanitary appliances. The method described above can also be used for such components, with the device for study being installed within the reverberant room and suitable arrangements made for the supply and disposal of water. In practice, however, it is not so much the noise from such components within the sanitary or kitchen accommodation or plant room that is of concern but rather the noise transmitted into adjacent rooms. Whatever measurements are done thus need to be converted eventually to what a receiver in a nearby room would be likely to hear. This idea is behind the method adopted in the international standard for laboratory tests on the noise emission from appliances (ISO 1980). It is based strongly on German studies that were carried out to establish methods for noise control in flats and houses, described by Fuchs (1983). The ISO standard is intended to facilitate comparison of noise from various commercial products made in different countries. The following notes indicate how it was developed.

Again a small test room is used and since the noise levels to be measured are somewhat higher than with pipework the room is not required to be reverberant. Sufficient absorbent materials are required, suitably disposed to ensure a diffuse sound field, to provide a reverberation time of 1–2 seconds varying by not more than ±10 per cent in the frequency range of 125–2000 Hz. Unlike the previous method, the principle here is to fix the piping that will convey water to the device to be tested on the *outside* of the test room. Noise generated by that device is then transmitted back along the piping to the room, excites the wall on which the piping is mounted, and causes airborne noise inside the room that is measured. The test layout is shown schematically in figure 10.13, with some basic requirements.

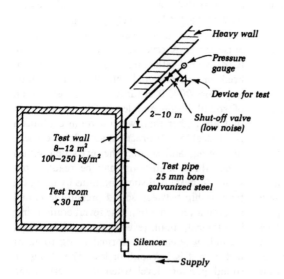

Figure 10.13   Schematic layout for noise testing based on ISO 3822/1

The test pipe is to be mounted rigidly, without insulation, on the test wall using four unequally spaced brackets. Any flow meter is to be mounted well away from this layout and a precision sound-level meter is to be used for noise measurement. The pressure and flow ranges suggested as relevant, at least for domestic situations, are respectively up to 5 bar and up to 2 l/s with water at a temperature up to 25 °C. Pressure regulators may need to be tested up to 10 bar.

The arrangement in figure 10.13 thus provides a basis for determining the airborne sound levels in a room adjacent to sanitary or kitchen accommodation. A problem in the use of such a method for comparative studies in different laboratories is to be sure that a given device will give similar results wherever it is tested. The originators made use of a special fitting for ensuring standardized conditions and as a basis for stating the noise level of the valve or other device studied. This fitting — termed the Installation Noise Standard (INS) — consists of a tubular pipe coupling about 50 mm long with an 18 mm bore inlet and fitted with an insert for noise generation. The latter incorporates a carefully made disc containing four 2.5 mm bore holes followed by a 5 mm orifice. For standardizing the whole apparatus the INS is fitted in the position provided for the device to be tested (figure 10.13) and discharges to the drain via 13 mm bore hose. Under these conditions cavitation occurs immediately downstream from the 5 mm orifice. Noise measurements in the test room provide a value for the INS, after which the device for study can be measured. Additional to this requirement, the facility is called upon to provide a background noise level in the test room at least 10 dB below that of the device under test, which implies a background level of less than 30 dBA or possibly 20 dBA for a very quiet device. The intrinsic noise level of the system, incorporating a low-noise-level valve at the point of test, must also be 10 dB below that of the device to be tested.

Against this background may be understood the way in which results are to be expressed. Two methods are suggested, one by means of a standard level difference, the other as a sound level. For the former the difference is given by:

$$D_S = L_S - L$$

where L_S is the average dBA level produced in the test room by the INS at a water pressure of 3 bar, and L is the corresponding value for the device under test. Instead of dBA values the average octave band sound pressure levels with mid-frequencies of 125–4000 Hz may be used; recent studies have shown that they must be used if the noise spectra of the device to be tested and the INS are substantially different, and this is reflected in the 1980 ISO revision. The result may also be given as a characteristic value for the appliance, L_{AP}, in a building, expressed by:

$$L_{AP} = L + (L_{SR} - L_S)$$

where L_{SR} is a reference level described below. Again the results may be in dBA or octave band sound pressure levels.

It is obvious that D_S is the smaller the noisier the device tested. As a guide to numerical values it may be noted that the intrinsic noise level for the INS is likely to be 65 dBA or more, and that the value of L_S could be expected to decrease from 62 dBA or thereabouts with the test wall weighing 100 kg/m^2 to perhaps 55 dBA with a weight of 250 kg/m^2, greater mass giving a greater measure of insulation. Thus with L equal to, say, 40 dBA, D_S could be 15 and with L equal to 45 dBA, D_S could be 10 dBA, assuming the heavier wall in each case.

The value of L_{AP} obviously increases the noisier the appliance. That L_{AP} is not the same as L may be explained by reference to the terms L_{SR} and L_S. Essentially L_{SR} is a 'standard' sound level assumed to originate in some device such as a noisy tap and heard in a room above or below and displaced sideways from the source room. L_S again is, by definition, a noise level in a room (the test room) away from the source. Thus L_{AP} is the value of L modified by the difference between these two quantities. German experience from measurements in actual buildings suggests L_{SR} might be taken as 45 dBA and the international standard indicates that each country could choose its own value for typical national construction. Assuming that $L_{SR} = 45$, as an example, and that $L_S = 55$, then L_{AP} — the characteristic value for the appliance in a building — is seen to be some 10 dBA less than L.

At first sight the foregoing method seems rather complicated, linking as it does what should be a straightforward determination of the noise level for a device with some arbitrary assumptions about the insulation afforded by differing building constructions. Further experience is needed to judge the merits of the method.

Notes on terms

Sound pressure The increase or decrease in pressure relative to atmospheric pressure caused by sound propagation. Pressures caused by speech are tiny, averaging around 1 microbar. The ear can respond to pressures varying from about 0.0002 microbar to 2000 microbars, i.e. a ratio of 10^7 to one.

Decibels The large ratio of sound pressure levels is shrunk to manageable size by use of a logarithmic scale and a sound level in decibels is given by the formula:

$$dB = 20 \log_{10}(P_1/P_2)$$

An increase in pressure level by 10 times is measured as a doubling in decibels. The usual reference level is taken as 0.0002 microbar and thus a level of 140 dB above this reference represents a sound pressure 10^7 times the reference level.

Frequency Noise generally involves a substantial number of cycles of overpressure and underpressure and the number of cycles per second defines the frequency of the sound. The ear can detect a broad range of frequencies from about 30 to 20 000 cycles per second (also called hertz, Hz). The pitch of a sound is heard by the ear on a logarithmic scale; a doubling of frequency means that the pitch goes up one octave.

Weighting The ear tends to ascribe different loudness to sounds of different frequency. Sound-level meters include weighting networks with responses to sound that vary according to frequency. The A-weighting (others are called B, C and D) discriminates against low-frequency sounds, as does the human ear, and decibels A-weighted — dBA — are the units used in this chapter.

Reverberation time This is the time required for sound in a room to decay by 60 dB. It is directly proportional to the volume of the room and inversely proportional to its total absorption.

11 Water Conservation

Much water used by industry or agriculture has little or no treatment before use; many processes hardly contaminate this water and hence it is often discharged untreated. In contrast water for domestic use normally requires treatment, and sewage is normally treated also, for reasons of public health. These treatment processes and the provision and maintenance of supply and sewerage systems add to the cost of the water. In many places, moreover, the domestic demand for water is increasing, leading to the development of new water resources and to an expansion of the capacity of treatment systems and pipe networks. Losses from mains and service reservoirs, plus losses from customers' own pipes, together with water drawn off for fire-fighting, street cleaning and the flushing of mains and services, mean that much water put into supply does not reach the customer. This 'unaccounted for water' amounted in the UK to over 20 per cent of that put into supply. Since 1995 considerable attention has been paid to the reduction of water leakage in the supply network. Current OFWAT figures, Environment Agency 2001, OFWAT 2001, indicate that there has been a steady improvement with total industry leakage figures dropping from 4980 Mlitres/day in 1995/6 to 3243 Mlitres/day in 2000/01.

With a view to economy, therefore, interest has grown both in the UK and abroad in the development and use of domestic equipment and practices that require less water. In the UK rain water collection and reuse are being studied with a view to applications such as WC flushing, clothes washing and garden use. Greywater reuse, i.e. waste water from buildings excluding WC discharges, also has possibilities. There is, naturally, a special interest in such matters in countries where water is scarce.

The present chapter reviews some of the methods tried in this field, and refers both to equipment that is well proven and to newer approaches. Where there is an energy use implication associated with water saving this is brought out, and in some examples water and energy saving may go together.

The chapter does not attempt to deal with legislative and administrative aspects of the subject. So far as cost effectiveness is concerned, any measures proposed should clearly be cheap to install and require little maintenance, and avoid additional energy use. The implications require to be assessed on a local basis and are not discussed here in detail, although a brief review is given in conclusion. Needless to say, such measures should not reduce the levels of hygiene considered necessary. There are presumably limits to the collaboration that can be expected from users in effecting water economy, depending on such factors as the availability of water, charging policy and so on.

Water closet specification and design and its impact on water conservation

During the seventeenth to nineteenth centuries many WC 'designs' were experimented with, however, significant development awaited the provision of reliable water supply and drainage connections to both dwellings and commercial buildings. By the late nineteenth century these prerequisites were in place and the WC suites available at the end of the Victorian era would be recognizable today. The invention of the trap seal and the introduction of a siphon initiated cistern flush established the design with which we are all familiar. Originally high flush volumes were considered necessary, 40 litres being used in the London Metropolitan

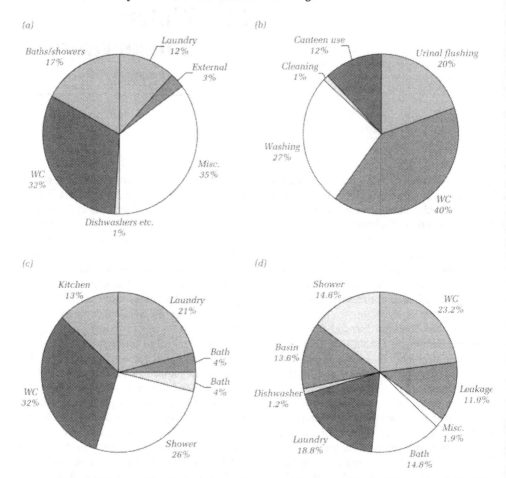

Figure 11.1 (a) UK domestic water usage data (Griggs and Shouler 1994) (b) UK commercial building 'domestic' water usage (Griggs and Shouler 1994) (c) Australian domestic water usage data (Cox 1997) (d) US domestic water use (AWWA Research Foundation 1999)

Water Board area in 1890. It was quickly realized that such profligate use of water could not be sustained by the water supply system and a reduction to 9 litres was agreed against the wishes of the UK sanitaryware manufacturers. In the USA higher volumes, at around 15 litres persisted. The reduction in UK flush volume at that time was supported by the results of the first WC drain line carry test undertaken in 1900 by the Sanitary Institute in London that showed that a 2 gallon WC flush would clear 95 per cent of the test solids (0.5 inch diameter balls) from the trap, depositing 21 per cent along the length of a 50 foot drain (Billington 1982). A very similar drainline carry test remains in use today in the American National Standards Institute code (ANSI 1990).

In order to understand the significance of WC design to the broader water conservation scene it is necessary to review water usage by the range of appliances found in both domestic and the domestic areas of commercial building use. Figure 11.1(a–d) illustrates the outcome of water use surveys in Australia, the UK and USA, adding to the picture given earlier in tables 1.1 and 1.2. It is immediately apparent that the WC is a major consumer of potable water. Also as the likely usage of WC appliances in both domestic and commercial buildings may be predicted with some accuracy, any savings brought about by good design reducing flush volume will be guaranteed and not subject to the vagaries of fashion or water supply pressure — as is the case with attempts

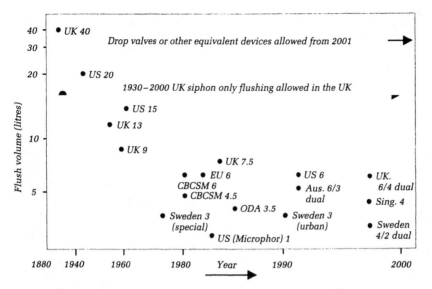

Figure 11.2   Reduction trends in WC flush volume — WC indentified by nation and flush volume, except experimental CBCSM and ODA bowls

Note: CBCSM – Confederation of British Ceramic Sanitaryware Manufacturers
ODA – Overseas Development Administration

to save water by the introduction of showers. Figure 11.2 illustrates the way in which flush volume has decreased from the Victorian high of 40 litres per flush to the UK norm from January 2001, enshrined in the Water Regulations, of 6 litres maximum with a dual flush upper volume of 4 litres.

The process that brought about the reduction in flush volume over the last 30 years has been controversial as it has necessitated the removal of barriers to trade to allow the free movement of innovative solutions to WC design problems. By around 1950 the cost and unnecessary water usage associated with high flush volumes had been fully recognized. However, little action was taken. By the 1970s UK water consumption was under scrutiny, exacerbated by the drought of 1976. The Building Research Establishment recommended a reduction to 6 litres per flush (Griggs and Shouler 1994), however, this recommendation was not put in place in the Water ByeLaws issued in 1988, although a minor adjustment down to 7.5 litres per flush from the 9 litres then current in England and Wales and 13 in Scotland was introduced. Parallel research funded initially by the Confederation of British Ceramic Sanitaryware Manufacturers (Uujamhan 1982) and later by the Overseas

Development Agency (Swaffield and Wakelin 1990, 1996) clearly showed that a 6 litre flush was achievable with a siphon operation and that volumes as low as 3 litres could operate successfully with drop valves to initiate the flush. Figure 11.3 illustrates the ODA funded WC design site tested in Botswana, Lesotho and Brazil.

European countries, and in particular the Scandinavian countries, appear to have been more aware of the importance of a reduction in flush volume much earlier. The Swedish authorities facilitated the introduction of a 3 litre WC as long ago as 1973, initially for use in weekend cottages with septic tank connections but from 1990 for use in urban environments provided drain clearance was assured due to the presence of sufficient joining flows or the provision of siphon tanks that periodically discharge volumes of stored appliance discharge into the network. The European Union nations in general accepted 6 litres as a norm from the mid-1980s. Similar research and action was underway in Australia where water shortages are part of the normal design criteria. Strong cultural links to the UK had, up to the 1980s, limited innovation in Australian WC design. However, as water conservation in the major cities became an issue

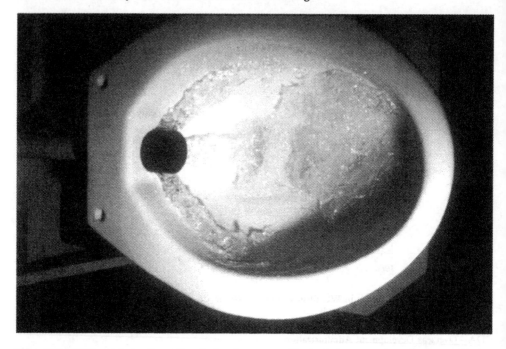

Figure 11.3   WC bowl design for a 4 litre flush volume as used in the Gaberone project (Courtesy of Heriot-Watt University)

there was a move away from British methodologies, both in terms of drainage system and WC design. The flush volume was recognized as the major target for reduction and progressive reductions were implemented as well as a move to dual flush. In parallel the normal mode of initiating the flush became the drop valve as opposed to the UK siphon. Dual flush had been tried in the UK during the 1970–86 period but had not been entirely successful. This failure was attributed to difficulties in operation of the optional reduced flush, these difficulties exacerbated by the need to restrict flushing to the siphon equipped cistern. The Australian approach was to standardize on the drop valve with separate buttons for full and half flush. This was an immediate success. The 6/3 litre dual flush regime is predicated on the assumption that most humans require to use a WC five times in a 24 hour period and that for four of these visits a reduced flush operation will be sufficient. Based on these figures Cox (1997) predicted the Australian water demand for WC flushing illustrated in figure 11.4. Similar calculations offered by Ifo, a major Swedish ceramic sanitaryware manufacturer, illustrate the reduction in annual water use by WC flushing for a family of

four as the dual flush volume is reduced, data to accompany the company's introduction of the first 4/2 litre dual flush WC (figure 11.5).

In the United States radical changes in WC flush volume were introduced as part of the 1992 Energy Act that passed through Congress with the flush volume issue almost unnoticed. Reductions to 6 litre flush volume, therefore, became law and provided both manufacturers and users with a major change to adapt to. The industry responded with a wide range of innovative solutions and despite current attempts to repeal the provisions of the act the efficiency of the designs available has improved immeasurably since the early 1990s. Among the most innovative solutions was the introduction of the pressure assisted cistern flush that allows WC operation down to 4 litres while retaining the large water area culturally expected by the national user. Figure 11.6 illustrates this device along with the more traditional drop valve and siphon. The pressurized tank mounted in the ceramic cistern allows the water supply pressure to compress trapped air above the water volume to the line pressure. When operated the flush thus has the line pressure as a driving force which results in

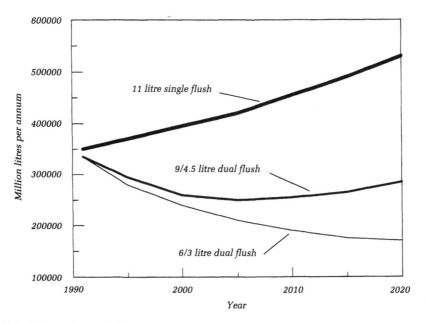

Figure 11.4   Estimated annual WC water consumption, Australia

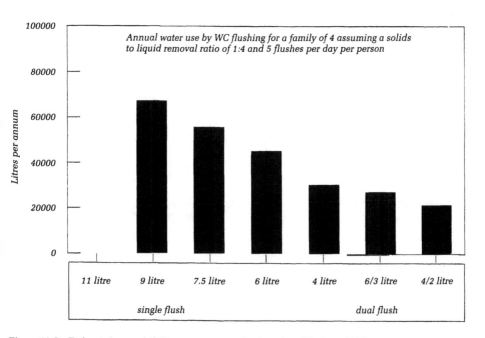

Figure 11.5   Estimated annual WC water consumption based on Ifo data, 1996

a high flowrate to the bowl and rather rapid attenuation of the discharge wave in the branch connection, see chapter 9. While ostensibly a 'new' idea, such pressurized tanks could be found in French buildings but without the camouflage of a ceramic cistern. Direct mains flushing is also allowable within US codes and again provides a strong flush to clear waste material from the bowl — although concerns do exist as to the safety of such installations in particular cases of poor maintenance — see chapter 9. In both these cases

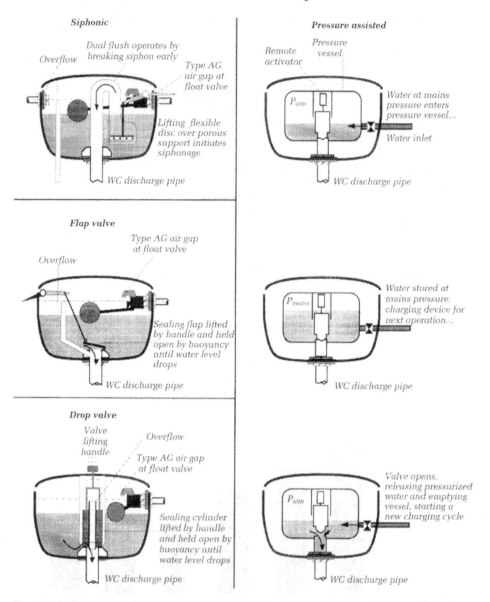

Figure 11.6 Schematic illustrations of various flushing solutions. From January 2001 internal overflows and siphonic, flap and drop valve dual flush allowed in the UK

noise may be a problem as both pressure cistern and direct mains valve can generate both acoustic and waterhammer problems. Site trials of the pressure cistern device in San Simeon, California showed that up to 40 per cent water savings could be achieved compared to the existing installations.

In the UK the replacement of the Water ByeLaws by the Water Regulations from July 1999, fully implemented from January 2001, offered an opportunity to introduce some of the water conservation innovations mentioned above. The reliance on the siphon, which since the 1930s was the only approved means of initiating a WC flush, limited the design of WCs and in particular the reduction in flush volume. In order to clear the pan with waste solids reasonably close to the front of the discharge wave requires a flow to the bowl approaching 1.6–1.8 litres/s. The operation of the siphon relies on lifting water over the weir to generate suction pressures that will drain the cistern. If the flush volume available is much below 6 litres there is insufficient water available for the column to accelerate to this flow rate before the cistern is drained. The drop valve, or pressure cistern with drop valve, does not suffer from this limitation as the maximum flowrate to the bowl is available almost immediately. Hence the monopoly of the siphon, introduced in the 1930s to ensure no water wastage, effectively blocked innovation and a reduction in water usage. The Water Regulations 1999 allow the siphon or any other equally effective discharge device to be used. In order to prevent excessive water leakage through the introduction of drop valves stringent testing is now part of the acceptance procedures for new products. This move will ensure that reduced flush volumes will continue to play the central role in reducing water consumption in UK domestic and commercial buildings.

To summarize the changes brought about by the Water Supply (Water Fittings) Regulations 1999 and the Water ByeLaws 2000, Scotland, that contribute directly to water conservation by WC modification:

(a) Both single and dual flush arrangements are now permitted, with the lesser flush of a dual flush device not to exceed two-thirds of the larger volume.

(b) A flush volume of not more than 6 litres is required from 1 January 2001, this being the maximum allowed for a single flush and the larger flush in a dual flush device.

(c) Along with these requirements there is permission to use alternative methods to deliver water to the WC bowl — a flushing cistern, a pressure flushing cistern or a pressure flushing valve, the latter not allowable in a house or where the flow of at least 1.2 l/s cannot be achieved.

(d) Flushing cisterns in the past have been required to be of the valveless siphonic type (the waste water preventer) but this requirement is removed. WC pans and cisterns have to conform to a test specification approved by the Regulator (see Preface), and cistern discharge to the pan may be by siphonic action or by the operation of a 'drop' or 'flap' valve, any of which

may be provided for a single or dual flush operation. The design of any dual flush device should ensure that the in the event of a failure the flush reverts to the full 6 litres.

(e) None of the above are retrospective. A WC suite installed before 1 July 1999 should have any cistern replacement to deliver a volume similar to its predecessor, either single or dual flush. A flushing cistern of the valveless type is required until January 2001, with a single flush not exceeding 7.5 litres, as mentioned previously.

New solutions to the problem of achieving a satisfactory WC operation at reduced volumes continue to be proposed. Currently US manufacturers are introducing pumped cistern discharge to the bowl while in the UK experimental work with a combination of air pressure and water volumes down to 1 litre is underway. Similarly both Scandinavian and Australian designs are under development that will further reduce the current dual flush volumes.

Reduced flush volume effects on drainage system design

Chapter 9 introduced the unsteady nature of drainage branch flows and the likely deposition of solids in transport. Clearly as flush volumes fall there will be an enhanced probability of deposition unless design accommodates these changes in specification. Reducing drain diameter in line with flush volume, particularly for isolated appliances or within a group of appliances discharging to a vertical stack has advantages.

However, while the effects of reduced flush volume on the self-cleansing nature of drainage systems has been rightly questioned, evidence does exist that it is possible to introduce low flush WCs, and other water saving appliances, where there is sufficient other flow in the network. Figure 11.7 illustrates a long-term investigation, 1985–1990, of low income housing water conservation in Gaberone, the capital of Botswana, where approximately 200 low flush volume WCs, set to 4 litres, were installed and compared to existing 10 litre appliances, (Swaffield and Galowin 1992). The drainage network in figure 11.7 was laid to UK

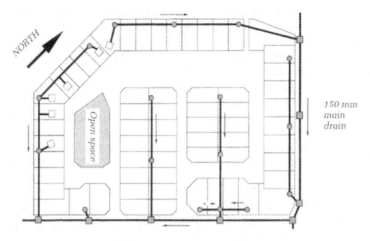

Layout of 63 low income Botswana Housing Corporation house
Plots, Gaborone West, fitted with 4 litre flush WCs. All branch
drains 100 mm diameter at 1/90 slope.
Note: pre-installation consumption similar for both groups.

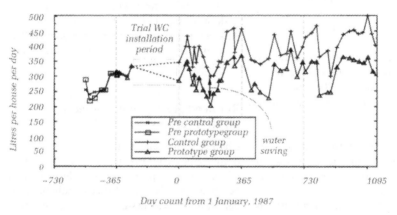

Figure 11.7   Water consumption of prototype 4 l flush installation compared with 10 l appliances over 3 years
in Gaberone project

standards and the drainage system was monitored for a 3 year period, the average water savings being approximately 25 per cent.

Similar testing supported the Australian decision to reduce flush volume from 11 to 6 litres and introduce the 6/3 litre dual flush option. In 1986 Brisbane City Council (Jeppesen 1988) initiated a series of site trials involving the monitoring of 'live' waste within 40 m of transparent 100 mm diameter drain laid at grades of 1/40 and 1/60 — the Australian norm for WC branches (AS/NZS 1996). These tests highlighted the fact that while the larger 11 litre flush would clear the drain on every operation, the lower 6/3 litre dual flush operation also cleared the drain despite the establishment of regular depositions then moved on by the following flush — a satisfactory mechanism as discussed in chapter 9. The study demonstrated the acceptability of the 6/3 litre solution which would lead to water saving of up to 60 per cent on WC operation. Follow-up testing at 223 sites in three towns in Western Australia together with trials in Melbourne and Brisbane confirmed the 6/3 solution which became standard in Australia from 1993 (Cummings et al. 2001).

In addition to the possibility of other appliance flows contributing to the transport of solids from low flush WCs in standard sized drains it must be remembered that matching the drain diameter to the WC flush is another design option. It has been shown, chapter 9, that reducing the drain diameter as flow volume falls is one way of maintaining flow depth and velocity and it is the relationship between these parameters and the solid in transport that determines drain self-cleansing. There is certainly a case for WC discharge pipe reduction to 75 mm diameter, as already demonstrated in Sweden, where the initial installations of 3 litre flush volume WCs called for 75 mm diameter drains set at a slope of 1/50.

In cases where low flush WC operation is considered to be detrimental to the drainage system, due to the increased likelihood of solid deposition, it is possible to include local flow increasers, either tipping tanks or siphons. The latter system is currently utilized in Sweden in conjunction with 3 litre WCs. Tipping tanks have a long history as a means of providing drain self-cleansing; in the UK this may be traced to the mid-19th century, as well as applications in irrigation and animal house cleansing. It will be appreciated that the tipping tank wave, while initially more severe, attenuates more rapidly than the siphon tank discharge which may therefore be more efficient in providing solid transport in long horizontal drains.

Hydraulic dependence of WC design

In the design of low flush volume WCs it is important to understand that the problem of efficient appliance operation is one of hydraulics; unfortunately the design of WCs has often been driven by aesthetic rather than hydraulic criteria. The design of the low flush volume WCs utilized in the ODA-funded Botswana study mentioned (Swaffield and Wakelin 1990) was based on previous ceramic appliance manufacturer funded research aimed at the development of a 6 litre close coupled WC in response to the UK drought conditions in 1976. This work, for the first time, identified the critical bowl design parameters, namely trap seal water volume, trap seal depth and discharge passage width, which, when related to flush volume, determined satisfactory performance for both solid and fluid waste removal. The design was also based on

quantifiable performance measurements, based on multiple ball discharge efficiency and residual fluid contamination measurements utilizing potassium permanganate as a trap water contaminant.

Water reuse and alternative WC solutions

Water conservation inevitably leads to alternative solutions to enhance or replace WCs. Table 11.1 summarizes some of these and many other innovations are in progress but not yet at a publishable stage. Vacuum systems have been available for several decades but were limited by the cost of maintenance and the reduced range of appliances, however, alternative vacuum systems utilizing collection tanks adjacent to a building have the advantage of allowing the full range of standard appliances to be used in the building while offering the advantages of vacuum transport across difficult sites. Similarly the use of combinations of pressurized air and water may be used to reduce actual flush volume, although care will need to be taken to ensure that other code requirements such as trap seal retention are fully considered. Reuse of rainwater and grey water is an attractive proposition as discussed later. The UK Water Regulations 1999 allow recycling but stress the importance of segregation of the possible rainwater, grey water and top-up potable water supplies that might be involved. These solutions will require careful consideration of backflow prevention and stringent application of water supply pipe colour coding. Initial trials by the Environment Agency has identified that considerable maintenance is required to ensure that the water storage is acceptable to the dwelling inhabitants. It would appear that such systems are more appropriate to either commercial buildings or communal apartment complexes where a facilities management team would be available to handle these issues.

All these considerations suggest that methods for achieving water economy, especially radical changes, should be thoroughly assessed both from a drainage and a water supply standpoint, including the overall economic case, any energy saving likely, hygiene considerations, and the likely user response, before they are introduced. This is true not only for

Table 11.1 Examples of closet installations

Description	Operation and maintenance	Water and energy use	Disposal of waste
A Water closet:		No electrical energy required for operation	Sewage works or septic tank
(i) Conventional 9 litre flush or more	—		
(ii) Conventional pan with dual flush, 6/3, 4/2 or less	Control requires careful design	Reduced flush volume	
(iii) Combined air/water operation	Trap seal retention	Ultra low flush volume	
(iv) Pressurized cistern	Pressure must be sufficient	Low, with high flow to bowl	
(v) Use of drop valves	Leakage prevention	Allows low flush volume	
B Composting closet:			
(i) Natural composting unit (may receive kitchen waste as well as closet contents). Substantial composting chamber required with vent	Process initiated in layer of peat, soil and organic waste, self-perpetuating thereafter; compost removal every few years	Water use — nil Electricity use — nil, except possibly for heater in cold climates	As fertilizer
(ii) Accelerated composting unit, designed with electric heating, mechanical ventilation and agitation. Smaller composting chamber, hence more easily incorporated into dwelling	As above, with compost removal once or twice a year	Water use — nil Electricity use — 100 to 200 W continuous	As fertilizer
C Closets based on thermal processes:			
(i) Incineration by gas or electric heater, initiated by seat closure. Flue for waste gases, fan assisted	Occasional removal of ash; skilled maintenance required	Water use — nil Gas and/or electricity required	Gas to atmosphere, ash for burial
(ii) Drying by air at about 70°C, enough to kill most bacteria present. Biological degradation less than with B(ii). Requires fan, heater and waste milling mechanism	Manual milling of waste at intervals, remove once or twice a year; skilled maintenance	Water use — nil Electricity use — 400 W	Bury in ground or incinerate
(ii) Freezing at −20°C in refrigerated box, when bacterial activity largely ceases. The waste is held in a plastics-coated paper bag	Removal of waste at about weekly intervals; maintenance as for refrigerators; supply of bags required	Water use — nil Electricity use — 40 W	Bury in ground or incinerate
D Chemical closets:			
Many available. The receiving liquor prevents bacteria from multiplying and contains a deodorant and colouring matter	Emptying of container at about weekly intervals, refilling with liquor. Well suited to temporary or mobile homes	Minimal water use	Empty into drains or soakaway if scale of use is small
E Packaging closet:			
Incorporation of waste into a plastics film to form sealed packages deposited into a sack	Removal of sack; renewal of plastics film and sack. Maintenance for feeder mechanism and heat sealing device	Water use — nil Minor electricity use for heat sealing	Incinerate

WC flushing, where the greatest potential for saving and the best economic case appear to lie, but also for the various other measures that are from time to time considered.

Urinals

On the basis of rounded estimates of the number of buildings of various types requiring urinal provision, numbers of male employees, and assumed scales of provision, it is possible to make a rough assessment of the quantity of water used for the flushing of urinals. This amounts to 1–2% of the water supplied by public undertakings in the UK. Measures such as the following can make a contribution to water economy in this context:

1. The traditional flushing quantity (4.5 l every 20 minutes) might be reduced for the small bowl type of urinal; indeed, some manufacturers recommend a smaller quantity, e.g. 3 l, and still smaller quantities may give an adequate flushing action. Maximum filling rates for cisterns are given in the 1999 Regulations.
2. Automatic flushing cisterns, normally left running continuously, might be turned off when urinals are not in use. Solenoid valves combined with time switches are available commercially for this application; UK by-laws required a control for new installations from 1989 and the 1999 Regulations continue this requirement. The manual control of flushing by means of a conventional flushing cistern might save small quantities of water where urinal use is limited but could increase consumption in some circumstances. The use of flushing valves, user controlled, may increase consumption, especially if they leak. The 1999 Regulations specify 1.5 l per bowl or position as the maximum delivery from such a valve.
3. Waterless urinals have been developed abroad, e.g. in Switzerland, as a stall-type urinal with surfaces treated with a water-repellent paint; the trap contains oil, some 13 mm in a 70 mm seal depth, floating on the seal surface and requiring to be replenished at intervals. This has been used a little in the UK, and bowl-type urinals of this type have also been tried with encouraging results.

Other equipment

Washing and bathing account for over a fifth of water used in buildings. Economy measures that may be considered include:

1. The provision of showers instead of baths in housing. Whereas the volume of water used in showering, assuming the shower is operating correctly, is likely to be some 20–30 l, the volume used for a bath is likely to be three or four times greater. Showers fitted with atomizing jets are likely to use less water than the conventional shower referred to above. Modern power showers, however, may not save water compared to a bath. Bidets may also offer some scope for economy.
2. The provision of spray taps in commercial and public premises. This practice has become established as a useful economy measure but has led to the need to clear the adjacent waste pipes more frequently of deposits.
3. Increasing water use is likely to stem from an increase in ownership of clothes and dish washing machines, especially of automatic machines. If total water use for clothes washing purposes is to be kept at around present levels the consumption per wash by automatics must be reduced substantially. Volumes for a 4 kg load have ranged from around 115 l to below 100 l, according to design.

Table 11.2 summarizes present domestic consumption, assuming a total use of 140 l/head/day and the distribution given in table 1.1, and ways in which consumption may change through economy measures and changes in the ownership and use of appliances.

Charging for water; metering

In the UK the traditional method of charging for water is based on rateable values of domestic property. Possible alternatives include a licence fee, charges based on household size or property type, and metering. The last gives people the opportunity to control their expenditure on water. It is contentious; for instance, whilst single-person households would be likely to pay less than on the current basis, large households would probably pay

Table 11.2 Domestic water use and possible changes

Activity	Present use (l/head/day)	Possible changes
WC flushing	43	Lower flushing volumes and radical design changes
Personal washing	36	Saving by conventional showers instead of baths Saving by atomizer showers More thermostatic control
Drinking, cooking and other	21	Small
Washing up	14	Increase from increased ownership of automatic dishwashers
Clothes washing	17	Increase from increased ownership of automatic washing machines
Other	9	Increase from increased use of automatic car washing; and garden use
Total	140	

more. And if metering forced a reduced use of water because of higher bills, that might affect health and hygiene. For a thorough discussion of the options in paying for water see Thackray (1992) and Foxon *et al.* (2000).

Whether the introduction of domestic meters would save water overall has been a subject for discussion over many years. Early work in the UK suggested that a saving in water use of 10–15 per cent was possible. Investigations at Malvern and Mansfield (Thackray *et al.* 1978) suggested, however, that metering may not effect a permanent reduction, and there was similar experience in trials in Stockholm. The cost of water to the consumer may not be sufficient to prompt action after an initial response.

In view of the proposed abolition of the rateable value system for charging, the water industry and Department of the Environment undertook extensive metering trials in the UK between about 1988 and 1992. The broad objective of the trials was to obtain information on how best to implement metering and to determine the effect on the consumption of water. One large-scale trial of about 50 000 properties (on the Isle of Wight) together with 11 each of about 1000 properties were put in hand (Gadbury and Hall 1989, Smith and Rogers 1990).

Early results (DOE/WO 1992) gave an estimated reduction in customer demand (excluding waste) on the Isle of Wight of 6 per cent in 1990–1991 while meters were being installed; in the small trials the comparable figure was 8 per cent. The report in 1993 (WSA 1993) gave an average fall in consumption of 11 per cent. Metering has encouraged customers to remedy leakage from their installations. Bills for over half of customers were lower, not least because of reduced consumption; but they were more than 25 per cent higher in nearly a quarter of the homes studied. It was acknowledged that the reduced bills were not necessarily realistic because costs were not being fully recovered on the tariff used. Other findings were that meter technology does not always permit accurate readings at low flows; some 20 per cent of meters ran slow or jammed in the trials, thus reducing bills falsely. Installing meters indoors (£165 per dwelling, 1990) was cheaper than outdoors (£200 per dwelling, 1990), but the costs of reading them were higher. The problem met in the earthing of electrical systems as a result of installing meters containing a good deal of plastics materials has been discussed by Bessey (1989). Details from early results in the Brookmans Park area trials were discussed by Russac *et al.* (1991).

It is likely that domestic metering will be increasingly used on a selective basis, especially in areas where there is a water shortage. In such cases the costs of installing, reading and maintaining meters would be accepted. Whether they effect a long-term saving in consumption remains to be seen.

Water is normally metered in industrial and commercial practice. Perhaps the most extensive investigations of water use practices and policies yet undertaken, involving over 3000 metered consumers in the Severn Trent region, was reported by Thackray and Archibald (1981). They provide a wide range of information useful for forecasting demand, for individual plants up to regional level. Amongst the data reported are those summarized in table 11.3, which are potentially useful parameters for building design. They note also an average

Table 11.3 Average water use in buildings (Thackray and Archibald 1981)

Type of building	Volume used	Standard error	Notes
Schools	75	11	Litres per head per day (including staff and pupils)
Hospitals	175	16	Litres per unit per day, 365 days per year (total units = total staff + number of beds)
Laundries	12	2	Litres per pound of washing
Retail shops	135	18	Litres per employee per day
Offices	62	9	Litres per employee per day
Hotels	763	67	Litres per employee per day, 365 days per year

Note: Based on 250 working days per year except where indicated

domestic-type use of water in manufacturing industry of 96 litres per head per day (based on 250 working days), not much less than the average domestic consumption in the Severn Trent region in 1980 of 115–120 litres per head per day.

Recycling of water

Reclamation and reuse of rainwater and greywater, i.e. wastewater from buildings excluding the WC component, has a long history in some parts of the world. Recent decades have seen much progress in this field, and many reviews and field studies have been reported. As examples of varied reuse schemes the following may be mentioned. From Namibia, direct reuse of wastewater for potable supply was reported (Haarhoff 1996); whilst the costs of treatment are greater than the usual raw water, they are less than incurred with long-distance transportation. Direct reuse for non-potable supply is commonplace in arid climates; Mills and Assano (1996) reviewed findings from many schemes in California. From Australia, Law (1996) described a first full-scale domestic

non-potable reuse application. In this case, treated and disinfected sewage was supplied via separate piping to houses in a new development, for WC use and an external tap. Many varied reuse schemes in Japan were reviewed by Aya (1994); and Maeda (1996) reported on a project in a Tokyo district, initially supplying $4000 \, \text{m}^3$/day of wastewater for WC flushing. This scheme is to be extended to $8000 \, \text{m}^3$/day in view of its success in this densely populated area. Recycling of sewage effluent via the river system for potable supply has long been practised in the UK, supported by epidemiological surveys and water quality analysis, e.g. DOE/WO (1990). Increasing water demand and climate change have encouraged an assessment and investigation of reuse possibilities at the scale of individual buildings and groups of buildings; and study of the conservation possible via redesign of sanitary appliances and practices. Investigations in this field have been pursued in the UK and abroad in three main areas: health and aesthetic concerns; the modelling of rainwater and wastewater collection and reuse systems; and on-site studies of water use and reuse.

Health oriented guidelines for reuse systems focused on faecal coliform as indicator organisms are given in a report of the Building Services Research and Information Association (Mustow et al. 1997). These were formed from a review of practice worldwide in rainwater and greywater reuse. Dixon et al. (1999a) support these and extend them, drawing on a risk analysis which distinguishes between multi-user and single family water reuse. 'Higher risk' is associated with the former, 'lower risk' with the latter. They discuss the application of a system for assigning scores for hazard from and exposure to greywater, showing how the risk score varies from source, through transit, in storage, and at use, taking bathwater as an example. Their proposed framework draws on this risk analysis and on the guidelines put forward by BSRIA and by WHO for reuse of partially retreated domestic sewage. They see the latter guidance as appropriate for design for greywater reuse in the industrial/commercial/municipal sector and in the domestic sector with multitransient occupancy, They are also seen as appropriate for reuse outdoors (garden use or exterior washing) for a single family resident occupancy. On the other hand, for a single family home to use greywater for WC flushing, they suggest that instead of a limit on the number of faecal coliform present there should be criteria controlling system design with the aim of ensuring that the following are kept to a minimum:

1. residence time of greywater in the system, to minimize bacterial growth;
2. human exposure to greywater;
3. odour.

In addition biofilm should be prevented, components should be clearly labelled, and a contract for maintenance of a greywater system after purchase would be an advantage.

Modelling supported by laboratory and/or field testing has been used in the investigation of quality changes in stored untreated greywater (Dixon et al. 1999b). They identify four main processes in operation: settlement of suspended solids, aerobic microbial growth, anaerobic release of soluble COD, and atmospheric reaeration. Storage is, of course, essential in a recycling system, and the above factors must be taken into account in sizing such storage in buildings. Their work confirms practical observations that greywater should not be stored for more than about 48 hours, but storage of around 24 hours can be acceptable.

Reuse of greywater within a day to flush WCs in a single family home is a simple application, probably not requiring treatment apart from filtering to ensure that the float valve does not become blocked and perhaps disinfecting to reduce odour. Hodges (1998) points out that investing in a designed treatment system to filter, store and disinfect greywater may make reuse more practicable. He lists the factors that require consideration including:

• the volumes of water available from the various appliances in a building;
• the contaminants present, which vary with the appliance;
• the use to be made of the greywater after treatment, whether for garden uses which have their own special needs, WC flush, or other application.

The report describes methods of storage, disinfection and filtration that are appropriate.

Modelling has also been used in studying the design of water reuse systems and their performance. The advantage of rainwater alone, i.e. without wastewater, is that it is suitable for a range of uses without extensive treatment. Drawing on available data on rainfall and on patterns of demand for water in buildings, modelling can predict performance in terms of roof area, demand, rainfall, and the volume stored. Performance is given in terms of water saving efficiency, a measure of how much potable water is saved in relation to overall demand. The approach may, of course, be extended to deal with the reuse of greywater and also with the reuse of rainwater and greywater in combination, an area that has not been widely studied.

The possibility of using rainwater for the flushing of WCs has been investigated in the UK by Fewkes and Frampton (1993). They used a computer to simulate the operation of a system consisting of a single cistern located at or preferably below ground level to collect rainwater from a roof, on the lines of a commercially available system shown in figure 11.8. Water for flushing is supplied under pressure from the accumulator at up to 4 bar. When there is insufficient rainwater, make-up water from the mains is admitted via a magnetic valve, the latter being activated by a float switch near the bottom of the cistern. To aid design, they sought to evaluate the optimum size of storage vessel to satisfy flushing demand for a range of conditions and with an eye to limiting the quantity of make-up water. Based on a Monte Carlo technique, they produced sets of performance curves, relating cistern size to roof areas with efficiency, i.e. the proportion of flushing demand satisfied by rainwater, varying from 50 to 99 per cent. It was concluded, for instance, that with a typical semi-detached house in Nottingham having a roof with a $40\,m^2$ catchment area, 70 per cent of WC flushing water could be supplied using a cistern of about 600 l capacity. The performance of this reuse system was subsequently monitored in field tests over 12 months. The data collected were used to refine the model, in particular to allow for the loss of some rainwater in the collection process. With an acceptable model achieved, dimensionless design curves were produced for the Nottingham area, enabling

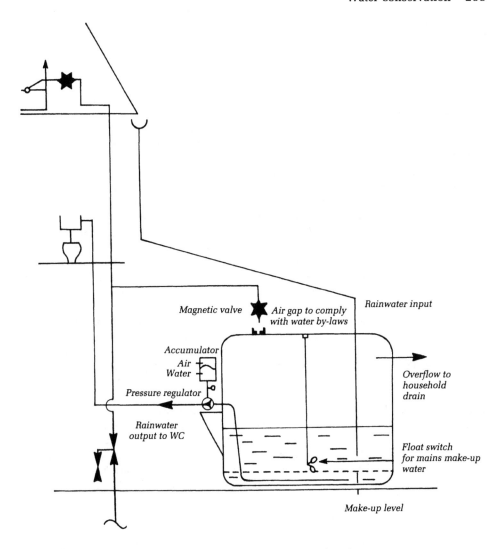

Source: Wilo-Rain Water Collector, Wilo Salmson Pumps Ltd, Ashlyn Road, Derby, UK

Figure 11.8 Example of rainwater collection system (see also Konig 2001)

storage capacity to be specified in terms of domestic roof area and demand patterns.

Dixon *et al.* (1999c) made use of a model similar to the above applied to the reuse of rainwater and greywater singly or in combination, drawing on data from the study of appliance usage reported by Butler (1993). Greywater was from bath, shower, washbasin and washing machine; the aim was to supply water for WC flushing. As background to considering such modelling, tables 1.1 and 1.2 indicate the scope in terms of quantities of water used for savings in WC flush by applying water from washing and laundry. Figures 1.2 and 1.3 are also of interest in showing the patterns of use of WCs and other appliances. For combined reuse their model results show that, whilst the occupancy of a dwelling influences the volume of storage required to achieve a given water saving efficiency, at least 90 per cent efficiency is achieved for all occupancies studied (1–5) when storage is from 100–2001. This efficiency level is also achieved when greywater only is used, the addition of rainwater thus bringing little benefit for these conditions. With smaller storage volumes, say 10–20 l, washbasin discharges supplemented by rainwater offer the chance of achieving savings in WC flush water, some 30–40 per cent efficiency with flush volumes of 9 l and 6 l. Efficiencies of 20–30 per cent may, however, be achieved with washbasin supply only, and the additional cost of incorporating rainwater into a simple reuse system using washbasin discharges may not, therefore, be justified. A small storage volume, moreover, implies that residence time for water in the system will be reduced, important as noted earlier in helping to maintain water quality. Simple schemes, such as the hand-basin toilet used in Japan, thus have their attractions particularly for domestic situations. Other interesting proprietary systems are also available, both in the UK and abroad, and may suit domestic situations or multi-occupancy. Studies such as those described above should help in their performance assessment, specification and design.

Data for such developments are also provided from full-scale studies undertaken by the Building Research Establishment in the UK. Water use was monitored extensively in a new development of 37 houses in Essex, which included 22 as controls, 12 with low-water use WCs and showers, and three with greywater reuse to flush WCs. Water savings per person of 10 per cent for WCs, baths and showers and 31 per cent for washbasins were reported in a brief outline (Building Research Establishment 2001). Studies reported by Surendran and Wheatley (1998) provide data from water monitoring undertaken in university buildings and halls of residence. Information is also given about the performance of a pilot plant for treatment to produce near potable quality from greywater. A full-scale plant has been installed in a hall of residence for study and demonstration purposes.

A review of existing reuse schemes in the UK (Diaper et al. 2001) describes experience from around 150 examples, mostly involving rainwater reuse in 75 single houses, eight groups of houses and 32 multi-occupancies. Twelve examples of combined greywater and rainwater reuse are mentioned, ten of which involve multi-occupancy. They conclude that guidelines for design and installation, for wide

dissemination, are needed; and that the reliability of small-scale systems and the design of medium-scale systems both need improvement. This should reduce excessive payback periods, the economic benefits of many schemes being poor. Education towards understanding the new approaches and the issues involved for users in accepting them is important.

Reuse of water might be accompanied by heat recovery. As an indication of the potential for heat recovery it may be noted that the average domestic consumption of hot water is in the region of 120 litres per household per day in the UK, delivered at about 55 °C. This uses about 9 GJ (gigajoules) per annum at the tap. In comparison the annual, average, primary energy consumption is in the region of 20 to 45 GJ per household for a reasonably well-insulated water heating system; this is about one-fifth of that for a reasonably well-insulated dwelling as a whole including space heating and other requirements. It might be assumed that perhaps one-half of the heat delivered at the tap (9 GJ) finds its way to the drains, and it is this fraction that might be recovered in a system to combine water and energy economy.

Heat recovery from stored waste water may be tackled either by conventional heat exchange apparatus or by a heat pump,* and with both there are various options. Some of these have been explored, not only for dwellings but also for public buildings of various kinds. Figure 11.9 illustrates the principle of a possible arrangement using a heat pump. The main components of the heat pump — the

---

* A heat pump is a device that can extract heat from its surroundings at one temperature (e.g. waste water at 25 °C) and pump this to a higher temperature (e.g. to produce water supply at 55 °C). Heat pumps work on the physical principle of evaporation and condensation as with most domestic refrigerators. With reference to figure 11.9, warm wastewater circulates round the evaporator containing the refrigerant, a liquid with low boiling point. The latter absorbs heat from the waste water and evaporates. The vapour flows to the compressor, has its pressure markedly increased, and in this state gives up its heat to the water in the cylinder via the condenser. The vapour has now become a liquid and passes back to the evaporator via an expansion valve which lowers its pressure. The amount of heat made available in this way can be two or three times that required to drive the compressor.

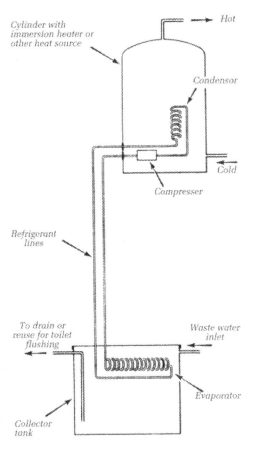

Figure 11.9 Diagram to illustrate the principle of heat recovery from wastewater using a heat pump

evaporator coil and the compressor and condenser coil — are conveniently separated as shown. The system is controlled by two thermostats, one in the cylinder and one in the tank. The control in the tank is set to prevent freezing of the water. A conventional heat source provides a back-up.

Autonomous house installations

In principle dwellings might be made fully autonomous in water, with little or no flush water input, by using some form of total recycling as in spacecraft. This is at present a very expensive option — it has been suggested that the cost of autonomous water supply and sewage disposal is about double that for central services at low population densities (three to four persons/hectare) and almost three times that for average urban densities. The

following notes indicate some of the radical designs that have been proposed, some aspects of which have been used in schemes for dwellings in areas remote from conventional water supplies. As indicated at the beginning of the chapter, water economy measures may affect or require some change in the habits or life style of occupants of buildings and, needless to say, that is likely to be true of autonomous systems. Regular supervision is likely to be needed if an autonomous system is to maintain purity of drinking water, and continue operating efficiently as a whole, whilst appropriate energy and chemicals will be needed in reliable supply.

Perhaps the simplest and cheapest autonomous system that has operated successfully, albeit for a short time with people actually living with the system, is that demonstrated by A Ortega and others in the Ecol house at McGill University, Montreal. Collected rainwater is stored and used for washing hands, showering and hair-washing. The washbasin is situated on top of the WC cistern so that the wash water can easily be used for WC flushing. It then mixes with the sewage and is fermented aerobically, aeration of the tank being achieved with a small aquarium pump. The water is then pumped by hand back up to the cistern and used for WC flushing again. Used shower water is collected and stored before being pumped up to a solar still where the distillate is collected and used for cooking and drinking. A member of the Ecol group and his wife lived in this house for 6 weeks, suggesting that with only slightly increased daily tasks, an ecological house makes a practical as well as an environmentally beneficial place to live.

The subject has been studied in the UK by, inter alia, Robert and Brenda Vale, their publication Vale (2000) bringing together the fruits of many years of experience. Sue Roaf (2001) has also provided a wide ranging review, with design guidance based on practical experience and case studies in various parts of the world. It may be said that to achieve autonomy or very substantial water economy, a good deal of equipment is needed, only some of which duplicates the normal hot and cold water system for a house with mains input. Installations require regular management and supervision, and steady supply of any necessary energy and chemicals. In a comprehensive account (Vale 1996), the authors describe the design, operation and costs-in-use of an autonomous house built in 1993. They find it satisfactory in terms of service provision and comfort but placing new responsibility on the occupants as managers of a resource.

In conclusion — the latter consideration, occupants required to be managers of a resource, is important for any new approaches to water conservation and autonomy more generally. A system may be too complex and demanding for a single family home. New approaches may be more appropriate in situations where a maintenance team is employed, such as in large apartment complexes or commercial developments.

12   Soil and Waste Drainage Underground

The introduction of water-carriage systems for soil and waste early in the nineteenth century involved installations both inside buildings and underground. Chapter 4 has given a brief historical introduction and figure 4.1 illustrates conditions that pertained even towards the end of the century. Pipelines needed underground are shown leaking due to poorly made joints or as a result of fractures, and with foul water liable to enter buildings and contaminate water supplies. The drains from buildings led into sewers which, especially in the first half of the century, were also of poor quality. Furthermore they were generally planned to take both sewage and rainwater — on what was known as the combined system — and in later years were provided with storm overflows to permit spillage to adjoining water courses on the occasion of storms. Pollution due to such overflows and also to the general leakiness of the sewers occurred.

From the latter part of the nineteenth century, and with the passing of such measures as the Public Health Act of 1875, there was a gradual improvement in the design and construction of drains and sewers. Brick sewers of better quality were constructed, and clay and cast iron pipe systems of better quality and also concrete, asbestos cement and pitch fibre pipes were introduced. Gradually more came to be known about the structural characteristics and performance of pipelines underground, and this together with improved quality control in pipe manufacture and better installation techniques led to significant improvements in underground installations. More recently, flexible joints have been introduced with various types of pipe, both in older and newer materials, to overcome the failures often associated with the fracture of rigid joints as a result of their movement in the ground.

In the 20th century, also, there has been a move away from combined systems of drainage and sewerage. With the construction of terraced houses in the early years of the century, for instance, there was the opportunity to introduce partially separate systems; the domestic sewage and rainwater from pipes at the back of the terrace discharged into one underground pipe, the rainwater including road drainage discharged into another. In later years sewage and rainwater were completely separated in some schemes, although at generally greater cost, and it is a common policy nowadays for new foul and rainwater installations to be separate. Many combined systems, of course, remain in use, most of which are old and will require renovation or replacement in the coming years.

A great deal has been written about the design, construction and maintenance of public sewers and it is not the intention to repeat this information here. The present chapter instead deals with soil and waste drainage underground between buildings and the public sewerage system. It thus considers the carriage of discharges from soil and waste pipe installations, such as those illustrated in chapter 4, from buildings into sewers. The prime concern here is with pipelines not more than 150 mm in diameter, and almost certainly under 300 mm diameter. The sewerage system, on the other hand, uses pipes ranging from 300 mm diameter to 2000 mm or more. In terms of installed length, the smaller pipes constitute by far the greater proportion of the underground system. The principles involved in designing and maintaining the installations required are considered and information is drawn from the chapters dealing with hydraulic loading (chapter 1) and fluid mechanics (chapters 8 and 9). Other publications describe practice in greater detail and some are listed in the Bibliography and Appendix 1.

Résumé of requirements

A comprehensive list of performance requirements for soil and waste pipe installations was given in chapter 4 and most of these apply to drainage underground. Thus the pipework must be capable of carrying the required flows with a minimum of blockages, be leaktight, durable, accessible for maintenance and capable of being tested. The system should be properly ventilated: it should be designed to avoid the escape of foul air near buildings except through appropriately sited vents, normally provided by the open tops of drainage stacks. It is perhaps hardly necessary to mention that drainage of this kind is almost invariably by gravity; sewage pumping or vacuum systems may be used mainly for specialized applications, and carry additional requirements relating to operation and maintenance. So far as the general requirements are concerned the following points are relevant:

Leaktightness — is important to contain the discharge and avoid polluting the surroundings, and to exclude groundwater since the latter may carry earth into the drains; leaks may also encourage the penetration of roots into drains.

Durability — involves resistance to breakage and crushing of pipes due to ground or moisture movement or possibly thermal effects, and also to erosion and chemical attack from inside (or outside). Joints are critical as regards these two general requirements, and some flexibility in joints is an important asset.

Accessibility — is necessary for inspection, testing and breaking up and clearing blockages. It should, in general, be possible to perform these functions without entering a building. Commonly used equipment for these purposes includes pipe fittings and chambers for access. The terms in common use are:

rodding eye — a pipe fitting, e.g. figure 12.5(a), accessible from ground level, having a sealed removable cover, to permit rodding along a drain in one direction, normally

downstream; usually constructed from the same piping system as the drain *access fitting* — as above but with rodding possible in more than one direction; may be incorporated as a sealed fitting within a chamber, e.g. figure 12.5(b) and may permit some debris to be removed

manhole — a chamber with a removable cover and large enough for a person to get into and be able to work reasonably freely, with breathing apparatus if necessary

inspection chamber — shallower than a manhole, with removable cover, to enable a person on the surface to inspect, test, clear blockages and remove debris.

General design aspects

Taking these general requirements as a starting point, the aim of the designer should be to limit costs by minimizing the amount of excavation needed, the number of manholes, the total length of drain and the number of connections to the main sewer. The significance of excavation for overall costs may be appreciated by the example in table 12.1 relating to a 150 mm clay sewer pipe laid on a granular bed at a depth of 3 m to invert under a light road.

It can be shown also that, for typical 150 mm diameter drains, the amount of earth removed in comparison with pipe volume is likely to be roughly in accordance with table 12.2.

Table 12.1 Example of distribution of costs in laying a sewer (Boden 1977)

Operation	Percentage of total cost
Provide pipe	7
Bed and lay pipe	17
Excavate and backfill	41
Reinstate carriageway	35

Table 12.2 Proportion of earth volume removed to pipe volume for 150 mm drain

Depth to invert (m)	Earth volume removed / Pipe volume
1.3	55
2.0	80
4.0	150
6.0	220

Minimizing excavation should thus be a primary consideration of the designer and we turn later to a research finding important in this context — that the gradient of a drain (which determines the amount of excavation needed) is not particularly significant as regards the risk of blockage. Flat gradients do not necessarily imply a high risk of blockage.

Some of the variants possible as regards number of manholes, total length of drain and numbers of connections to the main sewer are illustrated for domestic installations in figure 12.1. It is assumed that the sanitary appliances in each dwelling are grouped to discharge via one stack into the underground drainage system. The captions in the figure comment on significant features of each layout. It has to be borne in mind that the main sewer is likely to be several metres below ground level, but the 'private sewer' much shallower, with a minimum cover of about 0.75 m under gardens and 1.25 m under roads and paths. Layout 1 would normally be the most expensive and layouts 3 and 4 the least expensive to install.

Figure 12.1 brings out some general features as regards access. The basic principles are that every length of drain should be accessible for maintenance, and access points should be so situated as to permit rodding in either direction from the outside of the building; there should be access at changes of direction (when sharp and at about 90°), gradient or pipe diameter and at the head of each length of drain. Access may not be necessary at junctions and more gradual changes of direction where rodding of that location is feasible from elsewhere. Layouts 1–3 illustrate some conventional solutions. Layout 4 incorporates reduced access in a form which some field experience has shown to be satisfactory. One option here may involve access from within the building, either via a rodding eye on the stack or possibly via the WC pan. The rodding eye should, however, be located if possible to permit access via, for instance, a public access or dustbin area, rather than from within a dwelling.

Traditional installations were also fitted with traps known as interceptor or disconnecting traps, located at the outlet of the last manhole

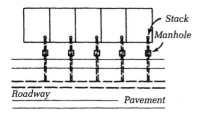

1. Traditional layout – Individual connections to main sewer with one manhole for each dwelling allowing rodding from within the curtilage. Interceptor traps commonly provided.

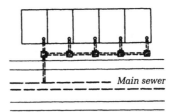

2. Layout with private sewer – single connection to main sewer with one manhole for each dwelling. Individual drains roddable from within the curtilage; rodding of private sewer may involve access to other properties

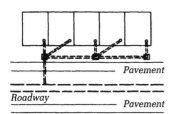

3. Layout with private sewer – number of manholes reduced, but each dwelling roddable from a manhole. Rodding of private sewer may involve access to other properties

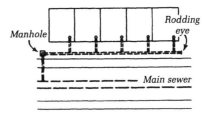

4. Layout with private sewer – one manhole at change of direction with rodding eye at head of private sewer. Access to private sewer is outside the curtilage of any property. Access for rodding also provided in the stacks

Figure 12.1   Diagrams to illustrate possible drainage layouts

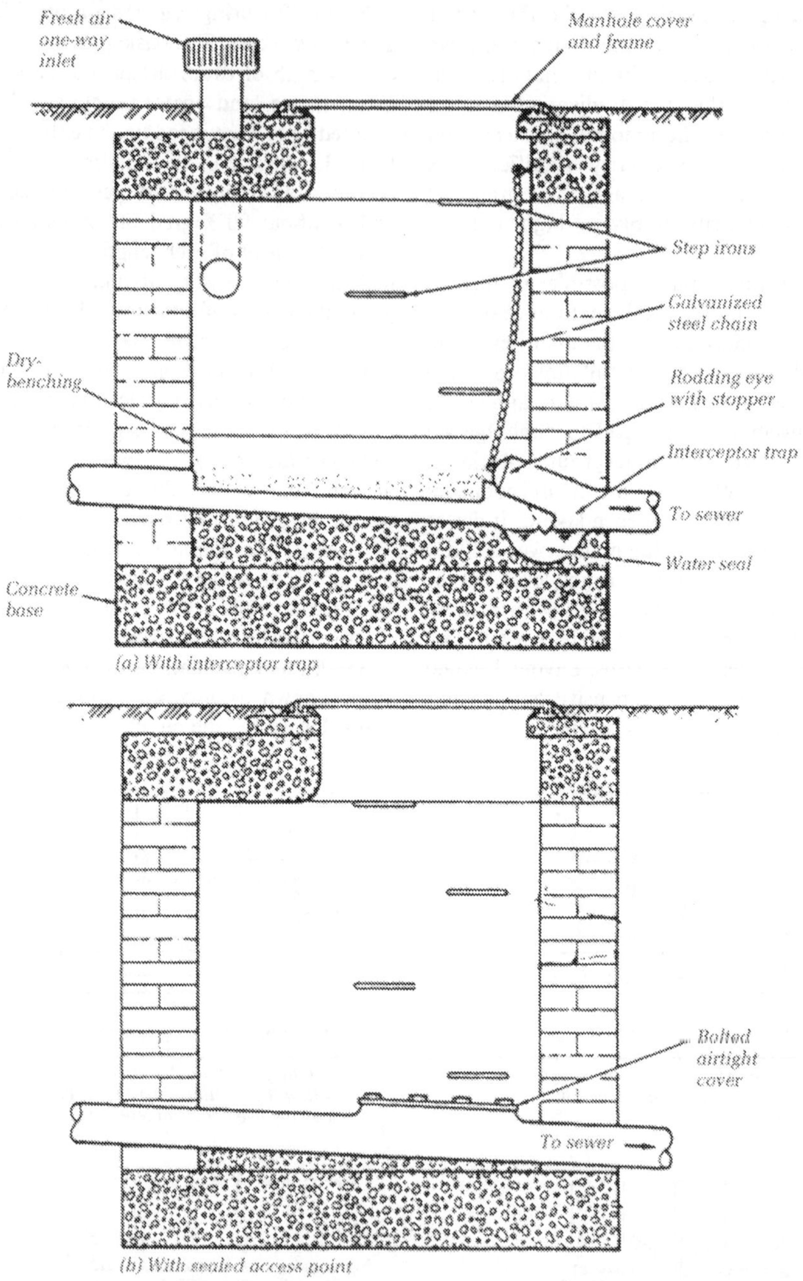

Figure 12.2   Diagrams of manholes (Building Research Establishment; Crown Copyright)

before the sewer, as in figure 12.2 (see also chapter 4). They were originally intended to exclude sewer gases, a factor now given much less weight, and they were found to help prevent rats entering the drains from sewers where they lived. The traps tended to block up readily, however, and required regular maintenance, with the disadvantages seen as outweighing the advantages (IPHE 1954). A recent survey (Griggs *et al.* 1993) shows, nevertheless, that they remain in widespread use; they are recommended in Building Regulations

(ADHI, 2002) in inspection chambers if rats are a problem. Alternatively a sealed arrangement, also illustrated in figure 12.2, may be used, eliminating the dry manhole space which rats find useful for turning around and nesting and as a refuge from water discharges; and they have no route out via cracks and gaps in brickwork. A sealed access unit may also be combined with an interceptor trap. Originally made in cast iron, sealed systems can now be obtained in a range of materials including plastics, clayware, stainless steel and glass.

Practice as regards gradients and hydraulic loading varies widely. On the whole it is unusual for more than 20 dwellings to be connected to a 100 mm drain, whilst this size of drain is unlikely to be laid flatter than 1 in 80. These limits have been associated with uncertainty over hydraulic conditions and a desire to limit inconvenience in the event of a blockage. It will be seen later, however, that extremes of 1 in 1200 and 1 in 7 occur with gradients in practice. Similar variations may be expected for other sizes of drains and the difficulties in generalizing are obvious. The following paragraphs are intended to provide a basis from which some broad conclusions may be drawn.

Flow in drains

Whereas sewers may be regarded as, on the whole, flowing fairly continuously, flow in the drains between buildings and sewers tends to be of a different character. Such drains serving large buildings may, of course, contain flow for a substantial proportion of the time and chapter 1 gives a method of assessing a peak flow. In many cases, however, the flow is highly intermittent with the drains empty for a substantial part of the time. Even when flowing they are likely to be carrying water at only a shallow depth. This may be readily seen from the example in figure 12.1 involving five dwellings connected to a drain (private sewer) leading to the main sewer. Table 4.2 suggests that the peak flow is likely to consist of the discharge of one WC, one basin and one sink only, perhaps 2 to 3 l/s, which occupies little of the cross-section of a 100 mm drain.

Under these conditions the depth of flow is greatest close to the point of connection of the stack to the underground drain. The flow then

Table 12.3 Numbers of 91 flushes needed to move 25 mm cubes over 12 m of well-made 100 mm clayware drain

Material	Gradient			
	1/40	1/80	1/120	Flat
Cube 0.9 SG	1	1	2	3
Cube 1.1 SG	1	1	2	5
Cube 1.6 SG	1	2	3	6
Newspaper	1	2	2	5

proceeds as a wave which gradually dies away along the drain, the decay of depth and velocity depending on such factors as the diameter, slope and roughness of the pipe. Solids sometimes are carried within such waves and, as these decay, tend to be deposited in the pipe, being moved further by the next flow of water. Solids tend to be carried a shorter distance initially if they do not emerge from the pan early in the flush wave. In the upper reaches of drains, therefore, the intermittent formation of stoppages — albeit temporary — is a characteristic feature of discharge. This is indicated also by the results of laboratory observations given in table 12.3. The open drain in these tests received water from a 91 washdown WC connected to it by a short vertical pipe and large-radius bend. More than one flush was in some cases required to carry a discharged solid the full 12 m length of the test section, although only one flush was required with well-made pitch fibre pipes at all gradients (Lillywhite and Webster 1979).

More flushes were needed to carry the solid the full length of the drain when joints were badly made or when a backfall was present. It was concluded that blockages were more likely to occur from badly made joints than from flat gradients.

These observations were independently corroborated by the work of Bokor (1982, chapter 9), who observed faecal and other waste transport in hospital horizontal drainage branches and by Cox (1997, chapter 11), who observed similar transport mechanisms during the Australian study leading to the introduction of the 6/3 litre dual flush standard for WCs. It is likely that repeated deposition and subsequent movement following the arrival of later flushes or other appliance discharges is the norm for drain operation.

While the severity of an obstruction or slope error might be expected to determine the effect of a particular defect, the location of that defect relative to the discharge entry point is also a determinant of its overall effect. As the discharge attenuates during its passage along the drain the relationship between the defect and the flow depth and velocity passing over or through it will depend upon the defect's position. Similarly the local drain slope will also affect the wave attenuation and the disruption caused by the defect. These issues are discussed in more detail in chapter 9 as part of the treatment of defective drains and their simulation utilizing the method of characteristics.

Observations of drainage in practice

Two studies (IPHE 1954 and Lillywhite and Webster 1979) have been reported of drainage in practice, and the following paragraphs draw on both of these in describing practical conditions. Both studies sought the cooperation of local authorities in obtaining the necessary information, and evidence from tenants was also collected in the more recent work. The latter also involved measurements *in situ* of the gradients of drains as installed, and a survey of conditions within some drains by means of a television camera and videotape recording. The systematic measurement and observation provide a sound basis of data for conclusions as regards design and operation. A further study (Griggs *et al.* 1993) dealt with rats in drainage.

Table 12.4 summarizes the results of observations on nearly 200 domestic clayware drains of 100 mm diameter, reporting the proportions involved in random blockages against the measured gradients expressed as ranges. The total sample involved is small and too much should not be read into the results, but this

Table 12.4 Gradient of 100 mm clayware drains and incidence of blockage

Approximate number of drains studied	Gradient	Percentage of drains blocking
25	1/500 to 1/71	4
140	1/70 to 1/33	11
25	1/32 to 1/20	3

information suggests that the incidence of blockages is not seriously influenced by gradient. It is interesting to note that a 100 mm drain of less than 1 in 1200 gradient was found to be working well with no history of blockages, being flushed satisfactorily over the short distance by WC discharge. At the other end of the scale there appeared to be no evidence of undue blockages in drains in Sheffield, for example, where the terrain can lead to gradients of 1 in 7 or thereabouts. A similar general conclusion as regards gradients was reached in the earlier work.

Drains subject to repeated blockages over 2 or 3 years were studied in detail as a separate exercise. These included drains serving domestic buildings and also a few serving hospitals, sports centres and club houses, some 70 cases in all. There was nothing unusual about the gradients observed in these cases since they mostly ranged between 1 in 30 and 1 in 100, the values commonly used in practice. As to hydraulic loading, calculations using discharge unit and hydraulic data indicated that at estimated peak flows the drains studied could be expected to be running about one-quarter to one-half full, not unusual, whilst the estimated flow velocities were somewhat greater than those generally accepted as 'self-cleansing' in sewers, 600–700 mm/s. In other words, there was nothing atypical about the gradients and estimated flows of the drains studied.

Careful investigation pointed to various causes of the repeated blockages and the conclusions are summarized in table 12.5. This lists features observed in the drains studied and which appeared to be the causes of the repeated blockages. Joints and manholes were particularly likely to cause trouble for the reasons given in the table. The joints concerned were the traditional mortar joints and it may be that the poor quality found was the result of a lack of skilled labour. Further enquiries were made in areas where modern systems of pipework and fittings with O-ring seals were installed, and these did not bring to light any examples of repeated blockage.

The study of rats in drains was pursued by means of a postal survey of local authorities. From the circulation of 392 questionnaires, 38 per cent replied and, of the latter, almost one-half reported a problem with rats in drains. They were not restricted to particular areas of

Table 12.5 Summary of features of drains of 100 and 150 mm size subject to repeated blockages

Feature observed	Percentage of sample, all gradients	Percentage of sample, gradients 1/40 to 1/100	Comments
Faulty joints	30	36	All traditional mortar joints (except for one badly made O-ring joint) either with misalignment of spigot and socket or mortar protruding into the bore. Faults appeared to be due to poor workmanship, not manufacturing errors
Broken pipes	10	9	All cases observed were in clayware. Ground subsidence or excessive loading were observed in several. One breakage led to root penetration
Backfalls	11	9	In the examples with backfall, congealed fat and grease was usually observed in the drains, and sometimes hard deposits of calcite
Deposits, roots present	24	24	Deposits of grease and fat were sometimes observed in drains without backfall, where the main loading was from sinks. Calcite from hard water was also seen. Irregularities, e.g. from poor joints, were sometimes a contributing factor. The presence of urine tended to increase deposits
Manhole faults	23	19	Several cast iron multi-branch junctions with branch and through connections at the same level and unequally loaded by the flow; hence cross-flow. The resulting deposits in the lightly loaded branch were eventually discharged into the main drain, becoming lodged on the rough iron surface. Also faults in brick manholes — rough or broken benching, and badly arranged branch inlets with excessive change of direction
Other features	2	3	Blocked interceptor trap
Total	100	100	

Note: Table based on the thorough investigation of 70 cases

the country like ports or markets; such factors as increases in fast food outlets, disposal of refuse in landfill sites, and sewers perhaps being less attractive because they receive more chemicals, may have created new habitats.

The survey reinforced the need for regular maintenance of interceptor traps. Rodding eye stoppers (figure 12.2) are, however, not always put back after rodding if an interceptor blocks frequently; and the breakage of clay stoppers, detachment of securing chains, and the rotting of cork stoppers present problems. Without a stopper rats can gain easy access into a manhole where they can nest and get outside. The findings pointed to the need for improved trap design to reduce blockages; secure robust stoppers to be readily removed and replaced; and improved designs of fresh air inlet valve to be robust and reliable.

Although sealed systems have been in use in the UK for a century or more, the survey brought few to light. When reported, sealed systems usually formed less than 10 per cent of drainage and sometimes were as low as 1 per cent. The report encourages their use by concluding that a completely sealed system, without conventional manholes, may prove to be as effective as an interceptor trap — in deterring rats — without the need for expensive maintenance. The report also draws attention to the role of systematic baiting of manholes where rats are a major problem.

The flow in building drainage networks, both above and below ground, has also been investigated by a range of international investigators. Chapter 11 has made reference to an extensive series of drain line transport observations undertaken to support the move to low flush volume WCs. Similarly a longer-term UK study funded by the Engineering and Physical Sciences Research Council investigated the effect of defective drains, in terms of obstructions and slope errors, on solid transport and wave attenuation. The results of this latter study were summarized in chapter 9 based on observations of drainage flow within the extensive glass drainage network mounted in the interfloor voids of Nottingham Teaching Hospital and the below-ground drainage necessary within London Underground stations (McDougall 1995). The results of this study are presented in tabular form in chapter 9, and in general supported and confirmed the findings of Lillywhite and Webster (1979).

Implications of practical and flow studies

The foregoing paragraphs suggest that it is very important for drainage installations to be well constructed, with good workmanship and with particular care as to joints. Flat gradients do not necessarily imply a high risk of blockages in these circumstances. The use of modern jointing systems and preformed access fittings and inspection chambers is likely to reduce trouble from blockages. Flow is intermittent and the layout should be designed to ensure that the flushing of the drains by normal discharges is as effective as possible. It is known that deposits can build up quickly when stacks and drains serve sinks only, and the connection of WCs to the piping, appropriately sized, can help to flush deposits away.

Backwater regions into which materials can meander should be avoided; for example, a little-used branch of a level invert junction may accumulate deposits from the main flow stream or from a heavily used branch directly opposite. The risk is greater when such branches are at nearly 90° to the main flow and where there are edges or rough surfaces that help to hold the deposits. Care should be devoted to the form and finish of benching in manholes, and to interceptor design. Sealed systems merit much greater attention.

The importance of good construction in drainage and sewerage practice has led to extensive research on the structural aspects of pipelines installed in the ground. Much has been written on the subject — see the reviews by Young (1977, 1978) and by CIRIA (1976). The findings from this work have led to the publication of practical design guides on the loading and bedding requirements of underground pipework and to much new material in the drainage and sewerage codes of practice and standards.

Establishing sizes and gradients

The minimum size of drains is no longer determined by Building Regulations. The 2000 issue requires simply in Schedule H1 that any system...'shall be adequate.' Good practice would suggest the following minimum values: 100 mm for pipes carrying soil and waste water and 75 mm for waste water only. For many simple installations these are the actual sizes used; pipes of 150 mm, 225 mm or perhaps 300 mm may be used for the larger installations. These are all nominal sizes and the actual internal bores vary a little according to the material.

The difficulties in establishing a rigorous method of sizing are considerable. The short section on flow in drains in this chapter, amplified by the account of the transport of materials given in chapter 9, indicate that flow conditions are normally far from steady and continuous. Flow is essentially intermittent and its assessment is complicated by the presence of solids which can be deposited at intervals and disturb the conditions. Flow being intermittent, the discharges from stacks do not normally build up into steady continuous flows until well into the sewerage section of the underground system. A further complication is that the surfaces of drains are likely to become rougher in the course of time as deposits and slime accumulate, and the bore likewise may reduce.

Whilst the methods outlined in chapter 9 have great potential, the shortage of information on these matters means that a rigorous method of hydraulic design is not available. Sizes and gradients can, however, be established by combining information from the discharge unit procedure described in chapter 1 with the hydraulic information in chapter 8. The former deals with the frequency of use aspect, the latter with the fluid-carrying capacity of the pipework. A final choice of size and gradient can then be made taking account of the sizes actually available, the topography, and the practical considerations set out earlier in this chapter.

Relevant data are given in tables 1.8 and 1.10 and figure 1.7 (discharge unit procedure) and figures 12.3 and 12.4 (flow capacity). The latter are based on the Colebrook–White

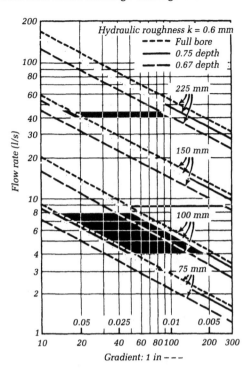

Figure 12.4 Approximate discharge capacity of drains with 0.6 mm hydraulic roughness

equation for steady flow in pipes given in chapter 8. In using this source it is necessary to assume a roughness for the pipe surface and to decide on the depth of flow for which the drains will be designed. Since air circulation for ventilation purposes is important it is usual to design for peak flow to be less than full bore. A proportional depth of 0.75 is commonly assumed and the data are given on this assumption in figures 12.3 and 12.4, with examples for 0.67 proportional depth and full-bore flow for comparison.

Smooth new pipes have a larger flow capacity than pipes that have become roughened in service. The degree of roughening that is likely to occur with drainage installations of this kind is not well established. The experimental work done, for example, by the Hydraulics Research Station (now Hydraulics Research Ltd) and Water Pollution Research Laboratory (now Water Research Centre) on the sliming and roughening of sewers shows that the process is affected by a variety of factors including the depth of flow. Figure 12.3 is for pipes having a roughness of 0.06 mm, and figure 12.4 has been calculated assuming that a

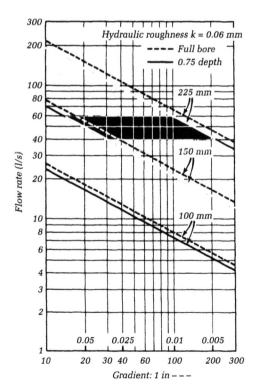

Figure 12.3 Approximate discharge capacity of drains with 0.06 mm hydraulic roughness

Table 12.6 Numbers of dwellings that may be connected to drains based on data in the text

Gradient	Pipe diameter (mm)	
	100	150
1 in 40	85	
80	35	320
150		190

roughness of 0.6 mm develops in practice. The latter is suggested as a guide for intermittently used drains in all pipe materials. Greater roughness, perhaps 1 to 2 mm, may be appropriate for drains subject to continuous flow and sometimes running full.

As an example of the application of this information consider domestic installations with drains of nominally 100 and 150 mm diameter, assumed to have a roughness of 0.6 mm, at gradients between 1 in 40 and 1 in 150. The numbers of dwellings that may be connected, based on this information and rounded to the nearest five, are given in table 12.6.

The calculations may be exemplified as follows. For a drain of 100 mm diameter at a gradient of 1 in 40 assumed to be flowing 0.75 full, the flow capacity is 9.2 l/s (figure 12.4). This is equivalent (figure 1.7) to about 1200

discharge units assuming mixed appliances including WCs. Taking each dwelling as possessing one WC, bath, basin and sink, the discharge unit value per dwelling is 14 (table 1.8). The discharge unit total of 1200 is, therefore, equivalent to 85 dwellings. The remaining figures in table 12.6 have a similar basis. As already explained, figure 12.4 on which the calculations have been based assumes that pipes will become roughened in service; the assumption of very rough pipes would give a smaller loading. Conversely, the assumption of a smooth surface without subsequent roughening would lead to substantially greater loadings. If it was possible to allow correctly for the attenuation of flow that occurs in such drains, a further increase in loading would be possible. Neglected in the foregoing is any estimation of secondary pressure losses (see chapter 8) due to bends and manholes. In view of the approximate nature of the calculations this additional refinement is hardly justified. It becomes more necessary for the design of sewers flowing more or less continuously at a substantial depth.

For the conditions specified, table 12.6 appears to be a reasonable numerical basis, possibly somewhat conservative. Even so, the loadings given are substantially greater than those used in many current designs. It was

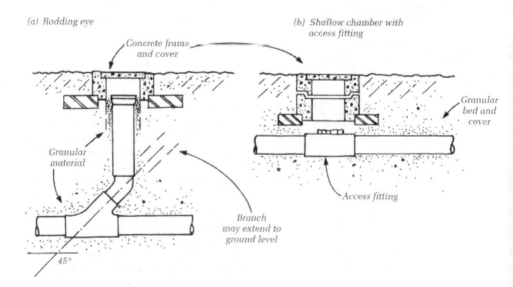

Figure 12.5   Diagrams to illustrate forms of access

noted earlier that more than 20 dwellings on a 100 mm drain is unusual. The loadings given in table 12.6 may well be beneficial in ensuring a more regular and effective flushing of the pipework. The 2002 edition of the Approved Document H1 related to the Building Regulations 2000, however, recommends 10 dwellings as maximum on a 100 mm drain.

Establishing the spacing and design of access

Whilst power-assisted drain clearance equipment is available, the clearing of drains manually by rodding is common and it is the latter that determines the spacing between access points. *BRE Digest* 292 (1984) reports tests to help establish the optimum spacing, in which measurements were made of the extent to which the force exerted by the operator was reduced by rods buckling and rubbing on hard edges and surfaces. Cane, steel and polypropylene rods applied through several forms of access formed the subject of the tests. It was concluded that the applied force was reduced by about 1 per cent per metre length of new, dry clay or PVC pipework of 100 and 150 mm diameter, with these three types of rod. A possible benefit from the smoothness of the polypropylene rods appeared to be cancelled out by their greater tendency to 'snake' as compared with the rougher, more rigid rods. Pipe diameter in this range had little effect, and results with a gradually curving pipeline were much the same as with a straight pipe, again due to the snaking effect; hence the simple conclusion stated above.

Within chambers, the changes in angular deflection of the rod appeared to be significant as did frictional losses on the sides and invert. Thus it was found that the bigger an inspection chamber or manhole the greater the reduction of force; for a chamber 1 m deep and 0.45 m long the loss was about 25 per cent, with the operator at ground level; with a 1.5 m depth and 1.2 m length the corresponding loss was 45 per cent, but it was very little when the operator stood within the chamber. Losses in rodding eyes and access fittings varied from 10 to 50 per cent depending on type and size; rodding branches at 45° to the drain run gave ready access for the rod and hence the best results.

Table 12.7 Maximum distances between access points, metres

| | To | | | | |
	Access fitting		Junction	Inspection chamber	Man hole
	(1)	(2)			
Start of external drain*	12	12	—	22	45
Rodding eye	22	22	22	45	45
Access fitting (1)	—	—	12	22	22
Access fitting (2)	—	—	22	45	45
Inspection chamber	22	45	22	45	45
Manhole	22	45	45	45	90

* At connection from stack or ground floor appliance outlet

Notes: With chambers constructed with access fittings, spacing should be as for the fittings alone. Spacings related to rodding eye assume an angle of not more than 45°, see figure 12.5. For access fitting dimensions, see table 12.8

Recommendations for spacing were made from these tests with the aim of ensuring that some 50 per cent of the applied force would remain for breaking up a blockage, a level considered sufficient for this purpose. They are summarized in table 12.7, and assume that rodding is possible from both ends of the drain. Table 12.8 gives commonly accepted dimensions for the forms of access under consideration. As further commentary on these tables, note that rodding eyes can serve their purpose over a wide range of depths to invert, and hence no limit is given. Access fittings, on the other hand, are limited to a depth of 0.6 m — about arm's length — to permit the removal of debris and the insertion of test plugs if needed.

The smaller of the two fittings listed in table 12.8 can offer the greatest resistance to rodding of any form of access, and this is reflected in table 12.7. The smaller inspection chamber in table 12.8 is also limited to a depth of 0.6 m, being large enough only to allow the arm to be inserted; the larger size permits entry of arm and shoulder and the invert may, therefore, be as deep as 1 m. This information has formed the basis for advice in BS 8301.

Table 12.8 Dimensions for forms of access

	Depth to invert (m)	Internal size Rectangular (mm)	Circular (mm)	Nominal cover size Rectangular (mm)	Circular (mm)
Rodding eye	—	—	100*	—	—
Access fitting (1)	0.6 max†	150 × 100	150	150 × 100	150
(2)	0.6 max†	225 × 100	—	225 × 100	—
Inspection chamber	0.6 max	—	190‡		min 190
	1.0 max	450 × 450	450	450 × 450	450§
Manhole	1.5 max	1200 × 750	1050	600 × 600	600
	Over 1.5	1200 × 750	1200	600 × 600	600

Notes: * Preferably same size as drain
† Except when situated in a chamber
‡ For drains up to 150 mm diameter
§ 430 for clayware and plastics chambers to ensure adequate support for cover and frame

Approved document H1 (2002) related to the Building Regulations (2000) emphasises the need for suitable means to clear blockages in drains. It expands on the kind of information given in tables 12.7 and 12.8.

13 Rainwater Drainage

Adequate means to remove rainwater is an essential requirement with most buildings. Roofs leaking, whatever the cause, and gutters overflowing, perhaps over long periods of time, can cause great damage to both structure and contents. The subject, however, is often one of the last items to be considered in a building project. This is unfortunate since advantages may accrue by considering the requirements at an early stage in planning and design. The subject should not be dealt with simply as a matter of detail as the project nears its end. Wider knowledge and application of the basic principles is a prerequisite for satisfactory and economical solutions.

Tradition has played a large part in the sizing and spacing of gutters, outlets and down pipes. There have been rules of thumb such as each $10\,m^2$ of roof requiring $650\,mm^2$ cross-sectional area of down pipe. The discharge capacity of a pipe sized on this basis will usually be excessive in relation to likely run-off. In any case, discharge may be restricted by the design of the entry at the top of the down pipe which is often critical as regards flow performance. It is important, therefore, to consider the design of roof, gutters, entries and down pipes, and even the underground drains, as a whole if the requirements of rainwater removal are to be met satisfactorily.

Research to improve understanding and provide design data has been limited. An American study (Beij 1934) of flow in gutters provided all the reliable information available for some years. In the 1950s the Building Research Station conducted tests on a full-scale mock-up, the results of which were applied in a *Digest* and later in the British Standard 6367. Rainfall data are, of course, necessary as a basis for design and the Meteorological Office contributed essential information. More recently that Office, Hydraulics Research Ltd and the Institute of Hydrology have undertaken practical and theoretical investigations reported in several publications including NERC (1975) and May (1982). The results of all this work have formed a sound basis for updating and extending codes and standards, including the development of European proposals under CEN. It is the intention here to set out some of the underlying principles and data that are fundamental to satisfactory performance and to illustrate their application to design.

While conventional open gutter and down-pipe rainwater systems still dominate the market, over the past 30 years increasing amounts of industrial and commercial roof area has been drained using siphonic roof drainage. Chapter 9 includes an introduction to the mode of operation of these systems that rely on full bore flow in the network linking gutter outlet to ground level sewer connections, together with the simulation of the unsteady flows generated by a storm. Siphonic systems will have significantly higher flow capacity than open gutter systems.

While siphonic systems offer advantages over the traditional approach they rely on the establishment of full bore flow in the network. This fully primed requirement implies that a siphonic system is only accurately designed for one storm intensity and minimum duration. Below this the system fails to prime and operates either as a free surface flow system or suffers oscillatory flow conditions as the priming is routinely broken by air ingress from the gutter outlets. Above the design storm condition overtopping of the gutters may occur if the rainwater flows are not safely re-directed.

In recent years many prestigious buildings have featured siphonic drainage, including the

Sydney Olympic Stadium and Hong Kong Airport as well as many large commercial and industrial buildings requiring large roof spans and ground cover. The design of successful siphonic rainwater drainage systems does require a higher degree of expertise than the traditional. Currently there is no UK standard for siphonic system design, however, when the design criteria are better understood there can be no doubt that this will become an attractive design choice for large building projects.

General considerations

Roofs and paved areas, being impervious to water, form small catchment areas and accumulate substantial quantities of water when the rainfall is heavy. This can cause flooding unless adequate drainage is provided and is, therefore, the problem with which design and installation are intended primarily to deal. An important factor is the time of concentration for the catchment area, i.e. the time taken for water to reach the outflow point from furthest away. With buildings, this distance is usually tens of metres, and the time of concentration is about 1 to 2 minutes. The type of rainfall of concern to the designer is that associated with summer storms since these can produce high intensities over short durations. It might be expected that melted snow should be allowed for in design but this is not necessary in the UK because the resulting run-off is less than the rates associated with summer storms.

Pitched roofs usually have eaves gutters fixed to a nominal fall, say 1 to 3 mm/m for domestic gutters, which discharge into down pipes. So-called flat roofs should have a slight fall to facilitate the movement of rainwater to a roof outlet or to a channel formed in the roof and leading to a gutter or an outlet. Valley gutters may be established between roofs, and parapet and boundary wall gutters may be necessary. The design of the outlets from the roof into down pipes is, as already mentioned, usually critical for satisfactory performance. A range of components and fittings is available for this purpose, and shape and curvature are important for effective discharge. Cast iron is a traditional material for gutters and down pipes; systems in plastics are now common, and systems are also available in asbestos cement, aluminium, galvanized mild steel and

enamelled steel, and wrought copper and zinc, each with particular features. Down pipes may discharge directly into the underground drain where the latter is for surface water only but must discharge into a trap en route to the drain if the latter also carried sewage.

The rate of flow of rainwater Q to be drained from an impermeable roof under steady conditions is given by the expression:

$$Q = rA \text{ l/s} \qquad (13.1)$$

where r is the rainfall intensity in $l/s/m^2$ and A is the effective roof area in m^2. Intensities are sometimes given in mm/h, converted to $l/s/m^2$ by dividing by 3600. Methods of determining intensities and areas for design purposes are given in the following sections.

Meteorological factors

Summer storms arise because large 'parcels' of air heated by the ground or sea become less dense than the surrounding air and rise through the atmosphere. This upflow forms a cloud and, through condensation, produces the rain which, for reasons associated with the atmospheric conditions, falls outside the base of the cloud rather than back through it. The rainfall intensities produced by a continuing process of this nature are found generally to be between 25 and 100 mm/h. An intensity of 200 mm/h for 5 minutes is reached occasionally, whilst the maximum possible 1 minute intensity in the UK is thought to be about 300 mm/h. Reports of storms in the USA and the tropics have indicated rates of 1000 mm/h for similar periods. Accurate measurement of these intensities and the statistical analysis and classification of the results are needed to produce design data.

Modern methods rely on extreme value analysis, a branch of statistics which deals with the distribution of rare events within a so-called parent population. This enables calculation of the probability that in a given year some event such as the amount of rain measured at a point over a specified time is not exceeded. With design rainfall the convention is to refer to the return period T of the intensity which, if T is not less than 5 years, can be defined approximately by the probability $(1 - 1/T)$ that the event will not be exceeded in

Figure 13.1 Return periods in years for events of 100 mm/h for 2 minutes

Table 13.1 A comparison of return periods in the north and south of the British Isles

2 minute intensity of rainfall (mm/h)	Approximate return period (years)	
	Edinburgh area	London area
75	10	0.5–1
100	50	2–3.5
150	500	10–15
225	Over 500	100–200

any given year. If, for instance, a 2 minute intensity has a 50 year return period, that value will be reached or exceeded on average once every 50 years. This does not preclude the possibility of two rare events occurring in a shorter space of time, because the above is on average.

These methods have been used by the Meteorological Office to calculate the exceedance thresholds for rainfall amounts for a wide range of conditions in the UK, the results being tabulated and mapped. One important finding from this work is the geographical variation in short, heavy intensities of rainfall, as given, for example, in figure 13.1. This is based on one of several similar maps for the UK covering the range 75–225 mm/h (BS 6367). It shows that the return periods for 100 mm/h over 2 minutes are much greater in the north than in the south of the British Isles. Table 13.1 further illustrates this for London and Edinburgh. The reason for this difference lies in the terrain and climatic differences. In the south the flat terrain and proximity to warm continental air combine to give conditions especially suitable for producing summer

storms. Conditions in the north are less suitable. The higher annual average rainfall in many parts of the north is not relevant to the subject under discussion.

Before giving some design data, it will be useful to refer to one more analytical device because design may often require an estimation of the risk of rainfall exceeding the chosen design value during the proposed lifetime of the structure or perhaps the contents. This may be done (Butler 1983) using the expression:

$$P = 1 - (1 - 1/T)^L \qquad (13.2)$$

where P is the probability of exceeding the chosen design rate (the risk factor), T is the return period in years and L is the expected lifetime of the structure in years. The expression holds provided T is not less than 5 years. Combining this with the rainfall data enables the required rate to be found.

A basis for the UK (BS 6367) has been:

Category 1   A design rate of 50 mm/h (0.014 l/s/m²) recommended for paved areas when ponding can be tolerated during and for a short period after a storm.

Category 2   A design rate of 75 mm/h (0.021 l/s/m²) recommended for sloping surfaces, and for flat surfaces where ponding is not tolerable. This is generally satisfactory for eaves gutters where overflow is not likely to occur inside and for other gutters where some risk to contents may be acceptable. It is the value in Approved Document H3, Building Regulations 2000. The risks associated with, for instance, valley gutters may sometimes point to the need for design values in categories 3 to 5 rather than 2.

Category 3   The category is based on a probability of 0.5, and with this value a return period of $T = 1.5L$ is found to satisfy the above equation for a range of structure lifetimes. Then supposing, for example, that L is considered to be 40 years, $T = 60$. Maps such as figure 13.1 give the appropriate design rate; from this example it can be seen that 100 mm/h (0.028 l/s/m^2) could be selected for some places in Scotland but a higher figure would be appropriate in most of England.

Category 4   The probability is 0.2 and correspondingly $T = 4.5L$. With $L = 40$ years, $T = 180$ and this more rigorous condition would give higher design rates.

Category 5   The probability approaches zero and a return period of 35 000 years is recommended; this is an empirical figure for the frequency encompassing all known events.

In some codes such different categories have been incorporated through a list of safety factors which may be applied according to circumstances including, for instance, valley gutters and buildings requiring exceptional protection. A minimum rainfall intensity related to national and local requirements is selected; this is multiplied by a safety factor dependent on the type of gutter and the situation (and see BSEN 12056).

Rain, of course, can be driven by the wind and, as discussed in the next section, it is necessary to allow for this in estimating the effective catchment area of a roof or paved area. Studies by the Meteorological Office have shown that the angle to the vertical of the fall of heavy rain varies somewhat between exposed and sheltered sites, with a median angle for all regions in the UK of 28°. This is close to the value recommended in BS 6367 i.e. 26° to the vertical or $\tan^{-1}\frac{1}{2}$. A different point is that wind blowing along a gutter may influence its flow capacity.

Run-off

As already mentioned, the effective catchment area of a sloping or vertical surface depends on the angle of descent of the rain. The shapes of roofs and buildings, and layout, affect local wind flow and may be important. As a general rule design should be for the wind direction that will make the rainwater flow a maximum. Figure 13.2 shows how the effective catchment areas of simple roofs depend on the angle of descent of the rain and hence on the wind. With valley gutters one side of the roof tends to be exposed, the other sheltered. Figure 13.3 illustrates a basis of assessing the catchment area for vertical surfaces which may be estimated from:

$$A = \frac{1}{2}\sqrt{(A_{v_1}^2 + A_{v_2}^2 - 2A_{v_1}A_{v_2}\cos\theta)} \qquad (13.3)$$

The walls are assumed to present the maximum vertical area to the rain. For a single wall, the effective catchment area could thus be taken as one-half of the exposed vertical area.

The air flow around buildings is often complex and the following points arise from this. Some of the rain approaching a wall is carried over or round it by the wind. Of rain reaching a wall, some may rebound, falling in a curtain of drops up to 150 mm away. Some moves upwards or sideways, collecting in reentrant angles and beneath overhangs; this can

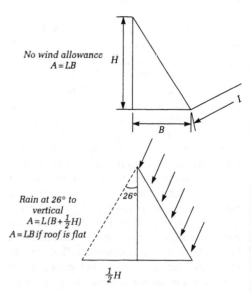

Figure 13.2   Examples of effective areas of roofs

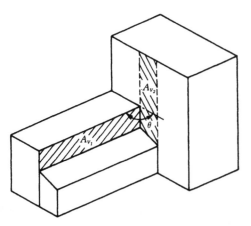

Figure 13.3 Catchment area for vertical surfaces

result in a substantial downflow in angles. Some water is shed from lintels or sills. Finally, some water may be stripped from walls by the strength of the wind. None of this is quantifiable at present and in designing wall drainage the total rain quantity approaching a wall may be used, usually ensuring a large safety margin.

Finally it should be noted that the form the jet of water takes as it leaves a roof is dependent on local detail. Figure 13.4 illustrates

this. Design to ensure a narrow jet is advantageous for efficient collection of the water.

Flow in gutters and outlets

The basic principles set out in chapter 8 are relevant but a special consideration applies with gutters: that the discharge increases in the direction of flow. The run-off from a roof into a gutter along which water is flowing has to be caused to move in the direction of that flow. This occurs through a transfer of momentum from the water already flowing along the gutter. The run-off thus exerts a retarding force on the main flow which causes the water to back up along the gutter; hence the maximum depth of flow occurs at the upstream end. This is independent of the frictional resistance of the gutter which normally causes only a small additional increase in the depth at the upstream end. Frictional resistance can become significant in gutters that are very long in relation to the depth of flow.

Drawing on the work of May (1982, 1984), consider a level rectangular gutter receiving water along its length, assuming that no frictional force occurs and that the discharge at the outlet end is not restricted in any way. Applying open-channel-flow theory it can be shown that the maximum possible discharge Q_c is given by:

$$Q_c = B\sqrt{g}(Y_u/\sqrt{3})^{3/2} \qquad (13.4)$$

where B is the breadth of the gutter, g is the acceleration due to gravity and Y_u is the depth at the upstream end. Under these conditions the depth at the downstream end is the critical depth (see chapter 8) given here by:

$$Y_c = Y_u/\sqrt{3} \qquad (13.5)$$

i.e. the downstream depth is about 60 per cent of the upstream depth.

From these results it can be concluded that the flow adjusts itself, depth decreasing towards the downstream end where the critical value occurs. The size of the outlet does not affect capacity provided it is large enough to allow free discharge. Furthermore Q_c is independent of the length of the gutter and, depending only on conditions at the ends, is not affected by a variation in the rate of inflow along the gutter.

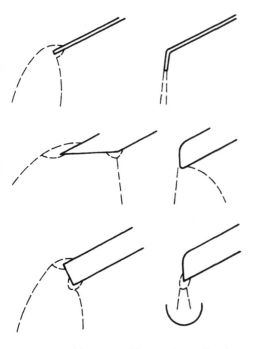

Figure 13.4 Diagrams to illustrate flow off various forms of roof edge

For half-round gutters a similar analysis can be made from which it can be shown that the maximum discharge is given by:

$$Q_c = 0.616\sqrt{gr^5} \qquad (13.6)$$

for a true half-round, and

$$Q_c = 0.438\sqrt{gr^5} \qquad (13.7)$$

for a nominal half-round, as in some components. Here r is the radius of cross-section, given by half the top width. In this case the downstream depth with free discharge is about 65 per cent of the upstream depth. Flow is just full at the upstream end and discharging freely at the downstream end. The simplifying assumptions behind these equations tend to overestimate the capacity of real gutters.

Figure 13.5 shows how gutter discharge may be expected to divide between outlets. The flow splits evenly, even if the inflow is not uniform, as in figure 13.5(1). The two lower diagrams show the effect of outlet position. Figure 13.5(2) shows an arrangement in which each section of the gutter requires a capacity of one-quarter of the total inflow. The lowest arrangement is less efficient with the gutter requiring a one-half capacity.

The free surface flow in an open gutter is another example of unsteady flow and has been included in the presentation of simulations in chapter 9, despite the fact that the rate of change of the flow conditions will be relatively slow, driven only by the changing intensity of a storm over its duration. However, the boundary conditions illustrated in figure 13.5 may be replicated, most notably the zero flowrate at the upstream termination of the gutter and some appropriate depth vs. flowrate expression at the gutter outlet. Chapter 9 illustrates the simulation techniques appropriate, while Swaffield, Escarameia and Campbell (1999) illustrates the degree of agreement possible between predicted and observed flow depth in a gutter during a rainstorm. Simulations also allow the possibility of including longitudinal as well as temporal variation of inflow during the storm duration, as may be the case as a result of a complex roof shape feeding into an edge gutter, a possibility not available through the use of existing steady state calculation techniques.

As might be expected, changes of direction reduce gutter capacity and this factor should also be considered in positioning outlets. Figure 13.6 shows how a sharp bend is likely to influence the discharge, a reasonable approximation for all normal types of gutter. An overriding factor in determining position may be building layout and design; nevertheless, the foregoing considerations should not be ignored.

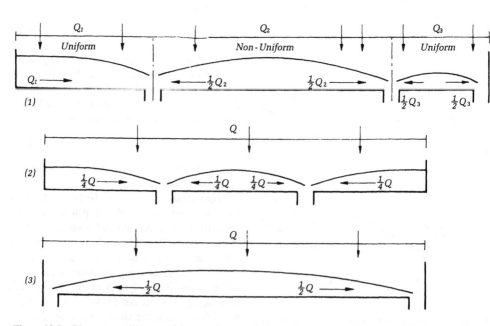

Figure 13.5 Diagrams to illustrate features of gutter flow

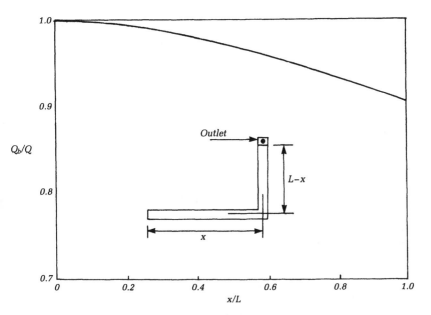

Figure 13.6   Effect of sharp bend on capacity of level eaves gutter

Free discharge, as assumed above, may not occur in practice. It depends on the capacity of the outlet which is a function of many factors. May (1982) reports extensive tests on a variety of outlets in rectangular and trapezoidal gutters and draws some general conclusions. As might be expected, flow at low depths is of a weir type, changing to orifice flow as the depth increases. With the former, and if unequal flows approach an outlet from two directions, depths are roughly the same on either side and fixed by the larger flow. Such flow in a box receiver is controlled by outlet size. With orifice flow, rectangular and circular outlets have closely similar coefficients, and depth is determined by the total discharge into the outlet. The complexity of the situation means that practical tests may often be an advantage to determine capacity. May (1997) encourages the greater use of box receivers to reduce the chance of flooding associated with the under-sizing of outlets.

Examples of design information

Data for eaves gutters are given in tables 13.2 to 13.5 to illustrate design information based on May (1982) and BS 6367. Gutters generally are assumed to discharge freely. Compared with the

Table 13.2   Flow capacities of level eaves gutters

Size (mm)	Capacity (l/s)	
	True half-round	Nominal half-round
75	0.38	0.27
100	0.78	0.55
115	1.11	0.78
125	1.37	0.96
150	2.16	1.52

Table 13.3   Reduction factors for long eaves gutters

l_g/Y_g	Reduction factor
50	1.0
100	0.93
150	0.86
200	0.80

Table 13.4   Reduction factors for corners in eaves gutters

Form of angle	Reduction factor	
	Angle below 2 m from outlet	Angle between 2 and 4 m from outlet
Sharp corner	0.80	0.90
Round corner	0.90	0.95

Table 13.5　Minimum outlet sizes for eaves gutters

Size of gutter	Outlet sharp (S) or rounded (R)	Throat diameter of outlet (mm)	
		When at gutter end	When not at end
75	S	50	50
	R	50	50
100	S	63	63
	R	50	50
115	S	63	75
	R	50	63
125	S	75	89
	R	63	75
150	S	89	100
	R	75	100

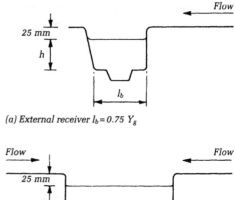

(a) External receiver $l_b = 0.75\ Y_g$

(b) Internal receiver $l_b = 1.5\ Y_g$

Figure 13.7　Diagrams to illustrate box receiver

theoretical method outlined above, and based on experiment, the ratio of downstream to upstream depth is assumed to be 5/9, i.e. 55 instead of 65 per cent, thus ensuring a safety margin. In this case the design capacity Q_{cd} is 0.73 of the theoretical capacity for true half-round gutters and 0.72 of Q_c for nominal half-round. Then equations (13.6) and (13.7) become:

$$Q_{cd} = 7.86 \times 10^{-6} W^{5/2} \qquad (13.8)$$

for true half-round and

$$Q_{cd} = 5.51 \times 10^{-6} W^{5/2} \qquad (13.9)$$

for a nominal half-round, where Q_{cd} is in l/s and W is the top width (size) of the gutter in mm. These have been used to calculate the capacities in table 13.2, the first two columns of which also appear in ADH3, 2002. These figures do not include an allowance for freeboard, but this is unnecessary provided any overflow can fall clear of the building. Anderson (1993) suggests that table 13.2 gives perhaps a 20 per cent underestimate of capacity, at least for short gutters.

Table 13.3 gives reduction factors on capacity to allow for friction in very long gutters, where l_g and Y_g are the overall length and depth of the gutter respectively. Table 13.4 gives reduction factors for corners, i.e. where gutters change direction; see also figure 13.6. Table 13.5 specifies minimum outlet sizes for standard eaves gutters to allow free discharge. It illustrates the effect that the shape of the outlet has on outlet capacity.

Whereas a rainwater pipe may be connected directly to the sole of a gutter, free discharge is encouraged by the use of a box receiver, examples of which, based on BS 6367, are given in figure 13.7. They may also enable a smaller rainwater pipe to be used. As a guide the minimum width of a box should not be less than the width of flow in the gutter at a depth equal to half the overall depth Y_g. Round-edged or tapered outlets, as shown, increase the capacity of a box; such an outlet acts as a weir at low depths, but as an orifice at greater depths.

A further consideration with valley gutters is that they permit walking along them. The top width should not be less than 500 mm for this purpose. With parapet and boundary wall gutters it should not be less than 300 mm for the same reason. Freeboard is intended to prevent overtopping and splashing.

Rainwater pipes for standard eaves gutters have the same nominal bore as the gutter outlets — table 13.5. For large gutters, such as valley or parapet types, tapered outlets may enable the size of rainwater pipe to be reduced. In general, outlet sizes are calculable by orifice or weir theory; owing to the great variety of layouts and shapes, capacities of outlets may have to be established by tests.

Down pipe sizes may be calculated from equation (8.11) in chapter 8 if required, or an equivalent using the more complex

Colebrook–White equation. For this purpose the flow might be taken as up to one-third full, whereas soil and waste stacks are designed to run less full than this. Flow, one-third full, is in the region of 50 per cent greater than one-quarter full, for which values are given in figure 5.12. Standard offsets and shoes in pipes do not significantly restrict the flow which tends towards full bore, with some air entrained. Drainage pipes receiving the flow from rainwater pipes may be sized in the usual way — see chapter 12. Document H3 associated with the 2000 Building Regulations recommends a minimum diameter of 75 mm. Such practical matters as inspection, clearing and maintenance are important.

The recommended design procedure thus follows the pattern:

Select design rainfall intensity.
Calculate the effective catchment area.
Determine the discharge of the gutter with free discharge.
Determine the number of outlets required and position them for efficient performance.
Check that outlets are large enough to allow free discharge.

If the last is found not to occur, or if the gutter is long in relation to its depth, further calculations may be necessary, or tests should be done.

European practice in this field varies. May (1982) provides a sound basis for considering any proposals for a general code, and for dealing with situations that are too complex for such a code.

14 Plastics and their Applications

Traditional materials retain a large share of the market for water, sanitary and waste services products, as indicated in earlier chapters, but the newer plastics materials are now well established in many applications. Extensive information is available in British Standards published in the past 20 years or so, arising in many aspects out of parallel work under the aegis of the International Organization for Standardization (ISO). This is contributing to the development of CEN standards. The following notes, which draw on such information, are intended as an introduction to some of the plastics used. For further reading, apart from the many standards and regulations, publications of the Institute of Materials contain much research and development information. Practical aspects are dealt with in the Institute of Plumbing's *Plumbing Engineering Services Design Guide*. *BRE Digest* 382 (1993) deals with new materials used in hot climates. Potential for development is discussed by Staniforth (1994).

Types of plastics

Plastics materials differ profoundly in chemical composition and structure from metals and ceramics and are themselves by no means uniform in structure and properties. These differences are reflected in physical properties such as density, strength, hardness, resistance to heat and pressure, ease of forming and working. The main advantages offered by plastics materials in general terms are: ready availability and competitive cost; ease of forming into complex shapes; tailoring for particular applications; ease of jointing; low density and hence lightweight components; and

good resistance to corrosion. On the other hand their coefficients of thermal expansion, propensity to mechanical damage, and susceptibility to the effects of heat, especially high temperatures and pressures in combination and, in the extreme, to fire, count against them for some applications.

Plastics are compounded of organic, i.e. carbon-based, polymers together with appropriate stabilizers, processing aids, fillers and pigments or other colorants. The main types of polymer or resin available are either thermoplastic or thermosetting, and those likely to be encountered in the present context are listed in table 14.1 together with abbreviations where defined. The majority are thermoplastic in nature, i.e. formed to shape by melting, and will soften and remelt if heated sufficiently; polyethylene, which is manufactured from ethylene under pressure assisted by the action of heat and catalysts, is one example. The more efficient the catalyst the lower the temperatures and pressures needed and, usually, the higher the density of the polyethylene produced. A polyethylene molecule has a long backbone, formed by linked carbon atoms, which may be essentially straight or branched in varying degrees. High-density polyethylene (PE-HD) has fewer, shorter branches than low-density (PE-LD), so that the molecules can pack together in a tighter, more orderly fashion. As a result PE-HD has the lower coefficient of thermal expansion and higher tensile strength and maximum working temperature. Polypropylene and polybutylene are normally manufactured by methods resembling those used for PE-HD and have highly ordered structures possessing relatively good heat resistance. Other polymers such as PVC-U, PS and

Table 14.1 Common names and abbreviations for some of the principal commercial plastics (TP denotes a thermoplastic, TS a thermosetting type)

Common name or abbreviation	Material	Abbreviation	(BS 3502)
ABS	Acrylonitrile–butadiene–styrene copolymer	ABS	(TP)
Acetal	Polyoxymethylene: polyformaldehyde	POM	(TP)
Acrylic	Methylmethacrylate polymer	PMMA	(TP)
Epoxy	Epoxide resin	EP	(TS)
EVA	Ethylene–vinyl acetate copolymer	EVAC	(TP)
GRP	Glass-fibre-reinforced plastic — usually based on UP or EP thermosetting resins	GRP	(TS)
Polycarbonate	Polycarbonate	PC	(TP)
Polyester	Unsaturated polyester — usually combined with styrene	UP	(TS)
Polyethylene or polythene	Polyethylene	PE	(TP)
	Low-density polyethylene	PE-LD	(TP)
	Medium-density polyethylene	PE-MD	(TP)
	High-density polyethylene	PE-HD	(TP)
Polypropylene	Polypropylene and copolymers in which propylene is the major constituent	PP	(TP)
Polystyrene	Polystyrene	PS	(TP)
Polyurethane	Polyurethane	PUR	(TS)
PTFE	Polytetrafluoroethylene	PTFE	(TP)
PVC (vinyl)	Polyvinyl choloride and copolymers in which vinyl chloride is the major constituent	PVC	(TP)
Unplasticized PVC	PVC without monomeric plasticizers	PVC-U	(TP)
Impact-modified PVC (PVC-MU)	PVC modified by blending with other suitable polymers, e.g. EVAC or CPVC	—	(TP)
Chlorinated PVC	PVC chemically modified by postchlorination	PVC-C	(TP)
Polybutene: polybutylene	Polybutene-1:	PB	(TP)
Cross-linked polythene	Polyethylene modified, for example, by radiation-induced cross-linking	PE-X	

PMMA have a more random distribution of substituent groups on the main polymer chain and their properties reflect this difference in structure.

The properties of a simple polymer may be modified by copolymerization, blending, or further chemical reaction. For example, a copolymer of acrylonitrile and styrene can be mechanically blended with a polybutadiene rubber to give ABS, a product of excellent impact resistance. Similarly PVC-U can be impact modified by blending, usually with a complex acrylic copolymer. Chlorination imparts improved heat resistance to PVC. Although PVC in many end uses is modified by the addition of plasticizers, this is not appropriate in the present context since excessively soft, heat-sensitive products would result, and leaching of plasticizer might occur on longterm contact with water. Thus the PVC referred to in several chapters is unplasticized.

By such means the properties of polymers can be tailored to meet specific requirements, but the cost may limit some practical applications.

Not all the polymers listed in table 14.1 possess continuous carbon backbones. In POM, for example, the backbone consists of alternate carbon and oxygen atoms. Others include in the chain characteristic groups of atoms which their names define, e.g. polyesters and polyurethanes. Acetals are thermoplastics having distinct, linear molecules. The polyurethanes and unsaturated polyesters are most commonly encountered in the present context, however, as rigid thermosetting materials. In practice they are supplied as prepolymers which must be mixed, together with catalysts for the UPs, immediately prior to forming to shape. The final structures are highly cross-linked — extended, three-dimensional networks — and do not melt on heating, though excessive heat can cause decomposition. Such products are generally characterized by relatively good heat resistance and hardness, but the properties can be varied by chemical composition.

Unsaturated polyesters have particularly complex structures. Flame-retardant characteristics can be imparted by introducing suitable additives. Since the cured resins are usually hard and brittle it is necessary for most purposes to provide additional reinforcement in the form of glass fibres, e.g. as random fibres in dough moulding compounds, or chopped strand mats. The type and amount of glass together with the resin-to-glass bond largely govern the strength and stiffness of the products, such as inspection chambers and settlement tanks.

Forming materials to shape

With thermoplastics the polymer is often supplied in the form of granules or powder; these are heated sufficiently to fuse together and flow as a viscous liquid to permit blending and forming to shape. Articles of fixed cross-section, such as pipes or gutters, are usually produced by extrusion. In this process the polymer melt is forced through a die of appropriate shape and dimensions. More complex shapes, such as pipe bends, traps, taps and fittings, are formed by injection of the melt into a mould. A sophisticated technology has developed to cope with the various problems which arise whilst permitting high production rates and standards of quality.

With acrylic baths a cast sheet of the specified thickness and colour is warmed sufficiently to allow it to be pressed and drawn by vacuum into a mould. A feature of such a process is that some areas of the moulding come out substantially thinner than the original sheet. To allow for this two types of construction may be specified in order to achieve the required rigidity.

Thermosetting resins such as UPs are normally formed at room temperature in the mould. Heat is usually evolved during curing which can cause problems in thick sections if uncontrolled. Even so it is often desirable to force the curing reaction to completion by application of heat (postcuring) in order to ensure subsequent dimensional stability and to minimize the content of volatile or leachable

residues. The latter factor is particularly important where the component is to be used in contact with potable water.

Various production methods are available for GRP, the choice depending on the type of product and scale of production. GRP pipes are normally made in individual lengths rather than by continuous extrusion.

Properties of plastics

Typical data for some of the more important properties of plastics used in buildings based on published information including *BRE Digest* 69 are given in table 14.2, but it will be appreciated that values may vary depending on composition and method of manufacture. A general appreciation of these and related characteristics is essential to the proper design and use of plastics components and systems.

The effect of heat on plastics can be particularly significant. The coefficients of thermal expansion are generally high, and the increase in length can be up to 10 times as much as for most metals for a given temperature rise. The strength properties of plastics, and especially of thermoplastics, are strongly temperature dependent so that it is usual to specify maximum working temperatures, related also to applied pressure. The temperature data in table 14.2 relate to normal operating conditions of services in buildings. The footnotes amplify this information for materials used for hot water (see later). Tensile strengths quoted are generally at 20 °C.

Moulded or extruded thermoplastics components sometimes shrink rather than expand when heated above a certain temperature. This phenomenon is called reversion and is a function both of the material and the manufacturing conditions. Standards for plastics pipework include reversion tests so that the effect can be controlled in practice. Thermo-plastics components should, in any case, be mounted so that they are protected from excessive radiant heat or contact with hot surfaces. The low thermal conductivity of plastics reduces the efficiency with which the effects of local heating can be dissipated. Frozen plastic pipes should be thawed out carefully; naked flames should not be used for this purpose. Low thermal conductivity is, however, an advantage for handles of hot

Table 14.2 Typical properties of plastics materials used in buildings

Material	Density (kg/m^3)	Linear expansion coefficient per °C	Maximum temperature recommended for continuous operation (°C)	Short-term tensile strength (MN/m^2)	Behaviour in fire
PE-LD	910	20×10^{-5}	60	7–16	Melts, burns like paraffin wax
PE-HD	945	14×10^{-5}	–	20–38	Melts, burns like paraffin wax
PP	900	11×10^{-5}	–	34	Melts, burns like paraffin wax
PVC-U	1395	5×10^{-5}	65	55	Melts, burns only with great difficulty
PVC-C	1560	6.5×10^{-5}	70 for 50 years	–	Requires a constantly applied flame to burn
POM	1410	8×10^{-5}	90	62	Softens, burns fairly readily
ABS	1060	7×10^{-5}	90*	40	Melts, burns readily
PMMA	1185	7×10^{-5}	80	70	Melts, burns readily
GRP	1600	$2\text{–}4 \times 10^{-5}$	90–150	100	Flammability controlled by choice of polyester resin
PB	920	13×10^{-5}	82 for 50 years 110 for short term	17	Melts, burns like paraffin wax
PE-X	938	14×10^{-5}	95	19–26	–

* A higher-temperature formulation is also available
Notes: PVC-C: Max. short term temp. 95 °C at reduced pressures
      PB: Max. operating temp. 110 °C, short term
      PE-X: Different pipe types for 95 °C depending on pressure, 6 bar or 10 bar; short term exceptional temp. 100 °C; softens at 133 °C

taps, baths and WC seats. Plastics materials also possess low electrical conductivities. Whilst this property can be put to good use, it means that plastics pipes are unsuitable for earthing purposes. Plastics components can build up high charges of static electricity on their surfaces.

The durability of a component depends both on the materials used to make it and on the environment and stresses to which it is exposed in use. Failures in practice are usually caused by exceptional short-term stresses — thermal, hydrostatic, or mechanical — due, for example, to impact or mishandling. In selecting a material for a particular application it is advisable to consider what the likely modes of failure are; what the consequences of failure would be; and how readily the failure can be repaired. The more accessible and exposed the component, the higher the risk of failure is likely to be — but usually the more easily it can be seen and made good. Concealed pipework may be less at risk — but more difficult to repair. Components may become more prone to failure as they age in use. PVC-U rainwater goods, for example, are exposed to the degradational effects of sunlight which causes embrittlement and makes the material more prone to impact damage. The consequences of such failure are unlikely to be serious in the short term and repairs are usually easily made. By contrast, plastics sanitary pipework is usually enclosed and so is exposed to a less severe environment. Failures, should they occur, may, however, go undetected for some time and be difficult to locate and repair.

Pipework for cold water supply

The main plastics used for conveying cold potable water are PE and PVC-U; ABS is specified for industrial uses. PB and PE-X are now specified for both cold and hot water and are dealt with in the next section. Standards (see Appendix 1), including codes and specifications of general requirements and for the installation of the pipes and fittings, are available. Major factors governing selection are wall thickness in relation to pipe diameter, maximum sustained hydrostatic pressure, and plastics type. Pressure ratings available range from 3 bar or less up to some 15 bar, in colour-coded form; projected lifetimes (50 years at 20 °C) are extrapolated from data obtained from pressure tests at higher temperatures. Pipes must be able to withstand high, short-term pressures and an increased wall thickness may be advantageous where such events are likely to be frequent.

Mechanical damage during handling or installation may weaken the material and accelerate failure under pressure. Care should be taken to avoid abrasion; crushing; failure to release twist from coils; sharp bending; inadequate support of horizontal runs; local heating effects; incorrect backfilling of buried pipes. Water pipes are unlikely to be exposed long-term to the embrittling effects of sunlight, but pipes may be pigmented with carbon black to provide protection. PE pipes, for example, with a density in the range 930–944 kg/m^3 (compare table 14.2) and a rating up to 12 bar at 20 °C, are available in blue (for use especially underground or in ducts with other services) and in black for above-ground use which requires greater resistance to light and weathering. Impact tests are included in standard specifications for PE and PVC-U pipes, the impact behaviour at low temperatures (0 °C) being particularly significant.

Joints present potential points of weakness, and have to be designed carefully in relation to the requirements of the system. Preferred methods of jointing vary with the material. PVC-U is amenable to solvent welding; the solvent must be allowed to evaporate before the joint is stressed. PE pipes are jointed either by mechanical (compression or flanged) fittings or by thermofusion. Both solvent welding and thermofusion make use of the thermoplastic character of the material. The processes require careful control and attention to manufacturers' instructions. Standards for jointing are available.

Pipework for hot water

The progressive softening and loss of tensile strength of thermoplastics as the temperature rises is a problem for possible hot water use. This reduces either the service life expectancy at a given pressure, or the maximum permissible working pressure compared with that at 20 °C. For example, at a maximum continuous service temperature for PVC-U piping of 60 °C the pressure rating is 40 per cent of that

corresponding to 20 °C. Allowance may also need to be made for any ambient temperature excess above 20 °C. Moreover, the effects of creep and distortion, reversion, and high coefficients of thermal expansion are exaggerated in this type of use. PE pipe, for instance, shows expansions of the order of 10 mm per metre when hot water passes through. Adverse effects are not just physical in origin; chemical degradation, e.g. oxidation of the anti-oxidant and hence the polymer, also occurs more rapidly the higher the temperature.

Plastics tried for this purpose include chlorinated polyvinyl chloride (PVC-C), nylons, polybutene 1, and cross-linked polyethylene (PE-X). Research and development, mainly by manufacturers but also trials by the Building Research Establishment, have been extensive and led to the publication of British Standards in 1990 for the use of PVC-C, PB and PE-X for hot as well as cold water and in heating installations in buildings. Fittings to go with the latter two pipes are made from several engineering plastics.

Pipes in thermosetting materials possess in principle the necessary strength and heat resistance for this use but have the major disadvantage that, once made, they cannot be formed into bends on site. GRP pipes are mainly used in large diameters for underground water supply or waste disposal applications. Thermoplastics are successfully employed to provide a protective insulating cover to metal pipes both in factory-applied and site-fitted designs.

Soil and waste pipes

This is a major use for plastics materials. A fundamental requirement is for components to be designed to withstand the effects of short-term heating, albeit at relatively low hydrostatic pressures, as hot waste passes through them. Standards include a 2500 cycle proving test for this purpose, based on some years' experience. Materials used include PVC-U; ABS; PVC-C; PP; PE-HD and styrene copolymer. Waste traps and connectors for horizontal outlet WC pans are widely used.

The durability of these materials may be affected by impurities in the waste water. As an example of this, PE is prone to environmental stress cracking promoted by domestic detergents. In order to meet requirements, PE components must pass a test of resistance to hot detergent solutions. In some applications waste pipes and fittings may be required to withstand the action of a wide variety of chemicals (see the standards available), whilst advice can be sought from manufacturers. Leak testing by smoke should be avoided with plastics pipework, especially PVC-U, PVC-MU and ABS, because the naphtha can have a detrimental effect on the material; rubber jointing components may also be affected.

One source of problems can be the over-tightening of threaded plastics connections. Weeping joints of this type are best cured by the use of unsintered PTFE tape rather than by excessive tightening. Parts of waste systems may be exposed to impact damage in practice. Whilst they are best protected if possible, impact-modified compositions such as ABS or PVC-MU could well be preferred for use in vulnerable situations. Proper support of pipework is also important, bearing in mind the variable loadings and thermal movements which can occur. There is a need for care in clearing pipe blockages in these systems.

Other applications

Cold water storage cisterns and WC cisterns are widely used in plastics, the former (also floats and float valves) usually in polyolefins, the latter in toughened polystyrene or ABS. Taps are now widely made either wholly or in part in plastics. These provide an example of small components of intricate design for which high-quality mouldings in relatively expensive plastics such as acetal, acrylics or polycarbonates find useful applications. Resistance to mechanical stress is an important performance factor here.

Rainwater goods in PVC-U, formulated to withstand weathering, are the norm for many situations. Overpainting of new components is not recommended since this is likely to cause embrittlement and hence susceptibility to impact damage.

Plastics components are widely used in buried pipeline applications. Major requirements here include provision for thermal expansion and contraction, especially in long, straight runs; proper bedding and backfilling of trenches; and also the possible long-term effects of ground contamination, as by gas oil, for example. The latter may, in particular, affect the long-term performance of elastomeric joint rings.

Appendix 1
Notes on Standards

As noted in the Preface, European Standards (Norms) including appropriate national provisions and adopted as British Standards will cover most of the fields dealt with in this book. Some are already available, others are expected in the future. The list below indicates the current position for selected standards related to the chapters here.

As an example, the first item below indicates some important features of the move towards British Standard, European Norms. The Standard, prepared over some 10 years, includes five parts. There were obstacles to achieving a fully harmonized Standard in view of the different methods and systems in use across Europe. A format of multiple design systems was eventually chosen. Part 2 includes four such systems, and there is the option for each country of including a supplement, termed National Annex, not subject to CEN scrutiny. The UK National Annex to Part 2 is essentially BS 5572, suitably edited. The well-established methods therein continue to be available to UK designers, therefore, along with other material including any other National Annexes. The National Annexes have advisory status (as with BS 5572 in the past as a Code of Practice) but may be written into contract documents (as was BS 5572) and become mandatory. Part 3 replaces BS 6367. Part 4 focuses principally on German practice. Part 5 relates to both Parts 2 and 3.

BS 5572 : 1994 Sanitary Pipework

> Replaced by BS EN 12056 : 2000 'Gravity drainage systems inside buildings'
> Part 1 General and performance requirements
> Part 2 Sanitary pipework – layout and calculation

Part 3 Roof drainage – layout and calculation
Part 4 Waste water lifting plants – layout and calculation
Part 5 Installation and testing, operation and maintenance.
The National Annex to Part 2 is BS 5572, suitably edited.

BS 6367 : 1983 Drainage of roofs and paved areas
Replaced by BS EN 12056 : 2000, Part 3 (see above)

BS 6465 : 1995 Part 1 Provision, selection and installation of sanitary appliances
Part 2 Space requirements for sanitary appliances
Current

BS 8301 : 1985 Building drainage
Replaced by BS EN 752 Parts 1–7 but remains current and forms the National Annex to the EN, suitably edited

BS 8313 : 1997 Accommodation of building services in ducts
Current

BS 5906 : 1987 Storage and on-site treatment of solid waste from buildings
Current

BS 5810 : 1979 Access for the disabled to buildings
Current

BS 6700 : 1997 Water installations for buildings
Current but overtaken by the Water Supply (Water Fittings) Regulations 1999 and Water Byelaws 2000, Scotland
EN 806 dealing with water installations is in preparation

BS 3502 Part 1 1991 Common names and abbreviations for plastics and rubbers
Current

BS 3505 : 1986 Specification for PVC-U pressure pipe for potable water
Replaced by BS EN 1452: 2000 Parts 1–5 but remains current

BS 3943 : 1979 Specification for plastics waste traps
Current

BS 3955 : 1986 Specification for electrical controls for household appliances
Partially replaced by BS EN 60730 and 61058 but remains current

BS 4201 : 1979 Specification for thermostats for gas-burning appliances
Replaced by BS EN 257 : 1992

BS 4514 : 2001 PVC-U soil and ventilating pipes
Current

BS 4576 : 1989 PVC-U rainwater goods
Partially replaced by BS EN 607, 1462 and 12200 but remains current

BS 4660 : 2000 Thermoplastics ancillary fittings for below ground gravity drainage
Current

BS 4991 : 1974 Specification for propylene copolymer pressure pipe
Current

BS 5254 : 1976 Specification for polypropylene waste pipe and fittings
Replaced by BS EN 1451 : 1:2000 but remains current

BS 5255 : 1989 Specification for thermoplastics waste pipe and fittings
Partially replaced by BS EN 1329, 1519, 1565, and 1566 but remains current

BS 5412 : 1996 Specification for low resistance taps
Current

BS 5955, various dates. Plastics pipework
Current

BS 6283 : 1991 Safety and control devices for use in hot water systems
Current

BS 6572 : 1985 Specification for blue polyethylene pipes for potable water (below ground use)
Current

BS 6730 Specification for black polyethylene pipes for potable water (above ground use)
Current

BS 6920 Suitability of non-metallic products for use in contact with water intended for human consumption
Current

BS 7206 : 1990 Specification for unvented hot water storage units and packages
Current

BS 7291 : 2 : 1990 Specification for PB, PE-X and PVC-C pipes and fittings (for hot and cold water)
Current

Appendix 2 Notes on Sizing Storage and Piping for Water Supply

A statistical approach to tank sizing

The demands made on a supply system are variable, depending on a range of parameters that might include climate, building population, usage and time of year. It is, however, possible to utilize statistical techniques that are already well established to allow the designer to set a satisfaction level for the system. While the binomial distribution may be used to address problems in 'discrete' appliance usage, as in chapter 1, the prediction of usage of a supply commodity, such as water, has to be dealt with via the normal distribution. Ideally this distribution is characterized by a variable that is continuous and has infinite upper and lower limits. Obviously the supply of hot water or gas does not conform, but the upper and lower limits possible are usually sufficiently far from the mean for a reasonable approximation to be made to a normal distribution.

For example, assume that the following data represent the usage of hot water in a process or dwelling in litres per day over a 40 day period:

362	412	234	317	365	298	333	341
284	376	298	394	342	368	253	451
234	292	366	321	372	359	286	311
299	322	377	344	401	381	349	328
345	367	267	329	318	362	349	305

These data may be represented by a mean, x_m, and a standard deviation, S_x, where:

$$x_m = \Sigma fx / \Sigma f$$

and

$$S_x = \sqrt{\left(\frac{\Sigma fx^2}{\Sigma f} - x_m^2\right)}$$

A frequency distribution may be drawn up from the data, as in table A2.1, and these factors determined. Hence, from the table:

$$x_m = 13\,670/40 = 341.75 \text{ litres/day}$$

and:

$$S_x = (4\,760\,800/40 - 341.75^2)^{0.5}$$
$$= 47.1 \text{ litres}$$

By reference to the normal distribution in figure A2.1 the supply to provide any satisfaction level may now be calculated. The shaded area under the curve in figure A2.1 represents the probability of a variable having a value less than or equal to $(x - x_m)/S_x = Z$. The total area under the curve is thus unity. The probability of a variable having a value less than or equal to Z is given by $\phi(Z)$; the relationship between Z and $\phi(Z)$ is found in statistical tables (table A2.2).

Table A2.1

Hot water consumption range (l/day)	Mid-point x	Frequency		
		f	fx	fx^2
230–260	245	2	490	120 050
260–290	275	3	825	226 875
290–320	305	8	2440	744 200
320–350	335	10	3350	1 122 250
350–380	365	9	3285	1 199 025
380–410	395	5	1975	780 125
410–440	425	2	850	361 250
440–470	455	1	455	207 025
		40	13 670	4 760 800

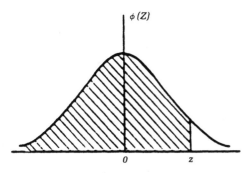

Figure A2.1   Normal distribution

Table A2.2   Extract from normal distribution

$\phi(Z)$	Z	$\phi(Z)$	Z
0.5	0.0	0.9	1.281
0.6	0.253	0.95	1.645
0.7	0.524	0.99	2.326
0.8	0.842	0.999	3.090

A satisfaction level may now be defined in terms of the number of occasions the user finds that the designed storage capacity is sufficient to meet the user's needs, e.g. a 99 per cent satisfaction level implies that on 1 occasion in 100 uses the storage was insufficient. If the mean and standard deviation defining a distribution are known, and if the distribution may be assumed to be normal, it follows that the value of the variable x corresponding to any designed user satisfaction, Z, may be defined as:

$$Z = (x - x_m)/S_x$$

or:

$$x = x_m + ZS_x$$

For example, if in the data given the design for the hot water use specified a 99 per cent satisfaction then the supply capacity required would be given by:

$$\text{storage capacity} = 341.75 + 2.326 \times 47.1$$
$$= 451.3 \text{ litres/day}$$

This figure may well be, as in this case, greater than the highest daily value recorded in the sample. The setting of a satisfaction level therefore has cost implications as it determines the storage tank capacity and/or the water

heating capacity and its response time. Lower satisfaction levels are cheaper and hence the choice also depends upon the ultimate user of the system or the importance of the application catered for.

While setting a satisfaction level will yield the necessary capacity for a particular time period, it does not size the storage tank necessary, nor the heating power necessary, in a hot water system. It must be remembered that the storage capacity may be continuously replenished, as with UK domestic water supply, and therefore the size of the storage tank will be considerably less than the satisfaction level capacity. The methodology described below is applicable to a wide range of storage situations, including domestic and cold water and rainwater reuse calculations (chapter 11), both these cases featuring a replenishing flow, and domestic hot water storage where the role of the replenishing flow is filled by the recovery time of the heater battery or immersion heater utilized.

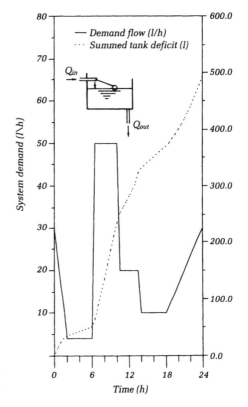

Figure A2.2   System demand and cumulative tank deficit over a 24 hour period

Assume that the cold water demand for a dwelling is characterized by the 24 hour profile illustrated in figure A2.2, and that the replenishing flow from the urban water supply is potentially constant, this inflow being controlled by the level in the storage tank. The storage capacity necessary then becomes the largest shortfall indicated by figure A2.3, plus whatever safety factor is added to ensure that water is not scavenged from the bottom of the tank, or in the case of hot water storage to allow for flow temperature stratification. Figure A2.3 was developed as for any time period, Δt, the balance between inflow, Q_{in}, and outflow, Q_{out}, results in a tank deficit, Vol_{tank} as follows:

$$(Q_{in} - Q_{out})\Delta t = Vol_{tank}$$

Subsequent summation of Vol_{tank} values at each time step will reach a maximum *negative* value representing the required minimum tank storage (Maver 1964).

Figure A2.3 illustrates the tank capacity required for three constant inflow rates of 20,

30 and 40 litres/hour. The inflow rate becomes zero if the tank is full. It will be seen that a 20 litres/hour inflow is insufficient as at 06.00 the tank is not full in readiness for the following day's demand, whereas 40 litres/hour would appear to be excessive. The predicted tank storage at 75 litres plus safety factor would appear reasonable with a 30 litre/hour inflow possible.

Sizing of distribution piping in a building

In the sizing of the distribution pipework within a building the usage pattern of the appliances becomes the deciding factor, necessitating the use of a demand unit method, as discussed in chapter 1. In order to determine the required delivery along any particular section of distribution pipework within the building it is necessary to sum the demand units for the appliances downstream served by that pipe. This summation may then be translated into a design flow by reference to the appropriate relationship between the unit total and flow. In the following example use is made of the CIBSE Book C procedure, to which figure 1.8 and table 1.11 in chapter 1 refer.

As an illustration consider the system in figure A2.4. In order to identify the index circuit, i.e the flow path with the lowest ratio of available head to total equivalent pipe length, approximated as the actual pipe length plus 30 per cent for the frictional resistance of bends, tees and so on, it will often be sufficient to assess the network by inspection and experience. In more complex networks comparable calculations should be undertaken for each possible route. An estimate for the extra equivalent length to be added for fittings should be made, which will inevitably be approximate. However, as the calculation will be dominated by the need to use standard pipe sizes, will include demand units and will be deemed satisfactory if the final result is that the available pressure exceeds that required, it is sufficient for this example. From figure A2.4 it would appear that the first floor is the determining route as this has the lowest available head to equivalent length ratio.

The equivalent length for the cold feed to first-floor level is $12 \times (1 + 0.3) = 15.6\,\text{m}$, and the available head is $4\,\text{m}$ or $39.24\,\text{kN/m}^2$. Hence

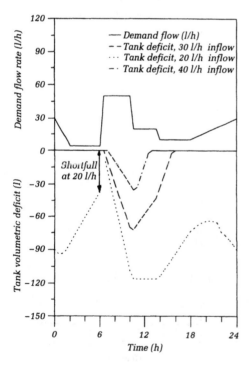

Figure A2.3 Example of cold water storage tank sizing for various inflow rates and a particular system demand

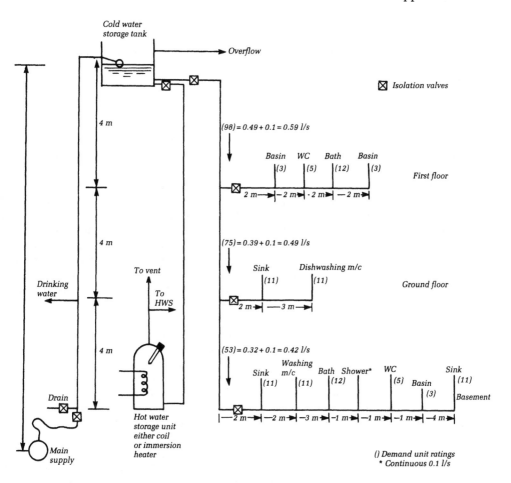

Figure A2.4   Schematic of cold water distribution network

the pressure loss per unit length for the first-floor index route is $39.24/15.6 = 2515\,\text{N/m}^2$. As the flow is greater in the cold feed than in the branch, it follows that the diameter of the cold feed will exceed that necessary in the branch. The water flow varies from 0.59 litres/s in the cold feed to 0.15 litres/s, based on a unit total of 23; in the first-floor branch, however, this is less than the minimum required for the bath, so this flow must be taken as 0.3 litres/s in the following calculation.

Suppose the water flow velocity in both pipes is limited to 1–1.5 m/s. Reference to table A2.3 for the required flow rates and the available pressure loss per unit length suggests diameters of 28 mm for the cold feed and 22 m for the branch. From table A2.3 it is now possible to check that these dimensions are

Table A2.3   Typical data linking pipe diameter, flow rate and diameter in copper tubes at $10\,^{\circ}\text{C}$, based on CIBSE Guide Book C

| Pressure loss/ unit length | Pipe diameter | | | |
| | 15 mm | 22 mm | 28 mm | 35 mm |
	Water mass flow rate (kg/s)			
600		0.101	0.297	0.601	1.09
650		0.106	0.311	0.629	1.14
					$v = 1.5$
700		0.110	0.324	0.656	1.19
750		0.115	0.337	0.682	1.23
800		0.119	0.350	0.708	1.28
850		0.123	0.362	0.732	1.32
					$v = 2.0$
900		0.127	0.374	0.757	1.37
950		0.131	0.386	0.780	1.41
1000		0.135	0.398	0.803	1.45
1100	$v = 1.0$	0.143	0.420	0.847	1.53
1200	$v = 1.50$	0.150	0.441	0.890	1.61

allowable within the available head provided by the storage tank by multiplying the pressure drop per unit length indicated by the total equivalent length for both the cold feed and the branch, i.e. $600 \times 4 \times (1 + 0.3) = 3120 \, \text{N/m}^2$ for the cold feed and $650 \times 8 \times (1 + 0.3) = 6760 \, \text{N/m}^2$ for the branch, a total of $9880 \, \text{N/m}^2$, considerably less than the total $39\,240 \, \text{N/m}^2$ available.

This disparity arises from keeping down the flow velocity in the cold feed and the branch. Increasing this flow velocity to 1.5–2.0 m/s — chapter 10 suggests a maximum of 3 m/s to limit noise — would have reduced the cold feed diameter to 22 mm and the branch diameter to 15 mm, raising the cold feed pressure loss per unit length to $2000 \, \text{N/m}^2$, and the branch pressure loss per unit length to approximately $4000 \, \text{N/m}^2$, thereby increasing the overall pressure required to $2000 \times 4 \times (1 + 0.3) + 4000 \times 8 \times (1 + 0.3) = 32\,000 \, \text{N/m}^2$, closer to that available.

This example demonstrates the limitations of the design method at low demand and emphasizes the importance of other design considerations, particularly noise. The linkage between pressure loss per unit length, flow velocity and pipe diameter has also been emphasized.

Hot water services

The choice of appropriate heating system depends upon the category of building and its occupation pattern. This choice includes the use of one or more instantaneous water heaters, storage systems utilizing either temperature-controlled immersion heaters or time-clock-controlled heat exchanger coils, in conjunction with a central heating system, or off-peak electrical supply to immersion heaters. Where the user requirements are known it is possible to determine the necessary heat input; however, it is more likely that these will be unknown and therefore some approximation to user needs will be necessary. BS 6700 recommends that for a dwelling with only one bathroom, bath use frequency will be no higher than once per 30 minutes, with additional hot water for kitchen use. With more than one bathroom BS 6700 recommends that bath use is assumed to be directly sequential. These figures can be used to give an approximate measure of heat load as the rating

required to raise that quantity of water to the required temperature is easily found.

In choosing instantaneous heating it must be remembered that the mass flow rate and temperature are linked as follows:

$$m_w = Q \, \text{kW} / [c(t_f - t_s)] \, \text{kg/s}$$

where c is the specific heat of water. The required flow rates may not be achievable at the temperatures desired. Storage systems, either immersion heater driven or intermittent, can supply the required flow rates but the storage capacity must be properly calculated.

Figure A2.5 illustrates a typical hot water heating and distribution network for a domestic-style application, in this case used in conjunction with a central heating system. Alternatively the water heating may be provided, in this example, by an immersion heater in the hot water storage tank. It will be appreciated that the effective head driving the hot water flow is the elevation of the high-level cold water storage tank, and this may have implications for the introduction of a shower facility in terms of maximum flow rate. The minimum capacity of hot water storage is illustrated in table 3.1. The sizing of hot water storage is illustrated in the same manner as the cold water storage already discussed. However, in this treatment it is necessary to 'replace' the inflow to the cold water tank from the mains supply with the rate of generation of hot water possible with the heating facility installed in the storage tank. It will be clear from figure A2.5 that the hot water tank is always full; however, it is the temperature of the stored water that is of concern.

For the 'standard' domestic dwelling the hot water storage is normally adequately handled by a 114 l tank (but see table 3.1), with either an immersion heater or a heat exchanger coil run from the central heating system. In other, more complex situations, or where off-peak heating is to be used, it is best to ascertain the likely hot water demand over the day and match this to the available heat input and storage capacity, as discussed below. Where detailed information is not available it is necessary to utilize data on the typical hot water volumes required for a range of conditions, e.g. table A2.4.

The rate at which water may be raised from the cold water mains supply temperature to the

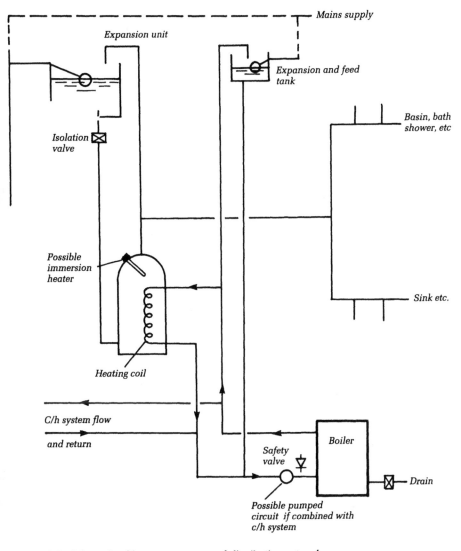

Mains supply

Expansion unit

Expansion and feed tank

Basin, bath shower, etc

Isolation valve

Possible immersion heater

Heating coil

Sink etc.

C/h system flow

and return

Safety valve

Boiler

Drain

Possible pumped circuit if combined with c/h system

Figure A2.5 Schematic of hot water system and distribution network

Table A2.4 Hot water consumption (figures in parentheses refer to total hot/cold volumes); see also table 1.8

Appliance	Water usage (l)	Flow rates (l/s)	Temperature (°C)
Wash basins	2–18 (5–20)	0.15	40–60
Shower	15 (40)	0.05–0.1	40
Bath	70 (110)	0.3	
Washing machine	70 (150)		
Sinks	5–15 (10–15)	0.2	60

required storage and use temperature within the system is known as the system recovery rate. This depends upon the rating of the installed immersion heater or coil. If Q kW is the rating of an immersion heater designed to raise the temperature of stored water through a difference $t_f - t_s$, then the mass of water, m kg, specific heat c kW/kg °C, raised in temperature in 1 hour is given by:

$$Q = mc(t_f - t_s)/3600$$

i.e. if $Q = 1\,\text{kW}$, $c = 4.2\,\text{kW/kg}\,°\text{C}$, $t_f = 60\,°\text{C}$ and $t_s = 10°\text{C}$, then $m = 17.1\,\text{l/h/kW}$.

Table A2.5 Storage and heating system capacity sizing for both continuous and intermittent heating regimes

Time	00	07	08	09	10	11	12	13	14	15	16	17	18	19	20	21	22	23	Appliance
		2	2	2	2	2	2	2				4	4	4	4	4	4	8	Casual
		4	20	4		8	20				20		8	8			12		Basin 4
			30	30	30	30	30			30		30	30				30		Sink 30
																			Sink 30
													30						Dishwasher 30
					40		40												Washing m/c 40
			60					60						120				60	Bath adult 60
		10	10										20						Shower 10
		16	122	38	72	40	92	62	0	30	20	34	92	132	4	4	46	68	Total litres (762 litres)

(Left bracket spanning the appliance rows: Water usage, litres per hour per appliance)

	00	07	08	09	10	11	12	13	14	15	16	17	18	19	20	21	22	23
Deficit at 6 kW (Max 102 litres/hour recovery)	0	0	20	0	0	0	0	0	0	0	0	0	0	30	0	0	0	0
Deficit at 3 kW (Max 51 litres/hour recovery)	0	0	71	58	79	68	109	120	69	48	17	0	41	122	75	28	23	40
Off-peak 10 kW from 14.00–16.00 (340 litres recovery max) + 02.00–06.00 (680 litres recovery max)	0	16	136	174	246	286	378	440	270	164	174	208	300	432	416	420	466	534
Off-peak 15 kW from 14.00–16.00 (510 litres recovery max) + 02.00–06.00 (1020 litres recovery max)	0	16	136	174	246	286	378	440	185	0	20	54	146	278	282	286	332	400

Note: If only one off-peak period is available, say 02.00–06.00, then the immersion heater rating would rise to at least 11.5 kW and the storage capacity to 800 litres

Thus if the daily usage of the system can be approximated, and the satisfaction level to be achieved determined, the size of the storage for a given immersion heater can be calculated, or alternatively the immersion heater rating for a given storage may be determined. Similarly it is possible to predict the effect on storage capacity of varying the heating control parameters, i.e. the use of off-peak electricity or time clock controls on a central heating system hot water control.

Table A2.5 illustrates the techniques available for sizing tank and heating system capacity. For the traditional immersion heater system it would appear from this example that a storage capacity of 120 litres would be sufficient with a 3 kW heater. The use of an off-peak supply increases the storage required to a minimum of 400 litres with a 15 kW heater or 550 litres for a 10 kW heater with two 'on' periods, or 800 litres for a 12 kW heater with one on period, depending on the availability of these off-peak supplies. Intermittent water heating from a central heating system may be handled in exactly the same way providing the coil heat transfer capacity and both the likely water usage and boiler 'on' times are known. It would be prudent to size the storage tank above these minimum values owing to the random nature of appliance usage, natural heat losses from the storage tank, even if minimized by insulation, and tank stratification. A 1.25 safety factor on tank capacity would not seem unreasonable, but oversizing is undesirable since adequate turnover of water helps to limit the growth of bacteria — see chapters 2 and 3.

Stratification

Stratification occurs in a hot water storage tank owing to the differing densities of the cold feed water, entering at the base of the tank, and the heated water that collects at the top of the tank and is normally drawn off with little mixing. To facilitate this effect it is important that hot water storage tanks are mounted with the longest axis vertical and should have a height to diameter ratio of not less than 2:1. To prevent a lowering of the stored water temperature by undue mixing with the incoming cold feed it is useful to position a baffle plate close to the cold water inlet to prevent jetting within the tank. An allowance for stratification should be made in tank sizing to allow the stored hot water volume, rather than the whole volume, to be equated to the recommended storage illustrated in preceding sections. A striving for economy can, however, conflict with water quality requirements. Temperatures maintained within 20–50 °C encourage the growth of bacteria — see chapters 2 and 3. This is an important consideration for design.

Example

The daily hot water use within a domestic environment may be represented by the appliance usage pattern shown in table A2.5 and may be used to determine both the storage and heating system capacity.

Heat losses from hot water pipes

Table A2.6 illustrates the recommended length of hot water supply pipes. Such pipes lose heat to their surroundings and, therefore, there can be an unsatisfying delay before hot water is provided at the appliance outlet. This may be countered by the provision of pumped recirculation, commonly found in large building applications; figure A2.6 illustrates this.

As shown in figure A2.6 flow may reach an open hot tap via the direct supply from the storage tank or via the recirculation network and pump, normally of smaller diameter. Generally the bulk of the tap flow is drawn from the storage tank directly as this route has the lower frictional resistance. However, when all the taps on the system are closed hot water recirculates to ensure that on opening the tap the user receives heated water. The sizing of the recirculation system requires a knowledge of the heat losses in the pipework and the

Table A2.6 Recommended maximum hot water pipe lengths from tank to tap

Pipe diameter (mm)		Maximum length (m)
Steel	Copper	
15	15	12
20	22	8
25	28	3
Spray taps		1

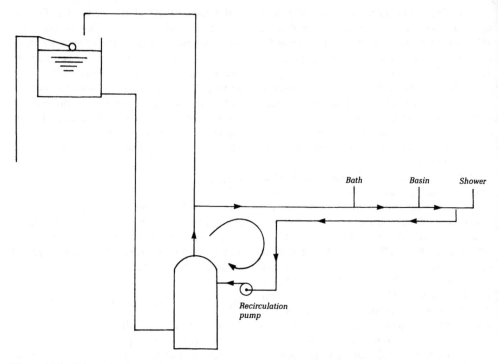

Figure A2.6 Schematic of a simple hot water system recirculation network designed to minimize heat loss from long pipes

temperature difference between the flow and return water at the storage tank. Normally a 10 °C drop is tolerated and if the heat loss expected is Q kW it follows that the required mass flow is

$$m_w = Q \text{ kW}/\Delta tc \text{ kg/s}$$

Once the required flow rate is known the pipe diameter may be calculated to yield an acceptable pressure drop around the recircula-tion circuit, utilizing techniques introduced earlier for supply pipe sizing. Table A2.7 illustrates the magnitude of the heat losses to be expected from insulated hot water pipes.

Table A2.7 Insulated pipe heat losses

Pipe diameter (mm)	15	22	28	35	42
Heat loss (W/m K)	0.19	0.23	0.25	0.29	0.32

Notation

A	Flow cross-sectional area, effective roof area in rainfall calculations, m^2	p	Pressure, N/m^2; probability of appliance discharge
A_c	Critical flow cross-sectional area, m^2	Q	Pipe, duct or partially filled conduit flow, m^3/s
B	Gutter flow width in rainfall calculations	q	Lateral inflow, m^3/s; probability that an appliance is not in use
C	Chézy coefficient for free surface flows	R_e	Reynolds number
C_2, C_3	Coefficients in St Venant equations of continuity and momentum	R_H	Hydraulic radius, hydraulic mean depth, m
c	Wave propagation velocity, m/s	r	Fraction of stack cross-section occupied by water annular film; rainfall intensity, l/s/m or mm/h
D	Pipe and trap diameter, m		
E	Specific energy in free surface flows, m; Young's modulus of elasticity, N/m^2	S_0	Pipe slope
e	Pipe wall thickness, m	S	Friction slope in free surface flows
f	Friction factor in Colebrook–White equation	S_p	Stack pressure, mm water gauge
G	Drain slope	T	Free surface width, m; interval between discharges
g	Acceleration due to gravity, m/s^2	T	Air temperature in stack, K; return period for rainstorms
H	Total head, flow depth in channel, m	T_c	Critical free surface flow width, m
H_0, H_1	Appliance trap seal depths, appliance and system sides, m	T_p	Pipe period
h	Flow depth, elevation, stack height, m	t	Time, s; annular film thickness in stack flow, m; duration of discharge, s
h_f	Frictional loss, m		
K	Bulk modulus of elasticity, N/m^2; boundary loss coefficient; junction depth coefficient	t'	Time in vertical stack treatment, s
K_s	Separation loss coefficient	V	Flow velocity in St Venant equations and mean free surface flow velocity, m/s
k	Roughness of conduit walls, m		
L	Pipe length, m; expected lifetime of building	V_t	Terminal water velocity in stack flow, m/s
L_{eqv}	Pipe length equivalent to a separation loss, m	v	Mean flow velocity; wind velocity, m/s
m	Hydraulic mean depth, A/P	W	Top width of rainwater gutters, m
n	Manning loss coefficient; number of pipe diameters representing L_{eqv}	W_p	Wind kinetic energy (pressure), mm water gauge
P	Wetted perimeter, m; probability of exceeding the expected design rainfall	x	Distance, m
		Y	Rainfall gutter flow depth, m

Z	Elevation relative to a datum; vertical distance to terminal conditions in stack flow, m	Suffix	
		A,B	Locations
		a	air
α	Pipe slope	air	Entrained airflow conditions
Δh_s	Separation loss, m	atm	Atmospheric conditions
Δp	Pressure difference or loss, N/m^2	c	Critical flow conditions, air core conditions in vertical stacks
Δp_{input}	Fan or pump input to Steady Flow Energy Equation, N/m^2		
		cd	Design capacity in rainwater gutters
Δp_{out}	Turbine extract in Steady Flow Energy Equation, N/m^2	exit	Exit boundary condition
		g	Gutter conditions
Δp_{pipe}	Friction and separation loss along a pipe, N/m^2	i	Pipe or section number
		j	Junction conditions
Δp_s	Separation loss, N/m^2	min	Minimum value
θ	Pipe slope	n	Normal flow conditions
λ	Darcy friction factor	sub	Subcritical flow conditions
μ	Coefficient of dynamic viscosity, kg/ms	super	Supercritical free surface flow
ν	Coefficient of kinematic viscosity, m^2/s	t	Terminal
ρ	Density, kg/m^3	u	Upper
τ	Shear stress, N/m^2	v	Vertical surface areas in rainwater runoff calculations
τ_0	Shear stress, N/m^2		
γ	Ratio of specific heats	w	Water

Bibliography

Ackers P 1969 Charts for the hydraulic design of channels and pipes. Hydraulics Research Paper 2, HMSO, London

Ackers P 1978 Urban drainage, the effects of sediment on performance and design criteria *Proc. Int. Conf. Urban Storm Drainage*, Southampton, April

Anderson J A 1993 Design of true half-round roof eaves gutters. *Building Services Engineering Research and Technology* 14(1): 17–21

Anon 1971 Non-domestic refuse: an investigation by Building Research Station. *Surveyor* April: 23–24

ANSI A112 19.6 1990 Hydraulic requirements for water closets and urinals – an American National Standard. *ASPE Journal of Engineered Plumbing* 1(2): 97–122 July

Arthur S and Swaffield J A 2001 Siphonic roof drainage: current understanding, *Urban Water* 3: 43–52

ASCE 1963 Report of task force on friction factors in open channels *J. Hyd. Div. ASCE* 89 HY2

AS/NZS 1996 National Plumbing and Drainage, Part 2.2: Sanitary plumbing and drainage-acceptable solutions. AS/NZS 3500.2.2: 1996 Sydney and Wellington: Standards Australia and Standards New Zealand

Auzou S 1967 Le bruit des equipments: les robinets. *Paris Cahiers du CSTB* 85: 739

AWWA 1999 Conserv 99. Water efficiency making cents in the next century. American Water Works Association Proceedings January 31-February 3 1999, Monterey California

Aya H L 1994 Modular membranes for self-contained reuse systems. *Water Quality International* IAWQ (4): 21

Ball E F 1974 Noise in water systems. *Plumbing* 2(5): 26–27, 41

Ball E F and Webster C J D 1976 Some measurements of water flow noise in copper and ABS pipes. *Building Services Engineer* 44(5): 33–40

Ballanco J A 1998 Investigation and analysis of violently fracturing water closets. *Technical Proceedings ASPE 1998 Convention*, Indianapolis, October 1998, pp. 109–150

Bartlett C L R 1981 Legionella pneumophila in hospital water systems. *Proc. Symp. on Hospital Water Supplies*, pp. 1–6, Institution of Public Health Engineers, London

Beij K H 1934 Flow in roof gutters. *Journal of Research, US National Bureau of Standards* 12(644): 192–213

Bessey S G 1989 Problems encountered in the Hotwells District small-scale metering trial. *Journal Institution of Water and Environmental Management* 3(6): 579–582

Billington N S and Roberts B M 1982 *Building Services Engineering, a review of its development*, Pergamon Press, Oxford

Boden J B 1977 Standards of site practice — implications for innovation in the design and construction of buried pipelines. *Proc. Water Research Centre Conf, Medmenham*

Bokor S D 1982 Correlation of laboratory and installed drainage system solid transport measurements. PhD Thesis, Dept. Building Technology, Brunel University, Uxbridge

Bowler R and Arthur S 1999 Siphonic roof rainwater drainage – design considerations. *CIBW62 Water Supply and Drainage Conference*, Edinburgh, Scotland, September

Bugliarello G 1976 *The impact of noise pollution*, Pergamon Press, Oxford

Building Research Establishment, Watford
Digest: 69 (1977) Durability and application of plastics
    81 (1967) Hospital sanitary services: some design and maintenance problems
    83 (1973) Plumbing with stainless steel
    248 (1981) Sanitary pipework: Part 1 Design basis
    249 (1981) Sanitary pipework: Part 2 Design of pipework
    292 (1984) Access to domestic underground drainage systems
    308 (1986) Domestic unvented hot-water systems
    337 (1988) Sound insulation: basic principles
    365 (1991) Soakaway design
    382 (1993) New materials in hot climates

Environmental Assessment Methods
BREEAM: 2/91 New superstores and supermarkets
    3/91 New homes
    1/93 New office designs
    4/93 Existing office buildings
    5/93 New industrial and warehousing units

Building Research Establishment 2001 Constructing the Future 9 : 4. BRE Watford

Burberry P and Griffiths T J 1962 Service engineering: hydraulics. *Architects' Journal* 21 November: 1185–1191

Butler AP 1983 Design rainfalls for large roof drainage systems: *Proc. Semin. on Drainage Design, Brunel University, Uxbridge*

Butler D 1991 A small-scale study of wastewater discharges from domestic appliances. *Journal of Institution of Water and Environmental Management* 5(2): 178–185

Butler D 1993 The influence of dwelling occupancy and day of the week on domestic appliance wastewater dis-

charges. *Building and Environment* 28(1): 73–79

Campbell D P and McLeod K D 2000 Detergency in soil and vent pipe systems. *Building Services Engineering Research and Technology* 21(1): 35–40

Campbell D P and McCleod K D 2001 Detergents in drainage systems for buildings. *Water Research* 18(1): 128–134

Carter R 1979 Vacuum waste problems overcome. *Surveyor* May: 16–18

Chakrabarti S P 1986 Studies on the development of economical drainage systems for multistorey buildings. National Bureau of Standards, Washington DC: funded publication by Amerind Publishing Co, New Delhi, TT 38-04-000

Chapman A 1948 A survey of noise in British homes. National Building Studies Technical Paper 2, HMSO, London

Chartered Institution of Building Services Engineers 1986, 2001 CIBSE Guide: Volumes A, B and C, London

Chartered Institution of Building Services Engineers 1991 Minimising the risk of legionnaires' disease. TM13, London

Coe A L 1978 *Water supply and plumbing practices in continental Europe*, Hutchinson Benham, London

Coggins C 1993 Research report analyses the initatives in Sheffield. *Waste Management* October: 18–20

Colbourne J S 1981 Influence of non-metallic materials and water fittings on the microbial quality of water in plumbing systems. *Proc. Symp. on Hospital Water Supplies*, pp. 1–15, Institution of Public Health Engineers, London

Colbourne J S and Dennis P J 1989 Ecology and survival of legionella pneumophila. *Journal of Institution Water and Environmental Management* 3(4): 345–350

Construction Industry Research and Information Association 1976 Report of Working Party on the design and construction of buried thin-walled pipes, London

Courtney R G 1976 A multinomial analysis of water demand. *Building and Environment* 11(3): 203–209

Courtney R G and Sexton D E 1973 Refuse collection from houses and flats by pipeline. CP4/73, *Building Research Establishment*, Watford

Cox D 1997 Designing with water. *The Society for Responsible Design, Newsletter* 48: 4.16

Crisp J and Sobolev A 1959 An investigation of the performance of lavatories using spray taps and of sanitary accommodation in an office building. *Journal of Institution of Water Engineers and Scientists* 3(6)

Cummings S J, Wright G B and Bonollo E 2001 Australian water closet developments and their role in sustainable sanitation. *Building Services Engineering Research and Technology* 22(1): 47–57

Dadswell J V 1990 Microbiological aspects of water quality and health. *Journal of Institution of Water and Environmental Management* 4(6): 515–519

Davidson P J and Courtney R G 1976 Revised scales for sanitary accommodation in offices. *Building and Environment* 11(1): 51–56

Davidson P J and Courtney R G 1980 A basis for revised scales for sanitary accommodation in schools. *Building Services Engineering Research and Technology* 1(1): 17–23

De Cuyper K 1993 Towards a standardised code of practice for drainage systems inside European buildings. *Proc. CIB W62 Symp. on Water Supply and Drainage for Buildings, Porto*, pp. 2–13

Department of the Environment 1972 Spaces in the home. *Design Bulletin* 24, HMSO, London

Department of the Environment 1974 Back-siphonage. HMSO, London

Department of the Environment 1989 The Water Supply (Water Quality) Regulations 1989. Statutory Instrument 1147, HMSO, London (Also Regulations issued in Scotland)

Department of the Environment 1991 Recycling of wastes. Waste Management Paper 28, HMSO, London

Department of the Environment 1992 The UK Environment. HMSO, London

Department of the Environment and National Water Council 1976 Domestic unvented hot water systems. Standing Committee Technical Report No. 3, DOE, London

Department of the Environment/Welsh Office 1990. Drinking Water 1990. HMSO, London

Department of the Environment/Welsh Office 1992 Using water wisely. HMSO, London

Department of the Environment and Welsh Office 1995 *Making waste work* HMSO, London

Department of Health and Department of Environment 1983 The bacteriological examination of water supplies. Report 71, HMSO, London

Diaper C *et al.* 2001 Water recycling technologies in the UK. *Journal of Institution of Water and Environmental Management* 15(4): 282–286

Dixon A M, Butler D and Fewkes A 1999 (a). Guidelines for greywater reuse: health issues. *Journal of Institution of Water and Environmental Management* 13(5): 322–326

Dixon A M, Butler D, Fewkes A and Robinson M 1999(b) Measurement and modelling of quality changes in stored untreated greywater. *Urban Water*, 1:293–306

Dixon A M, Butler D and Fewkes A 1999(c) Water saving potential of domestic water reuse systems using greywater and rainwater in combination. *Water Science and Technology* 39(5): 25–32

Douglas, Gasiorek and Swaffield 2000 Fluid Mechanics, Prentice-Hall, London

Ellis K V 1993 Legionellosis: a concise review. *Journal of the Institution of Water and Environmental Management* 7(4): 418–430

Environment Agency 2001 Leakage targets hit by foot and mouth. *Demand Management Bulletin* Issue 49 October: 3

European Community 1980 Council Directive 80/778 on the quality of water intended for human consumption. *Journal of European Communities* L229/11

Felstead R K 1974 An unusual water-contamination problem. *Hospital Engineering* October: 20–26

Fewkes A and Frampton D I 1993 The development of a computer model to evaluate the performance of a rainwater supplied WC flushing system. *Proc. CIB W62 Symp. on Water Supply and Drainage for Buildings, Porto*, pp. 117–132

Flenning D N 1977 Use of public washrooms in an enclosed suburban shopping plaza. D13R714, *Research Council of Canada*, Ottawa

Foxon T J *et al.* 2000 An assessment of water demand management options from a systems approach. *Journal of Institution of Water and Environmental Management* 14(3): 171–178

Fuchs H V 1983 Generation and control of noise in water supply installations Part 1: Fundamental Aspects. *Applied Acoustics* 16: 325–346

Fuchs H V 1993 Generation and control of noise in water supply installations Part 2: Sound Source Mechanisms. *Applied Acoustics* 38: 59–85

Gadbury D and Hall M J 1989 Metering trials for water supply. *Journal of Institution of Water and Environmental Management* 3(2): 182–186

Galbraith N S *et al.* 1987 Water and disease after Croydon: a review of water-borne and water-associated disease in the UK. *Journal of Institution of Water and Environmental Management* 1(1): 7–21

Gatt K 1993 An analysis of domestic dry weather flow in sewerage networks for the Maltese Islands. Unpublished MSc Dissertation, Imperial College, University of London

Gibson E 1978 The design of water storage systems for buildings. PhD Thesis, The University, Glasgow

Grace S 2000. RPZ valves uncovered. *Plumbing* May/June

Gray P G, Cartwright A and Parkin P 1958 Noise in three groups of flats with different floor insulations. National Building Studies Research Paper 27, HMSO, London

Griffiths T J 1962 The hydraulic design of drainage pipework for domestic and public buildings. *Proc. Public Works and Municipal Services Congr, London*

Griggs J C, Hall J and Shouler M C 1993 Rats in drainage systems. *Proc. of CIB W62 Symp. on Water Supply and Drainage for Buildings, Porto,* pp. 17–30

Griggs J C and Shouler M C 1994 An examination of water conservation measures. *Proc. of CIBW 62 Symp. on Water Supply and Drainage for Buildings, Brighton*

Griggs J C, Pitts N J, Hall J and Shouler M C 1997 Water conservation: a guide for design, installation and maintenance of low-flush WCs. IP8 (Parts 1 and 2) Building Research Establishment, Watford

Haarhoff J and Van der Merwe B 1996 Twenty five years of wastewater reclamation in Windhoek, Namibia. *Water Science and Technology* 33(10/11): 25

Hall M J, Hooper B D and Postle S M 1988 Domestic per capita water consumption in South West England. *Journal of Institution of Water and Environmental Management* 2(6): 626–631

Hanslin R 1993 Siphonic rainwater drainage system. *Proc. CIB W62 Symp. on Water Supply and Drainage for Buildings, Porto,* pp. 217–225

Hayes C R *et al.* 1997 Meeting standards for lead in drinking water. *Journal of Institution of Water and Environmental Management* 11(4): 257–263

Hayhurst D 2001 Installation benefits of pre-assembled space and water heating systems. *Plumbing* January/February

Health and Safety Executive 1992 Workplace (Health, Safety and Welfare) Regulations, London

HERU 1961 Study of the use of water outlets in wards. Hospital Engineering Research Unit, The University, Glasgow

Hodges D 1998 Safe use of household greywater. *Water and Environmental Management* 3(6), 15–16

Hope P and Papworth M U 1980 Fire main failure due to rapid priming of dry lines. *Proc. of the 3rd Int. Conf. on Pressure Surge,* BHRA, Cranfield

HSE 2001 'Legionnaires' Disease': The control of legionella bacteria in water systems. Guidance Note 70, Health and Safety Executive, UK

Hulsmann A D 1990 Particulate lead in water supplies. *Journal of Institution of Water and Environmental Management* 4(1): 19–25

Hunter R B 1940 Methods of estimating loads in plumbing systems. BMS 65 and BMS 79, National Bureau of Standards, Washington, DC

Institute of Plumbing 1988 Plumbing Engineering Services Design Guide. Hornchurch, UK

Institute of Plumbing 1990 Legionnaires' Disease: good practice guide for plumbers. Hornchurch, UK

Institute of Wastes Management 1984 Advice on storage and on-site treatment of household, commercial and industrial wastes. IWM 3, Northampton

Institution of Public Health Engineers 1954 Report of the Joint Committee on field research into drainage problems, London

International Organization for Standardization 1980 Laboratory tests on noise emission by appliances and equipment used in water supply installations, Part 1, Method of Measurement. ISO 3822/1, 1977(E), revised 1979/80, Geneva

Jack L B 1997 An investigation of the air pressure regime within building drainage vent systems. PhD thesis, Heriot-Watt University

Jack L B 2001 Developments in the definition of fluid traction forces within building drainage vent systems. *Building Services Engineering Research and Technology* 21(4): 266–273

Jackson G M and Leventhall H G 1975 Household appliance noise. *Applied Acoustics* 8: 101–118

Jeppesen B 1988 Investigation of reduced flushing volumes in sewer systems. Technical Dimensions, October, Brisbane, Brisbane City Council

König K W 2001 The Rainwater Technology Handbook, Wilo-Brain, Dortmund

Konen T P 1989 Water demand research: plumbing fixture requirements for buildings. *Journal of American Society of Plumbing Engineers* 4(2): 23–62

Law I B 1996 Rouse Hill – Australia's first full-scale domestic non-potable reuse application. *Water Science and Technology* 33(10/11): 71

Lawson J D, O'Neill I C and Graze H R 1963 Pressure surge in fire services in tall buildings. *Proc. 1st Australasian Conf. on Hydraulics and Fluid Mechanics*, Pergamon Press, pp. 353–368

Lillywhite M S T and Webster C J D 1979 Investigations of drain blockages and their implications for design. *Journal of Institution of Public Health Engineers* 7(2): 53–60

Lillywhite M S T and Wise A F E 1969 Towards a general method for the design of drainage systems in large buildings. *Journal of Institution of Public Health Engineers* 68(4) (reprinted as Building Research Establishment Current Paper 27/69)

Lister M 1960 Numerical solution of hyperbolic partial differential equations by the method of characteristics. In *Numerical methods for digital computers*, Ed. Ralston, Wiley, New York

Maeda M et al. 1996 Area wide use of reclaimed water in Tokyo. *Water Science and Technology* 33(10/11): 51

Marseille B 1965 Noise transmission in piping. *Heating and Ventilating Engineer* 38(455): 674–681

Martin B R 1992 The San Simeon Story. *CIBW62 Int. Symp. on Water Supply and Drainage for Buildings*, Washington DC September

Maver T W 1964 The design and performance of hot water storage calorifiers. *Journal of Institution of Heating and Ventilating Engineers* 32(12): 330–334

May R W P 1982 Design of gutters and gutter outlets. Report IT 205, Hydraulics Research Station, Wallingford

May R W P 1984 Hydraulic design of roof gutters. *Proceedings of the Institution of Civil Engineers* Part 2, 77: 479–489

May R W P 1997 The design of conventional and siphonic roof drainage systems. *Journal of Institution of Water and Environmental Management* 11(1): 56–60

McDougall J A 1995 Mathematical modelling of solid transport in defective building drainage systems. PhD thesis, Heriot-Watt University

McDougall J A and Swaffield J A 1994 Assessment of WC performance using computer-based prediction techniques. *Proc. of CIBW62 Symp. on Water Supply and Drainage for Buildings*, Brighton

McDougall J A and Swaffield J A 2000 Simulation of building drainage system operation under water conservation design criteria. *Building Services Engineering Research and Technology* 21(1): 41–52

Meers P D 1981 Bacterial multiplication in hospital water. *Proc. Symp. on Hospital Water Supplies*, 1–5, Institution of Public Health Engineers, London

Meikle J 1996 Home building. *Guardian* Newspaper, 4 June

Mein R G and Jones R F 1992 Determination of flow depths in roof gutters. *Proc. Int. Symp. Urban Stormwater Management*, Sydney, February, pp. 276–278

Millar D S 1978 Internal flow – a guide to losses in pipes and ducts, British Hydromechanics Research Association, Cranfield, Beds

Mills R A and Asano T 1996 A retrospective assessment of water reclamation projects. *Water Science and Technology* 33(10/11): 59

Mustow S *et al.* 1997 Water Conservation – implications of using recycled greywater and stored rainwater in the UK. Report 13034/1 BSRIA, Bracknell

National Bureau of Standards 1976 Noise control in multi-family dwellings. US Department of Housing and Urban Development, Washington, DC

National Institutes of Health 1936 Epidemic of amoebic dysentery — the Chicago outbreak of 1933. *NIH Bulletin* 166, Washington, DC

National Joint Utilities Group 1986 Recommended positioning of Utilities Mains and Plant for New Works. NJUG 7, London

Natural Environment Research Council 1975 The Flood Studies Report, NERC, London

OFWAT 2001 Leakage and the effective use of water. OFWAT 2000–2001 Report *www.ofwat.gov.uk*

Parkin P, Purkis H J and Scholes W E 1960 Measurements of sound insulation between dwellings. National Building Studies Research Paper 33, HMSO, London

Pink B J 1973 Building Research Establishment, Watford:
(i) A study of the water flow in vertical drainage stacks by means of a probe. Current Paper 36/73
(ii) Laboratory investigation of the discharge characteristics of sanitary appliances. Current Paper 37/73
(iii) The effect of stack length on the air flow in drainage stacks. Current Paper 38/73

Purdew R C 1969 Rationalisation of services in the home. *Proc. Symp. on Engineering in the Home*, pp. 56–63, Chartered Institution of Building Services Engineers, London

Roaf S 2001 Ecohouse, Architectural Press, Oxford

Rogers W L 1959 Noise and vibration in water piping systems. *ASHRAE Journal* 1(3): 83–86

Royal Commission on Environmental Pollution 1993 Incineration of waste. Report 17, Cmnd 2181, HMSO, London

Russac D A V, Rushton K R and Simpson R J 1991 Insights into domestic demand from a metering trial. *Journal of Institution of Water and Environmental Management* 5(3): 342–351

Sexton D E and Smith J T 1972 Studies of refuse compaction and incineration in multi-storey flats. *Public Cleansing* 62(12): 604–623

Shouler M C 1998 Household waste : storage provision and recycling. BR 356 Building Research Establishment, Watford

SLASH Group 1979 Maintenance Practice Manual: Section 6 — Plumbing Services. Scottish Local Authorities Special Housing Group, Edinburgh

Smith A L and Roger D V 1990 Isle of Wight water metering trial. *Journal of Institution of Water and Environmental Management* 4(5): 403–409

Smith J T 1976 An appraisal of the pneumatic refuse collection system at Lisson Green. CP58/76, Building Research Establishment, Watford

Sobolev A 1955 Failures of ballvalves and their remedies. *Journal of Institution of Water Engineers* 208–222

Sobolev A and Lloyd G J 1964 Trials of dual-flush cisterns. *Journal of Institution of Water Engineers* 18(1)

Soulsby P G 1994 The use of composting toilets for domestic purposes. *Proc. of CIBW62 Symp. on Water Supply and Drainage for Buildings, Brighton*

Spielvogel L G 1969 New concepts for sizing water storage systems. *Building Systems Design* 61(11): 75–80

Staniforth G 1994 A brief history of hot and cold water supply and plumbing systems, and their potential for development in the 21st century. *Proc. of CIBW62 Symp. on Water Supply and Drainage for Buildings, Brighton*

Stonemetz R E 1965 Liquid cavitation studies in circular pipe bends. TND3176, NASA

Strumpf F M 1967 Water installation units and their control. *IEE Conf. on Acoustic Noise and its Control*, London

Surendram S and Wheatley A D 1998 Greywater reclamation for non-potable reuse.

Journal of Institution of Water and Environmental Management 12(6): 406–413

Swaffield J A and Boldy A P 1993 *Pressure surge in pipe and duct systems*, Avebury Technical, Gower Press, Aldershot

Swaffield J A and Bridge S 1983 Applicability of the Colebrook–White formula to represent frictional losses in partially filled unsteady pipeflow. *J. Res.National Bureau of Standards* 88(6): 389–393, Nov/Dec

Swaffield and Campbell 1992 Air pressure transient propagation in building drainage vent systems including the influence of mechanical boundary conditions. *Building and Environment* 27(4): 456–467

Swaffield J A and Campbell D P 1995 Simulation of air pressure transient propagation in building drainage vent systems. *Building and Environment* 30(1): 115–127

Swaffield J A, Escarameia M and Campbell D P 1999 Unsteady roof gutter flow: development and application of a simulation. *Building Services Engineering Research and Technology* 20(1): 29–40

Swaffield J A and Galowin L S 1992 *The engineered design of building drainage systems*, Ashgate, Gower Press, Aldershot

Swaffield J A, McDougall J A and Campbell D P 1999 Drainage flow and solid transport simulation in defective building drainage networks. *Building Services Engineering Research and Technology* 20(2): 73–82

Swaffield J A and Wakelin R H M 1990 Low water use sanitation. *Proc. 2nd World Plumbing Conf.*, Barbican, London

Swaffield J A and Wakelin R H M 1996 Water conservation: the impact of design, development and site appraisal of low-volume flush toilet. In *Lowcost sewerage*, Ed. Mara D, Wiley, pp. 175–188

Teale T P 1881 *Dangers to health* (3rd Edition), J and A Churchill, London

Thackray J E 1992 Paying for water: policy options and their practical implications. *Journal of Institution of Water and Environmental Management* 6(4): 505–513

Thackray J E *et al.* 1978 The Malvern and Mansfield studies of domestic water usage. *Proceedings Institution of Civil Engineers* 1(64): 37–61

Thackray J E and Archibald G G 1981 The Seven-Trent studies of industrial water use. *Proceedings Institution of Civil Engineers* 1(70): 403–432

Uujamhan E J S 1982 Water conservation WC design: a study of the design parameters affecting WC performance. PhD thesis, Brunel University

Vale R and Vale B 1996 Water and sewage treatment in an autonomous house. *Plumbing* Jan/Feb: 15–19 (see also 1995 Nov/Dec)

Vale B and R 2000 *The new autonomous house*, Thames and Hudson, London

Walsh B 1993 The economics of waste compaction. *Waste Management* October: 21–23

Wassilieff C and Dravitzki V 1992 Prediction of noise from plumbing attached to light timber-framed walls. *Applied Acoustics* 37: 213–232

Water Regulations Advisory Scheme 2000 Type BA – verifiable backflow preventer with reduced pressure zone (RPZ valve). Information and Guidance Note 9–03–02, Newport

Water Services Association 1991 Water facts 1991. Water Services Association, London

Water Services Association 1993 Water meter trials: final report. Water Services Association, London

Webster C J D 1972 An investigation of the use of water outlets in multi-storey flats. *Building Services Engineer* 39(1): 215–233

Webster C J D 1979 Reducing demands for potable water. *Building Research and Practice* July/August

Wise A F E 1954 Self-siphonage in building drainage systems. *Proceedings of the Institution of Civil Engineers* 3(3) (Paper 5987): 789–808

Wise A F E 1957a Aerodynamics studies to aid drainage stack design. *Journal Institution of Public Health Engineers* 56(1): 48–64

Wise A F E 1957b *Drainage pipework in dwellings*, HMSO, London

Wise A F E and Croft J 1954 Investigation of single stack drainage for multi-storey flats. *Journal of Royal Sanitary Institute* 74(9): 797–826

World Health Organization 1984 Guidelines for drinking water quality. Geneva

Wright G B, Swaffield J A and Arthur S 2001 Investigation into the performance characteristics of multi-outlet siphonic rainwater systems. *CIBW62 Water Supply and Drainage Conference*, Portoroz, Slovenia, September

Wylie E B and Streeter V L 1983 *Fluid transients*, FEB Press, Ann Arbor, Michigan

Wyly R S 1964 *Investigation of hydraulics of horizontal drains in plumbing systems*, Monograph 86, National Bureau of Standards, Washington, DC

Wyly R S and Eaton H N 1961 *Capacity of stacks in sanitary drainage systems for buildings*, Monograph 31, National Bureau of Standards, Washington, DC

Young O C 1977 The structural design and laying of small underground drains of rigid materials. Supplementary Report 303, Transport and Road Research Laboratory, Crowthorne

Young O C 1978 The design and laying of small underground drains of flexible materials. Supplementary Report 375, Transport and Road Research Laboratory, Crowthorne

Index